C0-AVS-721

The Finite Element Method

Series in Computational and Physical Processes in Mechanics and Thermal Sciences

(Formerly the Series in Computational Methods in Mechanics and Thermal Sciences)

W. J. Minkowycz and E. M. Sparrow, Editors

Anderson, Tannehill and Pletcher, Computational Fluid Mechanics and Heat Transfer

Aziz and Na, Perturbation Methods in Heat Transfer

Baker, Finite Element Computational Fluid Mechanics

Baker, Finite Element Computational Fluid Mechanics, Second Edition

Beck, Cole, Haji-Sheikh and Litkouhi, Heat Conduction Using Green's Functions

Carey, Computational Grids

Comini, del Giudice and Nonino, Finite Element Analysis in Heat Transfer

Heinrich and Pepper, Intermediate Finite Element Method: Fluid Flow and Heat Transfer Application

Jaluria, Computer Methods for Engineering

Koenig, Modern Computational Methods

Patankar, Numerical Heat Transfer and Fluid Flow

Pepper and Heinrich, The Finite Element Method

Shih, Numerical Heat Transfer

Shyy, Udaykumar, Rao, and Smith, Computational Fluid Dynamics With Moving Boundaries

Tannehill, Anderson and Pletcher, Computational Fluid Mechanics and Heat Transfer, Second Edition

PROCEEDINGS

Chung, *Editor,* Finite Elements in Fluids

Chung, *Editor,* Numerical Modeling in Combustion

Haji-Sheikh, *Editor*, Integral Methods in Science and Engineering - 90

Shih, *Editor*, Numerical Properties and Methodologies in Heat Transfer

The Finite Element Method

Basic Concepts and Applications

SECOND EDITION

Darrell W. Pepper
University of Nevada
Las Vegas, Nevada

Juan C. Heinrich
University of New Mexico
Albuquerque, New Mexico

Published in 2006 by
CRC Press
Taylor & Francis Group
6000 Broken Sound Parkway NW, Suite 300
Boca Raton, FL 33487-2742

Printed in the United States of America on acid-free paper
10 9 8 7 6 5 4 3 2 1

International Standard Book Number-10: 1-59169-027-7 (Hardcover)
International Standard Book Number-13: 978-1-59169-027-6 (Hardcover)
Library of Congress Card Number 2005002971

Library of Congress Cataloging-in-Publication Data

Pepper, D. W. (Darrell W.)
The finite element method : basic concepts and applications / by Darrell W. Pepper and Juan C. Heinrich.-- 2nd ed.
p. cm. -- (Series in computational and physical processes in mechanics and thermal sciences)
Includes bibliographical references and index.
ISBN 1-59169-027-7
1. Finite element method. I. Heinrich, Juan C. II. Title. III. Series.

TA347.F5P46 2005
620'.001'51825--dc22 2005002971

Taylor & Francis Group
is the Academic Division of T&F Informa plc.

To our parents
Weldon and Marjorie Pepper
Carlos and Ruby Heinrich

Preface

It has been more than 13 years since the first edition of this book was published. Many changes in the art of finite element methodology have occurred since then. FORTRAN was the most prevalent programming language for scientific computing, and still exists in enhanced forms today. C/C++ has become one of the preferred choices for much of the computing performed on PCs using WINDOWS. As we progress into the 21st century, a new language is beginning to appear—JAVA—a platform-independent language that also runs on the Web.

The early finite element codes in the first edition were written in FORTRAN 77 and QuickBasic for graphical display under WINDOWS. FORTRAN still appears to be the preferred scientific language for the scientist–engineer, and so we felt that it was necessary to include source listings in FORTRAN. For this second edition, FORTRAN 95 versions of 1-D, 2-D, and 3-D codes are maintained on a Web site, located at femcodes.nscee.edu, along with several MATHCAD, MATLAB, and MAPLE algorithms. Interactive C/C++ and JAVA versions of the 2-D codes are also available from the Web. While the fundamental principles of the finite element method remain unchanged, applications of the method have continued to advance into new areas, including such fields as nanotechnology and biomedical.

This book is based on our experience in teaching the finite element method to both engineering students and experienced, practicing engineers in industry. Much of the material stems from the AIAA home study course and ASME short courses that we have given over the last 17 years and from the suggestions and recommendations of the many participants. There are many finite element books available in the literature today, and this book is among the multitude that continues to appear. When teaching finite element methodology to students, we always found that some alteration or simplification of much of the material was required before the concepts were grasped. Some of the confusion resulted from the mathematical "jargon" and deep theoretical aspects of the technique. We found that a much simpler approach was required before one can truly "appreciate" the detailed mathematical derivations and theory.

The primary intent of this book is to introduce the basic fundamentals of the finite element method in a clear, concise manner using simple examples. Much attention is paid to the development of the discrete set of algebraic equations, beginning with simple one-dimensional problems that can be solved by inspection, continuing to two- and three-dimensional elements, and ending with three chapters dealing with applications. The example problems are straightforward and can be worked out manually. The computer codes that accompany this text will be helpful for many of the exercises, especially multidimensional homework problems; however, almost any one of the commercially available finite element codes available today can be used for the problems. The FORTRAN source listings on the Web include simple 1-D and 2-D mesh generators and example data files. Descriptions of the codes are included in the help file on the Web (see Appendix E).

More exercises have been added to this revised edition, some of which involve working out the developments presented in the text in more detail. In addition, we have added executable files on the Web that can be run using COMSOL, a commercial finite element code that has been developed by COMSOL, Inc. (originally written to run under the MATLAB operating system). COMSOL is a very versatile finite element code that handles a wide variety of applications, including fluid flow, heat transfer, solid mechanics, and electrodynamics. This package runs on PCs and is easy to use.

Because many finite element books are slanted toward the structurally oriented engineer, the nonstructural engineer must sift through a considerable amount of uninteresting concepts and applications before finding a relevant problem area. We have found that most engineers are knowledgeable of the basic precepts of heat transfer and have, accordingly, directed the book toward heat flow and one degree of freedom (temperature). Many of the individuals whom we have instructed over the years have come from diverse backgrounds ranging from biology to nuclear physics; in nearly all cases, a simple generic approach focused on the transport and diffusion of heat (scalar transport) has allowed relatively easy mastering of finite elements.

We thank our colleagues and former short-course and home-study participants who greatly contributed to the presentation of the material in this second edition of our introductory text. We especially thank Taylor & Francis for its helpful comments and editorial assistance, and Professor W.J. Minkowycz and Professor E.M. Sparrow for their continued patience in reading and for their suggestions for revising the manuscript. We also express our thanks to Mrs. Jeannie Pepper for her Herculean efforts in preparing the manuscript and the graphical images in this second edition.

Darrell W. Pepper
Juan C. Heinrich

Table of Contents

CHAPTER

1

INTRODUCTION

1.1 BACKGROUND

The finite element method is a numerical technique that gives approximate solutions to differential equations that model problems arising in physics and engineering. As in simple finite difference schemes, the finite element method requires a problem defined in geometrical space (or domain), to be subdivided into a finite number of smaller regions (a mesh).

In finite difference methods in the past, the mesh consisted of rows and columns of orthogonal lines (in computational space—a requirement now handled through coordinate transformations and unstructured mesh generators); in finite elements, each subdivision is unique and need not be orthogonal. For example, triangles or quadrilaterals can be used in two dimensions, and tetrahedra or hexahedra in three dimensions. Over each finite element, the unknown variables (e.g., temperature, velocity, etc.) are approximated using known functions; these functions can be linear or higher-order polynomial expansions in terms of the geometrical locations (nodes) used to define the finite element shape. In contrast to finite difference procedures (conventional finite difference procedures, as opposed to the finite volume method, which is integrated), the governing equations in the finite element method are integrated over each finite element and the contributions summed ("assembled") over the entire problem domain. As a consequence of this procedure, a set of finite linear equations is obtained in terms of the set of unknown parameters over the elements. Solutions of these equations are achieved using linear algebra techniques.

1.2 SHORT HISTORY

The history of the finite element method is particularly interesting, especially because the method has only been in existence since the mid-1950s. The early work on numerical solution of boundary-valued problems can be traced to the use of finite difference schemes; Southwell (1946) discusses the use of such methods in his book published in the mid-1940s. The beginnings of the finite element method actually stem from these early numerical methods and the frustration associated with attempting to use finite difference methods on more difficult, geometrically irregular problems (Roache, 1972).

Beginning in the mid-1950s, efforts to solve continuum problems in elasticity using small, discrete "elements" to describe the overall behavior of simple elastic bars began to appear. Argyris (1954) and Turner et al. (1956) were the first to publish use of such techniques for the aircraft industry. Actual coining of the term *finite element* appeared in a paper by Clough (1960).

The early use of finite elements lay in the application of such techniques for structurally related problems. However, others soon recognized the versatility of the method and its underlying rich mathematical basis for application in nonstructural areas. Zienkiewicz and Cheung (1965) were among the first to apply the finite element method to field problems (e.g., heat conduction, irrotational fluid flow, etc.) involving solution of Laplace and Poisson equations. Much of the early work on nonlinear problems can be found in Oden (1972). Efforts to model heat transfer problems with complex boundaries are discussed in Huebner (1975); a comprehensive three-dimensional finite element model for heat conduction is described by Heuser (1972). Early application of the finite element technique to viscous fluid flow is given in Baker (1971).

Since these early works, rapid growth in usage of the method has continued since the mid-1970s. Numerous articles and texts have been published, and new applications appear routinely in the literature. Excellent reviews and descriptions of the method can be found in some of the earlier texts by Finlayson (1972), Desai (1979), Becker et al. (1981), Baker (1983), Fletcher (1984), Reddy (1984), Segerlind (1984), Bickford (1990), and Zienkiewicz and Taylor (1989). A vigorous mathematical discussion is given in the text by Johnson (1987), and programming the finite element method is described by Smith (1982). A short monograph on development of the finite element method is given by Owen and Hinton (1980).

The underlying mathematical basis of the finite element method first lies with the classical Rayleigh–Ritz and variational calculus procedures introduced by Rayleigh (1877) and Ritz (1909). These theories provided the reasons the finite element method worked well for the class of problems in which variational statements could be obtained (e.g., linear diffusion problems). However, as interest expanded in applying the finite element method to more types of problems, the use of classical theory to describe such problems became limited and could not be applied (this is particularly evident in fluid-related problems).

Extension of the mathematical basis to nonlinear and nonstructural problems was achieved through the method of weighted residuals, originally conceived by Galerkin (1915) in the early 20th century. The method of weighted residuals was found to provide the ideal theoretical basis for a much wider basis of problems as opposed to the Rayleigh–Ritz method. Basically, the method requires the governing differential equation to be multiplied by a set of predetermined weights and the resulting product integrated over space; this integral is required to vanish. Technically, Galerkin's method is a subset of the general weighted residuals procedure, as various types of weights can be utilized; in the case of Galerkin's method, the weights are chosen to be the same as the functions used to define the unknown variables.

Galerkin and Rayleigh–Ritz approximations yield identical results whenever a proper variational statement exists and the same basis functions are used. By using constant weights instead of functions, the weighted residual method yields the finite volume technique. A more vigorous description of the method of weighted residuals can be found in Finlayson (1972). Recent descriptions of the method are discussed in Chandrupatla and Belegundu (2002), Liu and Quek (2003), Hollig (2003), Bohn and Garboczi (2003), Hutton (2004), Solin et al. (2004), Reddy (2004), Becker (2004), and Ern and Guermond (2004).

Most practitioners of the finite element method now employ Galerkin's method to establish the approximations to the governing equations. The underlying theme in this book likewise follows Galerkin's method. The simplicity and richness of the method pays for itself as the user progresses into more complicated and demanding types of problems. Once this fundamental concept is grasped, application of the finite element method unfolds quickly.

1.3 ORIENTATION

This book is designed to serve as a simple introductory text and self-explanatory guide to the finite element method. Beginning with the concept of one-dimensional heat transfer (which is relatively easy to follow), the book progresses through two-dimensional elements to three-dimensional elements, ultimately ending with a discussion on various applications, including fluid flow. Particular emphasis is placed on the development of the one-dimensional element. All the principles and formulation of the finite element method can be found in the class of one-dimensional elements; extrapolation to two and three dimensions is straightforward.

Each chapter contains a set of example problems and some exercises that can be verified manually. In most cases, the exact solution is obtained from either inspection or an analytical equation. By concentrating on example problems, the manner and procedure for defining and organizing the requisite initial and boundary condition data for a specific problem become apparent. In the first few examples, the solutions are apparent; as the succeeding problems become progressively more involved, more input data must be provided.

For those problems requiring more extensive calculational effort (which becomes quickly discovered when dealing with matrices), a set of computer codes is available on the Web (see femcodes.nscee.edu). These source codes are written in FORTRAN 95 (including several simple MATLAB, MATHCAD, and MAPLE routines) and run on WINDOWS-based PCs. The purpose of these codes is to illustrate simple finite element programming and to provide the reader with a set of programs that will assist in solving the examples and most of the exercises. The computer codes are fairly generic and have been written with the intention of instruction and ease of use. The reader may modify and optimize them as desired. The reader is advised to read the file called README.DOC on the Web for a discussion of the codes and execution procedures. Two additional codes are also available; one is written in C/C++ and the other is written in JAVA. Both permit two-dimensional heat transfer calculations with mesh refinement (n-adaptation) to be run in real time under WINDOWS and on the Web. These two codes include simple pre- and post-processing of meshes and results.

A set of self-executing files that use COMSOL, a finite element code developed by COMSOL Inc., is also included. COMSOL is a fairly recent commercial finite element package, originally written to run with MATLAB, which is easy to use yet handles a wide variety of problems. The software can be used to solve one-, two-, and three-dimensional problems in structural analysis, heat transfer, fluid flow, and electrodynamics, and employs a rather elaborate, but easy to use mesh generator. The software also permits mesh adaptation (an upper-end capability in recent commercial finite element packages that allows local mesh refinement in regions of steep gradients and high activity).

A discussion of the method of weighted residuals is given in Chapter 2. This chapter provides the underlying mathematical basis of the Galerkin procedure that is basic to the finite element method. Chapter 3 serves as the actual beginning of the finite element method, utilizing the one-dimensional element—in fact, the entire framework of the method is presented in this chapter. Reinforcement of the basic concepts is achieved in Chapters 4 through 6 as the reader progresses through the class of two-dimensional elements. In Chapter 7, simple three-dimensional elements are discussed, utilizing a single-element heat conduction problem with various boundary conditions, including radiation. Chapter 8 describes applications to solid mechanics and the role of multiple degrees of freedom (e.g., displacement in x and y) with example problems in two dimensions. Chapter 9 discusses applications to convective transport, using examples from potential flow and species dispersion. In Chapter 10, the reader is introduced to viscous fluid flow and the nonlinear equations of fluid motion for an incompressible fluid. COMSOL is particularly effective at solving fluid flow problems. The advanced book by Heinrich and Pepper (1999) discusses fluid flow in greater detail, and includes a two-dimensional penalty approach code and source listing for incompressible fluid flow.

The finite element method has essentially become the *de facto* standard for numerical approximation of the partial differential equations that define structural engineering, and is becoming widely accepted for a multitude of other engineering and scientific problems. Many of the commercial computer codes currently used today are finite element based—even the finite volume computational fluid dynamics codes sold commercially employ mesh generators based on finite element unstructured mesh generation. It is the intent of this text to provide the reader with sufficient information and knowledge to begin application of the finite element method, with the hope of instilling an interest in advancing the state of the art in more advanced studies.

Since the first edition of this book, there has been a proliferation of commercial codes for the finite element method, including many that are applicable to a wide range of problems. Recent introduction of the finite element method through generalized mathematical solvers, such as MATHCAD, MAPLE, and MATLAB (and the supplementary software COMSOL), have helped to spread the training and use of the method. The development of the finite element method using these mathematical symbolic systems is described in Pintur (1998), Portela and Charafi (2002), and Kattan (2003). A computer-based finite element analysis software package that runs on PCs and MacIntosh computers is VisualFEA/CBT, developed by Intuition Software (2002). This package permits up to 3000 nodes and runs structural, heat conduction, and seepage analysis problems.

1.4 CLOSURE

There are some interesting Web sites that describe finite element methods and have codes that can be downloaded. It is recommended that the reader visit the following Web sites for more detailed information regarding commercial solvers and mathematical software packages:

http://www.ncacm.unlv.edu
http://www.cfd-online.com/Resources/topics.html#fem
http://femcodes.nscee.edu

Web sites have a tendency to change locations and addresses over time. Performing a Google search on the subject "finite elements" will generate numerous Web sites as well, many connected to universities and institutions around the world.

REFERENCES

Argyris, J. H. (1954). *Recent Advances in Matrix Methods of Structural Analysis*. Elmsford, N.Y.: Pergamon Press.

Baker, A. J. (1971). "A Finite Element Computational Theory for the Mechanics and Thermodynamics of a Viscous Compressible Multi-Species Fluid." Bell Aerospace Research Dept. 9500-920200.

Baker, A. J. (1983). *Finite Element Computational Fluid Mechanics*. Washington, D.C.: Hemisphere.

Becker, A. A. (2004). *An Introductory Guide to Finite Element Analysis*. New York: ASME Press.

Becker, E. G., Carey, G. F., and Oden, J. T. (1981). *Finite Elements: An Introduction*, Vol. I. Englewood Cliffs, N.J.: Prentice-Hall.

Bickford, W. B. (1990). *A First Course in the Finite Element Method*. Homewood, Ill.: Richard D. Irwin.

Bohn, R. B. and Garboczi, E. J. (2003). User Manual for Finite Element Difference Programs: A Parallel Version of NISTIR 6269, NIST Internal Report 6997.

Chandrupatla, T. R. and Belegundu, A. D. (2002). *Introduction to Finite Elements in Engineering*. Upper Saddle River, N.J.: Prentice-Hall.

Clough, R. W. (1960). "The Finite Element Method in Plane Stress Analysis." *Proc. 2nd Conf. Electronic Computations*. Pittsburgh, Pa.: ASCE, pp. 345–378.

Desai, C. S. (1979). *Elementary Finite Element Method*. Englewood Cliffs, N.J.: Prentice-Hall.

Ern, A. and Guermond, J.-L. (2004). *Theory and Practice of Finite Elements*. New York: Springer-Verlag.

FEMLAB 3 (2004). *User's Manual*. Burlington, Mass.: COMSOL, Inc.

Finlayson, B. A. (1972). *The Method of Weighted Residuals and Variational Principles*. New York: Academic Press.

Fletcher, C. A. J. (1984). *Computational Galerkin Methods*. New York: Springer-Verlag.

Galerkin, B. G. (1915). "Series Occurring in Some Problems of Elastic Stability of Rods and Plates." *Eng. Bull.* 19:897–908.

Heinrich, J. C. and Pepper, D. W. (1999). *Intermediate Finite Element Method: Fluid Flow and Heat Transfer Applications*. Philadelphia: Taylor & Francis.

Heuser, J. (1972). "Finite Element Method for Thermal Analysis." NASA Technical Note TN-D-7274. Greenbelt, Md.: Goddard Space Flight Center.

Hollig, K. (2003). *Finite Elements with B-Splines*. Philadelphia: Society of Industrial and Applied Mathematics.

Huebner, K. H. (1975). *Finite Element Method for Engineers*. New York: John Wiley & Sons.

Hutton, D. V. (2004). *Fundamentals of Finite Element Analysis*. Boston: McGraw-Hill.

Intuition Software (2002). *VisualFEA/CBT*, ver. 1.0. New York: John Wiley & Sons.

Johnson, C. (1987). *Numerical Solution of Partial Differential Equations by the Finite Element Method*. Cambridge: Cambridge University Press.

Kattan, P. I. (2003). *MATLAB Guide to Finite Elements, An Interactive Approach*. Berlin: Springer-Verlag.

Liu, G. R. and Quek, S. S. (2003). *The Finite Element Method: A Practical Course*. Boston: Butterworth-Heinemann.

MAPLE 9.5 (2004). *Learning Guide*. Waterloo, Canada: Maplesoft, Waterloo Maple, Inc.

MATHCAD 11 (2002). *User's Guide*. Cambridge, Mass.: Mathsoft Engineering & Education, Inc.

MATLAB & SIMULINK 14 (2004). *Installation Guide*. Natick, Mass.: The MathWorks.

Oden, J. T. (1972). *Finite Elements of Nonlinear Continua*. New York: McGraw-Hill.

Owen, D. R. J. and Hinton, E. (1980). *A Simple Guide for Finite Elements*. Swansea, U.K.: Pineridge Press.

Pintur, D. A. (1998). *Finite Element Beginnings*. Cambridge, Mass.: MathSoft, Inc.

Portela, A. and Charafi, A. (2002). *Finite Elements Using Maple, A Symbolic Programming Approach*. Berlin: Springer-Verlag.

Rayleigh, J. W. S. (1877). *Theory of Sound*. 1st rev. ed. New York: Dover.

Reddy, J. N. (1984). *An Introduction to the Finite Element Method*. New York: McGraw-Hill.

Reddy, J. N. (2004). *An Introduction to Nonlinear Finite Element Analysis*. Oxford: Oxford University Press.

Ritz, W. (1909). "Uber eine neue Methode zur Lïsung Gewisses Variations-Probleme der mathematischen Physik." *J. Reine Angew. Math.* 135:1–61.

Roache, P. J. (1972). *Computational Fluid Mechanics*. Albuquerque, N.M.: Hermosa.

Segerlind, L. J. (1984). *Applied Finite Element Analysis*. New York: John Wiley & Sons.

Smith, I. M. (1982). *Programming the Finite Element Method*. New York: John Wiley & Sons.

Solin, P., Segeth, K., and Dolezel, I. (2004). *Higher-Order Finite Element Methods*. Boca Raton, Fla.: Chapman & Hall/CRC Press.

Southwell, R. V. (1946). *Relaxation Methods in Theoretical Physics*. London: Clarendon Press.

Turner, M., Clough, R. W., Martin, H., and Topp, L. (1956). "Stiffness and Deflection of Complex Structures." *J. Aero Sci.* 23:805–823.

Zienkiewicz, O. C. and Cheung, Y. K. (1965). "Finite Elements in the Solution of Field Problems." *Engineer* 220:507–510.

Zienkiewicz, O. C. and Taylor, R. L. (1989). *The Finite Element Method*, 4th ed. New York: McGraw-Hill.

CHAPTER

2

THE METHOD OF WEIGHTED RESIDUALS AND GALERKIN APPROXIMATIONS

2.1 BACKGROUND

This chapter introduces the basic concepts needed to develop approximations to the solution of differential equations that will ultimately lead to the numerical algorithm known as the finite element method. We first establish the weighted residuals form of the governing differential equations and give a general method leading to the so-called "weak" formulation, which can be used to obtain finite element approximation of just about any kind of differential equation. The second step is to introduce the concept of shape functions and the Galerkin approximation to the integral form of the governing differential equation.

The theory of the finite element method is found in the variational calculus, and its mathematical basis allowed it to be developed in a very short time and become the powerful tool for engineers that it is today. However, this also created the misconception that a strong mathematical background is essential to understand the finite element method. Here, we show that this is indeed not the case and that all of the finite element methodology can be developed utilizing the theorems of advanced calculus and basic physical principles.

2.2 CLASSICAL SOLUTIONS

As a model problem, we consider determining the conduction of heat on a slender homogeneous metal wire of length L with constant cross section. Assume that the left end is exposed to a prescribed heat flux, q, the right end is held at a constant temperature, $T = T_L$, and the length of the rod is surrounded by insulating material. The situation is shown in Figure 2.1. We further assume that we can run an electrical current through the wire that will act as an internal heat source of magnitude Q.

Using Fourier's law, we can easily write the differential equation that governs the distribution of temperature across the rod. This is

$$-K\frac{d^2T}{dx^2} = Q, \qquad 0 < x < L \tag{2.1}$$

where x is the length coordinate, K is the thermal conductivity of the material (assumed constant), and Q is the internal heat generation per unit volume. The boundary conditions associated with the problem are

$$-K\frac{dT}{dx} = q, \qquad \text{for } x = 0 \tag{2.2}$$

$$T = T_L, \qquad \text{for } x = L \tag{2.3}$$

When $q > 0$, the direction of heat flow is into the rod at $x = 0$, which accounts for the negative sign in Eq. (2.2).

The solution to Eq. (2.1) with boundary conditions (2.2) and (2.3) can be found by direct integration if we assume that Q is an integrable function, and is given by

$$T(x) = T_L + \frac{q}{K}(L - x) + \frac{1}{K}\int_x^L \left(\int_0^y Q(z)dz \right) dy \tag{2.4}$$

If Q is constant, Eq. (2.4) reduces to

$$T(x) = T_L + \frac{q}{K}(L - x) + \frac{Q}{2K}(L^2 - x^2) \tag{2.5}$$

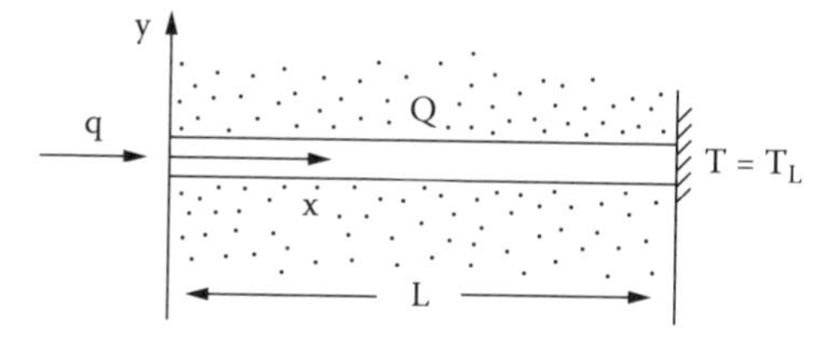

Figure 2.1 Conduction of heat in a rod of length L

Later we use Eq. (2.4) as a benchmark to compare solutions obtained with the finite element procedure. This example is simple and has a unique solution. More difficult problems do not lend themselves to easy analytical solutions if those can be obtained at all; therefore, it becomes crucial to fully understand the behavior of numerical solutions in simple problems to be able to interpret numerical approximations to more complex problems appropriately.

2.3 THE "WEAK" STATEMENT

There are basically two procedures that are normally used to formulate and solve equations of the form of Eq. (2.1) using finite elements. These are known as the Rayleigh–Ritz and the Galerkin methods. Other lesser-utilized methods are based on collocation, constant weights, and least-square techniques. All these procedures are subsets of the method of weighted residuals. For the motivated readers, a description of the Rayleigh–Ritz method can be found in the book of Reddy (1984). For other methods, the reader is referred to Fletcher (1984).

Regardless of which procedure we use, the first step is to define a partition or grid in the interval $0 \le x \le L$, consisting of a finite number of non-overlapping subintervals that cover the whole domain, as illustrated in Figure 2.2, where each subinterval is called an "element." We denote each element by e_k,

$$e_k = \{x : x_k \le x \le x_{k+1}\} \tag{2.6}$$

and the end points of x_k of each interval will be called "nodes." Over each of these elements, the distribution of temperature will then be approximated using known, predetermined functions of the independent variable x, denoted by $\phi_j(x)$, and corresponding unknown parameters a_j. Accordingly, we define an element as a subinterval e_k, together with a prescribed set of functions ϕ_j and an equal number of parameters a_j such that, if the parameters a_j are known, an approximation to the temperature field $T(x)$ is also known over the entire subinterval.

Over the whole domain $0 \le x \le \mathrm{L}$, we can then write

$$T(x) \cong a_1\phi_1(x) + a_2\phi_2(x) + \cdots + a_{n+1}\phi_{n+1}(x) \tag{2.7}$$

The functions $\phi_i(x)$ are called "shape functions" (this is explained later). We write expression (2.7) in summation notation and use an equal sign, i.e.,

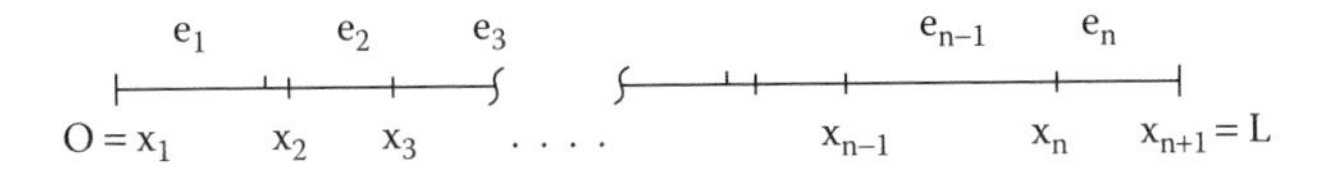

Figure 2.2 Partition of the domain into a finite element grid

$$T(x) = \sum_{i=1}^{n+1} a_i \phi_i(x) \tag{2.8}$$

The function $T(x)$ is chosen so that Eq. (2.3) is always satisfied, although this is not apparent yet. As we incorporate expression (2.8) into the latter steps of the procedure, the reasons for such an approximation will become evident.

As already mentioned, we utilize the Galerkin form of the weighted residuals procedure to formulate the finite element method. The Rayleigh–Ritz procedure is more desirable from the mathematical point of view because it allows us to develop a full mathematical theory of approximation and convergence. However, it has the disadvantage of becoming very difficult (and sometimes impossible) to apply in complex fluid flow and heat transfer problems, especially when advection due to a flow field is encountered. We discuss the Rayleigh–Ritz method further in Chapter 8. Details on the application of the procedure are not given here. These can be found in Huebner et al. (1995), Zienkiewicz and Taylor (1989), and Dawe (1984). The Galerkin method, on the other hand, is simple to implement and is guaranteed to yield a procedure to integrate the governing equations even when the Rayleigh–Ritz method cannot be applied.

When the solution to a problem is approximated using an expression of the form (2.8), in general we cannot obtain the true solution to the differential equation. Consequently, if we replace our approximate solution in the left-hand side of Eq. (2.1), we will not obtain an identity, but some "residual" function associated with the error in the approximation. We can define this error as

$$R(T, x) \equiv -K \frac{d^2T}{dx^2} - Q \tag{2.9}$$

Here T is the approximation to the true solution T^*. It follows that

$$R(T^*, x) \equiv 0 \tag{2.10}$$

However, for any $T \neq T^*$, we cannot force the residual to vanish at every point x, no matter how small we make the grid or long the series expansion. The idea of the weighted residuals method is that we can multiply the residual by a weighting function and force the integral of the weighted expression to vanish, i.e.,

$$\int_0^L W(x) R(T, x) dx = 0 \tag{2.11}$$

where $W(x)$ is the weighting function. Choosing different weighting functions and replacing each of them in (2.11), we can then generate a system of linear equations in the unknown parameters a_i that will determine an approximation T of the form of the finite series given in Eq. (2.8). This will satisfy the differential equation in an "average" or "integral" sense. The type of weighting function chosen depends on the type of weighted residual technique selected. In the Galerkin procedure, the weights are set equal to the shape functions $\phi(x)$, i.e.,

$$W_i(x) = \phi_i(x) \tag{2.12}$$

and because the number of unknown parameters a_j is equal to the number of shape functions ϕ_j, a system of linear algebraic equations with the same number of equations as unknowns will be generated. The existence and uniqueness of the solution to such a system of equations is guaranteed if the boundary conditions associated with the differential equation are correctly posed and imposed, as shown later.

This method turns out to be particularly advantageous in problems with irregular geometries in higher dimensions and nonlinear problems, and automatically yields a system of equations with the same number of equations as unknowns. The question of how to define the shape functions $\phi_i(x)$ is where the technique of finite element methodology really begins. We restrict ourselves almost exclusively to the use of simple (linear, quadratic, and cubic) interpolation; higher-order (and transcendental) approximations can also be used but at the expense of the added complexity, computational time, and storage requirements. The use of elementary linear and quadratic functions will confirm that the finite element concept is elegantly simple and yet extremely powerful.

We now wish to evaluate the left-hand side integrals in Eq. (2.11) using our proposed shape functions $\phi(x)$ as weights $W(x)$. Thus, the Galerkin procedure gives

$$\int_0^L \phi(x)\left[-K\frac{d^2T}{dx^2} - Q\right]dx = 0 \tag{2.13}$$

and we must find an appropriate form for the function $\phi_i(x)$ in Eq. (2.8). Since the temperature distribution must be a continuous function of x, the simplest way to approximate it would be to use piecewise polynomial interpolation over each element; in particular, piecewise linear approximation provides the simplest approximation with a continuous function and is very appealing, see Figure 2.3.

Unfortunately, the first derivatives of such functions are not continuous at the element ends and, hence, second derivatives do not exist there; furthermore, the second derivative of T would vanish inside each element. However, to require the second-order derivatives to exist everywhere is too restrictive. This would prevent us from

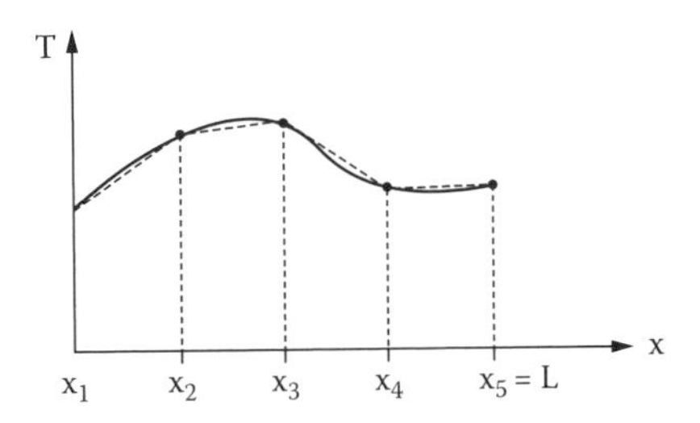

Figure 2.3 Piecewise linear approximation to a temperature field

being able to deal with many physical situations of great interest, such as the presence of a point source of heat of unit strength in the rod. In this case, Eq. (2.1) becomes

$$-K\frac{d^2T}{dx^2} = \delta(x - x_s) \tag{2.14}$$

where δ is the Dirac delta function and is zero everywhere except at $x = x_s$, the location of the source, where it is undetermined. Evidently, the second derivative of T does not exist at $x = x_s$; however, Eq. (2.14) has a well known solution, given by

$$T(x) = \begin{cases} -\frac{1}{K}\left[q(x-L) + (x_s - L)\right] + T_L & 0 \le x \le x_s \\ -\frac{1}{K}q(x-L)(x-L) + T_L & x_s \le x \le L \end{cases} \tag{2.15}$$

Fortunately, this difficulty can be readily resolved by an application of the principle of integration by parts to the second derivative term in Eq. (2.13), i.e.,

$$\int_0^L \phi(x)\left[-K\frac{d^2T}{dx^2}\right]dx = \int_0^L K\frac{d\phi}{dx}\frac{dT}{dx}dx - K\phi\frac{dT}{dx}\Big|_0^L \tag{2.16}$$

and Eq. (2.13) can be rewritten as

$$\int_0^L K\frac{d\phi}{dx}\frac{dT}{dx}dx - \int_0^L \phi Q dx - K\phi\frac{dT}{dx}\Big|_0^L = 0 \tag{2.17}$$

which is a "weak" form of our problem since it contains only the first derivative of the solution $T(x)$, whereas Eq. (2.13) contains the second derivative. The differentiation requirement on the function $T(x)$ has been weakened, hence the name "weak statement." Note that no approximations have been made yet; i.e., nothing has been lost in the formulation. On the other hand, simple piecewise linear approximations are now rendered plausible.

EXAMPLE 2.1

Let us consider Eq. (2.1) with boundary conditions

$$T(0) = T_0$$

and the convective boundary condition

$$-K\frac{dT}{dx}\Big|_{x=L} = h(T - T_\infty)$$

which was previously discussed in Section 2.2. Here, h is the convective heat transfer constant and T_∞ is a reference temperature.

The weighted residuals formulation, Eq. (2.17), remains unchanged for this problem since it was derived using Eq. (2.1) alone and is independent of the given boundary conditions. As is justified later in Chapter 4, whenever the temperature is given at a boundary point, we set the heat flux term at that point equal to zero by requiring that the weighting function vanish there. In this case, we set $\phi(x) = 0$ and, and using the convective boundary condition in the other boundary terms, Eq. (2.17) takes the form:

$$\int_0^L K\frac{d\phi}{dx}\frac{dT}{dx}dx - \int_0^L \phi Q dx + \phi(L)h(T_L - T_\infty) = 0$$

which is the weighted residuals formulation. Later, we further require that the weighting function be equal to one at boundary points where a flux is given. In this case, $\phi(L) = 1$, so that the weighting functions will never appear explicitly in the boundary terms. Let us now return to our basic problem given by Eqs. (2.1) through (2.3).

Utilizing Eq. (2.8) for each weighting function $\phi_i(x)$, we can write Eq. (2.17) in the form:

$$\sum_{j=1}^{n+1} K\left[\int_0^L \frac{d\phi_i}{dx}\frac{d\phi_j}{dx}dx\right]a_j - \int_0^L \phi_i Q dx + \phi_i\left[-K\frac{dT}{dx}\right]\Bigg|_{x=0}^{x=L} = 0 \qquad (2.18)$$

$$i = 1, 2, \ldots, n+1$$

Notice that the flux $-K(dT/dx)$ at $x = 0$ is automatically incorporated into the integral form. Also, the integrals in (2.18) can be readily calculated once the shape functions ϕ_i have been chosen.

The advantage of the Galerkin form (2.18) of the weak statement is that only a finite number of parameters a_i, $i = 1, \ldots, n + 1$, remains to be determined, as opposed to the value of $T(x)$ *at every point* x on $0 \leq x < L$ that is required by Eq. (2.1) or (2.11). At this point then, consideration of the exact solution is abandoned in favor of a compatible approximation.

To illustrate the above procedure, let us divide the interval into two equal-length segments, with a node placed at both ends of each segment. Thus, three nodes define the temperature field throughout the thickness of the slab; i.e., node 1 is placed at $x = x_1 = 0$, node 2 at $x = x_2 = L/2$, and node 3 at $x = x_3 = L$, as shown in Figure 2.4.

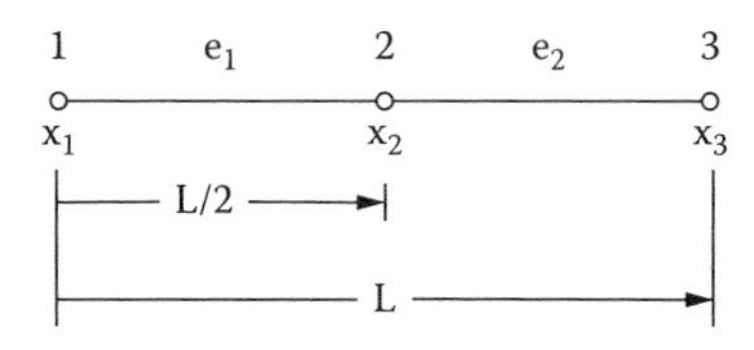

Figure 2.4 Segment discretization

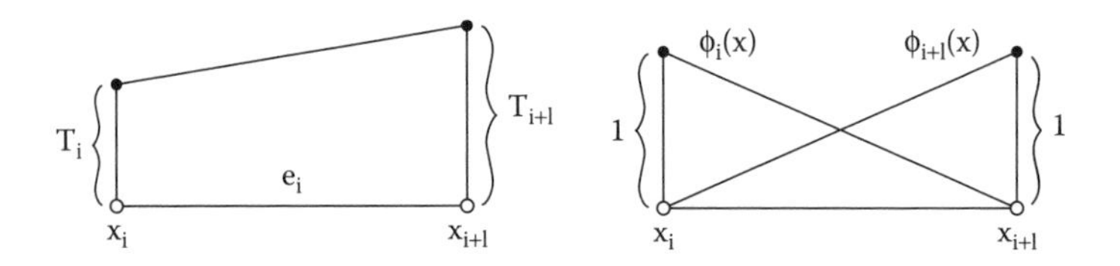

Figure 2.5 Linear shape functions over an element

If we assume that the variation of $\phi(x)$ between nodes is linear over each element, e_i, we can then conveniently express the dependent variable T in the form:

$$T(x) = \phi_i(x)a_i + \phi_{i+1}(x)a_{i+1} \qquad x_i \le x \le x_{i+1} \tag{2.19}$$

If the functions ϕ_1; are chosen so that $\phi_1(x_i) = 1$ and $\phi_i(x_{i+1}) = 0$, and vice versa $\phi_{i+1}(x_i) = 0$ and $\phi_{i+1}(x_{i+1}) = 1$, then the functions ϕ_i are given by

$$\begin{aligned} \phi_i(x) &= \frac{x_{i+1} - x}{x_{i+1} - x_i} \\ \phi_{i+1}(x) &= \frac{x - x_i}{x_{i+1} - x_i} \end{aligned} \tag{2.20}$$

and the parametric a_i become the nodal temperatures, i.e., $a_i = T(x_i) = T_i$, as depicted in Figure 2.5. The derivatives of the shape functions are

$$\begin{aligned} \frac{d\phi_i}{dx} &= -\frac{1}{x_{i+1} - x_i} \\ \frac{d\phi_{i+1}}{dx} &= \frac{1}{x_{i+1} - x_i} \end{aligned} \tag{2.21}$$

Equations (2.19) through (2.21) can be more succinctly written in matrix form as

$$T(x) = \varphi \mathbf{a} \tag{2.22}$$

where

$$\varphi = [\phi_i \quad \phi_{i+1}] \tag{2.23}$$

$$\mathbf{a} = \begin{bmatrix} a_i \\ a_{a+1} \end{bmatrix} \tag{2.24}$$

and hence

$$\frac{dT}{dx} = \frac{d}{dx}\varphi \mathbf{a} = \begin{bmatrix} \frac{d\phi_i}{dx} & \frac{d\phi_{i+1}}{dx} \end{bmatrix} \begin{bmatrix} a_i \\ a_{i+1} \end{bmatrix} \tag{2.25}$$

where

$$\frac{d}{dx}\varphi = \left[-\frac{1}{x_{i+1} - x_i} \quad \frac{1}{x_{i+1} - x_i} \right] \tag{2.26}$$

We can now apply this to our first element in Eq. (2.18),

$$\begin{aligned} &\sum_{j=1}^{2} K \int_0^{L/2} \left[\frac{d\phi_1}{dx} \frac{d\phi_j}{dx} dx \right] a_j - \int_0^{L/2} \phi_1(x) Q \, dx - K\phi_1(x) \frac{dt}{dx} \Big|_{x=0}^{x=L/2} = 0 \\ &\sum_{j=1}^{2} K \int_0^{L/2} \left[\frac{d\phi_2}{dx} \frac{d\phi_j}{dx} dx \right] a_j - \int_0^{L/2} \phi_2(x) Q \, dx - K\phi_2(x) \frac{dt}{dx} \Big|_{x=0}^{x=L/2} = 0 \end{aligned} \tag{2.27}$$

where these relations have been set to zero for convenience. Equations (2.27) can be written using (2.22) through (2.26) as

$$K \left[\int_0^{L/2} \left[\frac{d}{dx}\varphi \right]^T \left[\frac{d}{dx}\varphi \right] dx \right] \mathbf{a} - Q \int_0^{L/2} \varphi^T dx - \begin{bmatrix} q \\ 0 \end{bmatrix} = 0 \tag{2.28*}$$

Replacing

$$\varphi = \left[1 - \frac{2x}{L} \quad \frac{2x}{L} \right]$$

and

$$\frac{d}{dx}\varphi = \left[-\frac{2}{L} \quad \frac{2}{L} \right]$$

and integrating, Eq. (2.28) becomes

$$\frac{2K}{L} \begin{bmatrix} 1 & -1 \\ -1 & 1 \end{bmatrix} \begin{bmatrix} a_1 \\ a_2 \end{bmatrix} - \frac{QL}{4} \begin{bmatrix} 1 \\ 1 \end{bmatrix} - \begin{bmatrix} q \\ 0 \end{bmatrix} = \begin{bmatrix} 0 \\ 0 \end{bmatrix} \tag{2.29}$$

Performing a similar evaluation for element 2 (Exercise 2.3), we obtain

$$\frac{2K}{L} \begin{bmatrix} 1 & -1 \\ -1 & 1 \end{bmatrix} \begin{bmatrix} a_2 \\ a_3 \end{bmatrix} - \frac{QL}{4} \begin{bmatrix} 1 \\ 1 \end{bmatrix} - \begin{bmatrix} 0 \\ 0 \end{bmatrix} = \begin{bmatrix} 0 \\ 0 \end{bmatrix} \tag{2.30}$$

* It is not readily apparent why the second component of the vector $\begin{bmatrix} q \\ 0 \end{bmatrix}$ is zero, since $\phi_2(L/2) = 1$. This is clarified later. We can, for now, assume that if no flux is specified at a node, then it is zero at that node.

Expressions (2.29) and (2.30) are then combined using the "assembly" process, which, in this case, consists of adding up the second equation in Eq. (2.29) and the first equation in Eq. (2.30), since both correspond to the same weighting function $\phi_2(x)$. This process, described in the next chapter, yields the system

$$\frac{2K}{L}\begin{bmatrix} 1 & -1 & 0 \\ -1 & 2 & -1 \\ 0 & -1 & 1 \end{bmatrix}\begin{bmatrix} a_1 \\ a_2 \\ a_3 \end{bmatrix} = \begin{bmatrix} q \\ 0 \\ 0 \end{bmatrix} + \frac{QL}{4}\begin{bmatrix} 1 \\ 2 \\ 1 \end{bmatrix} \tag{2.31}$$

Because we know that $a_3 = T_L$, the system reduces to two equations in the unknowns a_1 and a_2, i.e.,

$$\begin{aligned} a_1 - a_2 &= \frac{qL}{2K} + \frac{QL^2}{8K} \\ -a_1 + a_2 &= \frac{QL^2}{4K} + T_L \end{aligned} \tag{2.32}$$

and the solution is

$$\begin{aligned} a_1 &= \frac{qL}{K} + \frac{QL^2}{2K} + T_L \\ a_2 &= \frac{qL}{2K} + \frac{3QL^2}{8K} + T_L \\ a_3 &= T_L \end{aligned} \tag{2.33}$$

EXAMPLE 2.2

There are many possibilities from which to choose the shape functions $\phi_i(x)$ in Eq. (2.7). Let us consider the solution of the differential equation

$$-\frac{d^2u}{dx^2} = 1, \qquad 0 < x < 1$$

with boundary conditions

$$u(0) = u(1) = 0$$

We can try, for example, an approximation of the form

$$u(x) = a_1 \sin \pi x$$

which satisfies the prescribed boundary conditions. The Galerkin weighted residuals formulation becomes

$$\pi^2 \int_0^1 (\cos \pi x)^2 dx \, a_1 - \int_0^1 \sin \pi x dx = 0$$

This results in

$$a_1 = \frac{4}{\pi^3}$$

At $x = 1/2$, the exact solution is $u(1/2) = 1/8 = 0.125$. The approximation obtained above, on the other hand, gives $u(1/2) \sim 0.129$, or approximately 3% error.

2.4 CLOSURE

The underlying principle of the finite element method resides in the method of weighted residuals. The two most commonly used procedures are the Rayleigh–Ritz and Galerkin methods. The Rayleigh–Ritz method is based on calculus of variations; however, the method is difficult to use on complicated equations. The Galerkin method is simple to use and is guaranteed to yield a compatible approximation to the governing differential equation.

In the Galerkin method, the dependent variable is expressed by means of a finite series approximation in which the "shape" of the solution is assumed known, and it depends on a finite number of parameters to be determined. Replaced in the governing differential equation, the approximation generates a residual function, which is multiplied by weighting functions and is required to be orthogonal to the weighting functions in the integrated sense, i.e.,

$$\int W(x) R(T, x) dx = 0$$

where $R(T, x)$ is the residual error function (the function obtained when the approximation to the exact solution T^* is replaced in the differential equation) and $W(x)$ is the weight. A set of linear algebraic equations can be generated from these expressions that allows us to determine the unknown parameters and hence an approximation to the solution. Reduction of second derivative terms using integration-by-parts yields the "weak" statement formulation. Application of the weak statement formulation produces a general algorithm extendable to a wide class of problems.

EXERCISES

2.1 Fill in the details leading to Eq. (2.17) and use it to find the weighted residuals formulation of Example 2.1.

2.2 Use Eq. (2.17) to find the weighted residuals formulation of Eqs. (2.1) through (2.3) in terms of the heat flux q.

2.3 Obtain expression (2.30).

2.4 Consider the equation:

$$\frac{d^2u}{dx^2} + u + x = 0, \qquad 0 < x < 1$$

with $u(0) = u(1) = 0$. Assume an approximation for $u(x) = a_1\phi_1(x)$ with $\phi_1(x) = x(1 - x)$. The residual is

$$R(u, x) = -2a_1 + a_1[x(1 - x)] + x$$

and the weight $W_1(x) = \phi_1(x) = x(1 - x)$. Find the solution from the following integral $\int_0^1 W(x)R(u, x)dx = 0$

2.5 Use the weak statement formulation to find solutions to the equation

$$\frac{d^2u}{dx^2} = 1, \qquad 0 < x < 1$$

$$u(0) = 0, \qquad \frac{du}{dx}\bigg|_{x=1} = 1$$

(a) Using linear interpolation functions, as explained in Section 2.3, use (i) one element and (ii) two elements.

(b) Using simple polynomial functions that satisfy the boundary conditions at the left-hand side, i.e.,
(i) $u(x) = a_1x$, hence $\phi_1(x) = W_1(x) = x$
(ii) $u(x) = a_1x + a_2x^2$, hence $\phi_1(x) = W_1(x) = x$, $\phi_2(x) = W_2(x) = x^2$.

(c) Using circular functions that satisfy the boundary conditions at the left-hand side, i.e.,
(i) $u(x) = a_1 \sin(\pi/2Lx)$
(ii) $u(x) = a_1 \sin(\pi/2Lx) + a_2 \sin(3\pi/2Lx)$

(d) Find the analytical solution and compare with results from (a), (b), and (c).

2.6 Subdivide the interval $0 \le x \le L$ into 10 linear elements, construct the element equations for four consecutive elements starting with element four, and show that none of the parameters a_i is related to more than two additional parameters in the equations. The only ones involved being a_{i-1} and a_{i+1} (as a consequence of this, the final system of linear equations will involve a tri-diagonal matrix, which is easy to solve).

2.7 The Rayleigh–Ritz formulation for the problem of solving Eq. (2.1) with boundary conditions (2.2) and (2.3) can be stated as: Minimize the functional

$$F(T) = \int_0^L \left\{ \frac{K}{2}\left[\frac{dT}{dx}\right]^2 - QT \right\} dx - Tq\,\big|_{x=0}$$

over all functions $T(x)$ with square integrable first derivatives that satisfy Eq. (2.3) at $x = L$. Approximate $T(x)$ using two linear elements as we did before and replace into Eq. (2.34) to obtain $F(T) \cong F(a_1, a_2, a_3)$. Now minimize $F(a_1, a_2, a_3)$ as a function of three variables. Show that the final system of equations is identical to Eq. (2.31); thus, in this case, the Galerkin method and the Rayleigh–Ritz method are equivalent.

2.8 Let us consider once more Exercise 2.3 above. Assume that $u(x) = a_1 x$ and $W_1(x) = x$, but this time use Eq. (2.11) and proceed as in Exercise 2.2 to obtain a solution. Show that this solution is identical to that obtained in Exercise 2.3(b)(ii), and hence, the process of integration by parts does not introduce any changes.

2.9 In Example 2.2, first find the exact analytical solution; then calculate two more Galerkin approximations using

(a) $u(x) = a_1 \sin \pi x + a_2 \sin 2\pi x$

(b) $u(x) = a_1 \sin \pi x + a_2 \sin 2\pi x + a_3 \sin 3\pi x$

Sketch the three different approximations for a few points and discuss the results.

2.10 Derive Eq. (2.32) from Eq. (2.31).

2.11 Fill in the details in Example 2.2 and verify the solution.

2.12 Show that if the expressions for (i) given by Eq. (2.20) are used in Eq. (2.19), then $a_i = T_i$.

2.13 Find the weak formulation of Eq. (2.1) with the boundary condition

$$-k\left.\frac{dT}{D_x}\right|_{x=0} = h\left(T - T_\infty\right) \text{ and -k} \left.\frac{dT}{D_x}\right|_{x=1} = q$$

2.14 Using two linear elements solve the equations in Exercise 2.13 with $L = 1$ m, $k = 200$ w/mK, $Q = 100$ W/m^3, $h = 150$ W/m^2 K, $T_\infty = 100$°C.

REFERENCES

Dawe, D. J. (1984). *Matrix and Finite Element Displacement Analysis of Structures*. Oxford: Clarendon Press.

Fletcher, C. A. J. (1984). *Computational Galerkin Methods*. Berlin: Springer-Verlag.

Huebner, K. H., Thornton, E. A., and Byrom, T. G. (1995). *The Finite Element Method for Engineers*, 3rd ed. New York: Wiley-Interscience.

Reddy, J. N. (1984). *An Introduction to the Finite Element Method*. New York: McGraw-Hill.

Zienkiewicz, O. C. and Taylor, R. L. (1989). *The Finite Element Method*, 4th ed. Maidenhead, U.K.: McGraw-Hill.

CHAPTER

3

THE FINITE ELEMENT METHOD IN ONE DIMENSION

3.1 BACKGROUND

The finite element method has its formal basis in the Galerkin procedure of weighted residuals. As discussed in Chapter 2, the implementation of the Galerkin procedure, coupled with a piecewise polynomial representation for the dependent variable, yields a set of algebraic relations obtained over a series of segments that discretize the problem domain. The approximation functions are defined locally over the segments, called *finite elements*, and any (every!) function, no matter how simple, can be used in the approximation, even a polynomial of first degree.

The first step in the finite element method is to divide the region of interest into subregions, or elements, which contain nodal points, and define the type of "shape" at each end of an element. A quadratic element has a third node placed at the midpoint of an element; a cubic element consists of four nodes, spaced at intervals. The collection of elements and nodal points is termed the "finite element mesh." The type of shape functions used in the procedure is directly linked to the type of mesh chosen. We begin our discussion with the various forms of the shape functions.

3.2 SHAPE FUNCTIONS

In Chapter 2, we dealt with piecewise polynomial shape functions $\phi_i(x)$ for the one-dimensional heat conduction problem, although in a rather loose way. Now we formalize the process of piecewise polynomial interpolation and show that it is more convenient to work with individual elements and local functions applicable only to that particular element.

3.2.1 Linear Elements

We begin defining a grid, in $0 \le x \le L$, of elements that are not necessarily of equal size. Over each of the elements we then define linear shape functions, as was done in Eq. (2.20); these appear as depicted in Figure 3.1.

This is the *global* representation of the finite element grid and shape functions in a coordinate system that allows us to describe the geometry of our problem. If the mesh consists of n elements, it has $n+1$ nodes with coordinates $x_1, ..., x_{n+1}$, as shown in Figure 3.1. The element domains are given by

$$e_i \; : \{x | x_i \le x \le x_{i+1}\} \qquad i = 1, 2, ..., n \tag{3.1}$$

The shape functions associated with each node are denoted by $N_i(x)$ such that $N_i(x_i) = 1$ and $N_i(x_j) = 0$ if $j \ne i$, and are given by

$$N_1(x) = \begin{cases} \dfrac{x_2 - x}{h_1} & x_1 \le x \le x_2 \\ 0 & \text{otherwise} \end{cases} \tag{3.2}$$

$$N_i(x) = \begin{cases} \dfrac{x - x_{i-1}}{h_{i-1}} & x_{i-1} \le x \le x_i \\ \dfrac{x_{i+1} - x}{h_i} & x_i \le x \le x_{i+1} \qquad i = 2, 3, ..., n \\ 0 & \text{otherwise} \end{cases} \tag{3.3}$$

$$N_{n+1}(x) = \begin{cases} \dfrac{x - x_n}{h_n} & x_n \le x \le x_{n+1} \\ 0 & \text{otherwise} \end{cases} \tag{3.4}$$

where

$$h_i = x_{i+1} - x_i \qquad i = 1, 2, ..., n \tag{3.5}$$

We can now define $T(x)$ as

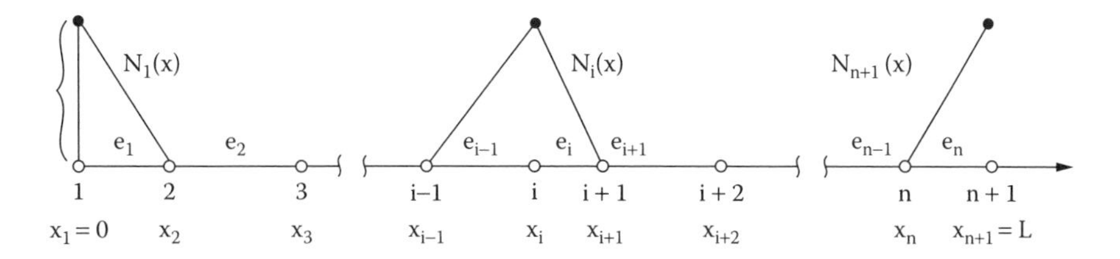

Figure 3.1 Shape functions in a general grid of piecewise linear elements

$$T\left(x\right) = N_1(x)T_1 + N_2(x)T_2 + \ldots + N_{n+1}(x)T_{n+1} = \sum_{i=1}^{n+1} N_i(x)T_i \tag{3.6}$$

where T_i denotes the value of T at x_i, i.e., at node i, and $T(x)$ is linear in between nodes. This can be simply observed for a one-element mesh, i.e., $n = 1$. If $T(x)$ is linear between the nodes, it will be of the form $T(x) = \alpha_1 + \alpha_2 x$. Furthermore, $T(x_1) = T_1$ and $T(x_2) = T_2$; hence

$$T_1 = \alpha_1 + \alpha_2 x_1 \tag{3.7}$$

and

$$T_2 = \alpha_1 + \alpha_2 x_2 \tag{3.8}$$

Solving for α_1 and α_2

$$\alpha_1 = \frac{T_1 x_2 - T_2 x_1}{h_1} \tag{3.9}$$

$$\alpha_2 = \frac{T_2 - T_1}{h_1} \tag{3.10}$$

Thus,

$$T(x) = \frac{T_1 x_2 - T_2 x_1}{h_1} + \frac{T_2 - T_1}{h_1} x \tag{3.11}$$

and, rearranging, we get

$$T(x) = \left[\frac{x_2 - x}{h_1}\right] T_1 + \left[\frac{x - x_1}{h_1}\right] T_2 \tag{3.12}$$

which is the same expression as Eq. (3.6) with $n = 1$. Here we have used a different notation than in Eq. (2.8). In fact, comparing Eq. (2.8) with Eq. (3.6) we have $\phi_i(x) = N_i(x)$ and $\alpha_i = T_i$. Also, a boundary condition of the form (2.3) is imposed simply by replacing the parameters, in this case T_{n+1}, by the given value of the temperature at the node.

It will be convenient in practice to work with each element in a *local* coordinate system associated with the particular element under consideration. If we isolate an element of length $h^{(e)}$ in the local coordinate system, we will use the following notation:

$$T^{(e)}(x) = N_1^{(e)}(x)T_1^{(e)} + N_2^{(e)}(x)T_2^{(e)} \tag{3.13}$$

where (e) refers to an element, and

$$N_1^{(e)}(x) = \left[1 - \frac{x}{h^{(e)}}\right] \tag{3.14}$$

$$N_2^{(e)}(x) = \frac{x}{h^{(e)}} \tag{3.15}$$

as shown in Figure 3.2. It follows that

$$\frac{dT^{(e)}}{dx} = \begin{bmatrix} \dfrac{dN_1^{(e)}}{dx} & \dfrac{dN_2^{(e)}}{dx} \end{bmatrix} \begin{bmatrix} T_1^{(e)} \\ T_2^{(e)} \end{bmatrix} = \begin{bmatrix} -\dfrac{1}{h^{(e)}} & \dfrac{1}{h^{(e)}} \end{bmatrix} \begin{bmatrix} T_1^{(e)} \\ T_2^{(e)} \end{bmatrix} \tag{3.16}$$

We will also need a relation between the local and global node numbering, as illustrated in Figure 3.3. But before we get further in the solution process, we introduce some higher-order elements.

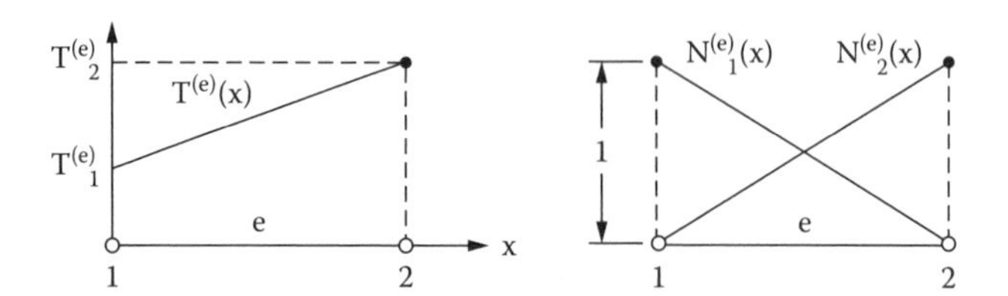

Figure 3.2 Local node numbering, temperature representation, and local shape functions

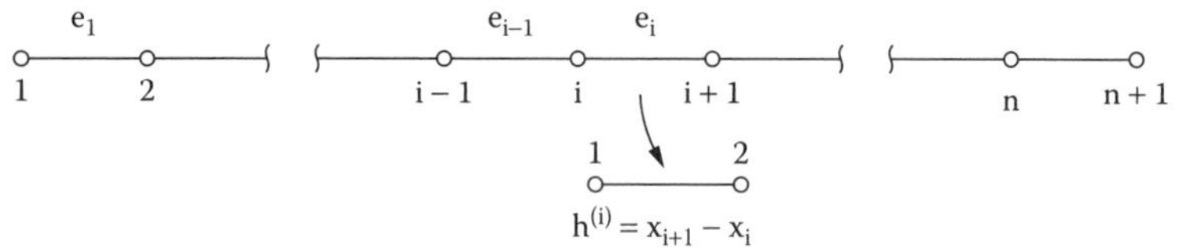

Figure 3.3 Relation between local and global numbering

3.2.2 Quadratic Elements

From the point of view of approximating a function by interpolation, we can see from Figure 3.4 that we could do better if we use parabolic arcs over each element rather than linear segments. Over each element, the functions $T^{(e)}(x)$ will be quadratic in x and therefore of the form

$$T^{(e)}(x) = \alpha_1 + \alpha_2 x + \alpha_3 x^2 \tag{3.17}$$

where we have three parameters to determine; hence, to require interpolation at the ends of the elements does not suffice to determine the function. To obtain a third relation, we introduce one more node in the middle of the element, and we also require interpolation of the function at that node. Figure 3.5 shows the resulting configuration for the local coordinate system.

The functions $N_i^{(e)}(x)$ are obtained from the requirement that

$$\left.\begin{aligned} T^{(e)}(0) &= \alpha_1 = T_1^{(e)} \\ T^{(e)}(h/2) &= \alpha_1 + \frac{\alpha_1 h^{(e)}}{2} + \frac{\alpha_3 (h^{(e)})^2}{4} = T_2^{(e)} \\ T^{(e)}(h) &= \alpha_1 + \alpha_2 h^{(e)} + \alpha_3 (h^{(e)})^2 = T_3^{(e)} \end{aligned}\right\} \tag{3.18}$$

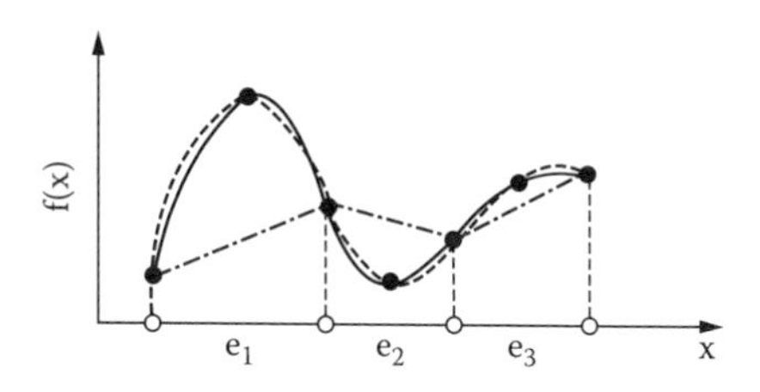

Figure 3.4 Piecewise linear vs. piecewise quadratic interpolation

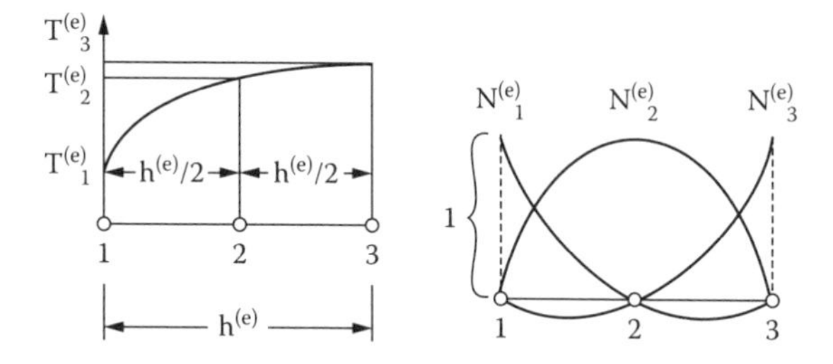

Figure 3.5 Quadratic element and shape functions in local coordinate system

The solution of this system of equations is given by

$$\alpha_1 = T_1^{(e)}$$

$$\alpha_2 = \frac{1}{h^{(e)}}\left(-3T_1 + 4T_2 - T_3\right)$$

$$\alpha_3 = \frac{2}{\left(h^{(e)}\right)^2}\left(T_1 - 2T_2 + T_3\right)$$

Substituting into (3.17) and rearranging, we obtain

$$T^{(e)}(x) = \left[1 - 3\left[\frac{x}{h^{(e)}}\right] + 2\left[\frac{x}{h^{(e)}}\right]^2\right]T_1 + 4\left[\frac{x}{h^{(e)}}\right]\left[1 - \frac{x}{h^{(e)}}\right]T_2$$

$$+\left[\frac{x}{h^{(e)}}\right]\left[2\left[\frac{x}{h^{(e)}}\right] - 1\right]T_3$$

Hence, the shape functions are

$$\left.\begin{aligned} N_1^{(e)}(x) &= 1 - 3\frac{x}{h^{(e)}} + 2\left[\frac{x}{h^{(e)}}\right]^2 \\ N_2^{(e)}(x) &= 4\frac{x}{h^{(e)}}\left[1 - \frac{x}{h^{(e)}}\right] \\ N_3^{(e)}(x) &= \frac{x}{h^{(e)}}\left[2\frac{x}{h^{(e)}} - 1\right] \end{aligned}\right\} \tag{3.19}$$

The notation used in this case to relate the global system to the local system of reference is shown in Figure 3.6 for a mesh consisting of n quadratic elements. The ith element in this case is defined as (from Eq. 3.1)

$$e_i = \{x \mid x_{2i-1} \le x \le x_{2i+1}\} \tag{3.20}$$

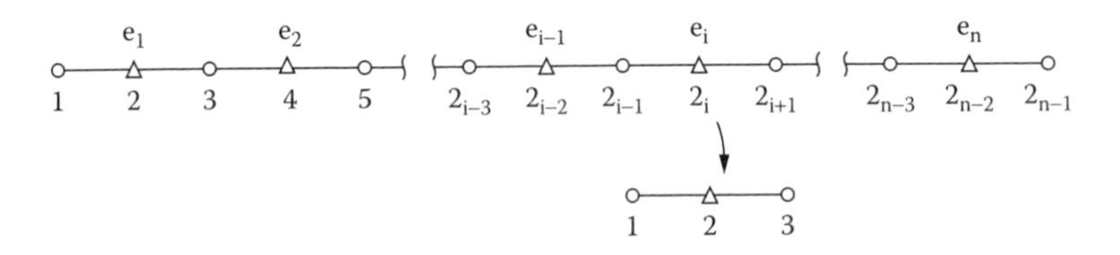

Figure 3.6 Relation between local and global coordinate systems. The triangles denote nodes interior to each element

and the element length is given by

$$h^{(i)} = x_{2i+1} - x_{2i-1} \tag{3.21}$$

The derivatives of the function $T^{(e)}(x)$ can now be obtained from

$$T^{(e)}(x) = N_1^{(e)}T_1^{(e)} + N_2^{(e)}(x)T_2^{(e)} + N_3^{(e)}(x)T_3^{(e)} = \left[N_1^{(e)}\ N_2^{(e)}\ N_3^{(e)}\right]\begin{bmatrix} T_1^{(e)} \\ T_2^{(e)} \\ T_3^{(e)} \end{bmatrix} \tag{3.22}$$

and are given by

$$\frac{dT^{(e)}}{dx} = \left[\frac{1}{h^{(e)}}\left(4\frac{x}{h^{(e)}} - 3\right)\ \ \frac{4}{h^{(e)}}\left(1 - 2\frac{x}{h^{(e)}}\right)\ \ \frac{1}{h^{(e)}}\left(4\frac{x}{h^{(e)}} - 1\right)\right]\begin{bmatrix} T_1^{(e)} \\ T_2^{(e)} \\ T_3^{(e)} \end{bmatrix} \tag{3.23}$$

which are no longer constant over the element.

3.2.3 Cubic Elements

We can proceed to even higher orders of approximation. The next level is given by the cubic functions. In this case, we have

$$T^{(e)}(x) = \alpha_1 + \alpha_2 x + \alpha_3 x^2 + \alpha_4 x^3 \tag{3.24}$$

where four nodes are required on each element, located equally spaced at $x = 0$, $h^{(e)}/3$, $2h^{(e)}/3$, and $h^{(e)}$. In this element, the shape functions are given by

$$\left.\begin{aligned} N_1^{(e)}(x) &= \left(1 - \frac{3x}{h^{(e)}}\right)\left(1 - \frac{3x}{2h^{(e)}}\right)\left(1 - \frac{x}{h^{(e)}}\right) \\ N_2^{(e)}(x) &= \frac{9x}{h^{(e)}}\left(1 - \frac{3x}{2h^{(e)}}\right)\left(1 - \frac{x}{h^{(e)}}\right) \\ N_3^{(e)}(x) &= -\frac{9x}{2h^{(e)}}\left(1 - \frac{3x}{h^{(e)}}\right)\left(1 - \frac{x}{h^{(e)}}\right) \\ N_4^{(e)}(x) &= \frac{x}{h^{(e)}}\left(1 - \frac{3x}{h^{(e)}}\right)\left(1 - \frac{3x}{2h^{(e)}}\right) \end{aligned}\right\} \tag{3.25}$$

The details on this element will be left to Exercise 3.2.

The following points should be noticed with regard to the elements defined above:

1. Even though the derivatives of quadratic and cubic elements are functions of the independent variable x, they will not be continuous at the inter-element nodes. An example of this is given by Exercise 3.5. The type of interpolation used here is known as Lagrangian, and it only guarantees continuity of the function across inter-element boundaries. The elements are known as C^0 elements, where the zero superindex means that only derivatives of order zero are continuous, i.e., the function.
2. Evidently, even higher-order elements, e.g., quartics, quintics, etc., can be constructed by adding more interpolation nodes in an element. Actually, we can construct elements that also interpolate derivatives at the nodes. The simplest such elements are the cubic Hermites that interpolate the function and its first derivative at the two nodes located at the ends of the element. These are C^1 elements because the first derivative will now be continuous everywhere in the domain. Even more sophisticated elements can be constructed. In fact, there is virtually no limitation to the degree of complexity or predetermined element behavior that can be attained. However, one has to keep in mind that the more sophisticated the element, the more computationally expensive it will be. In fact, cubic elements already become prohibitively expensive in multidimensional calculations and are seldom used.
3. The element interpolation functions considered above have the property that $N_i^{(e)}(x_j) = \delta_{ij}$, where δ_{ij} is the Kronecker delta function, i.e.,

$$\delta_{ij} = \begin{cases} 1 & \text{if } i = j \\ 0 & \text{if } i \neq j \end{cases} \tag{3.26}$$

   and x_j are nodal coordinates. This adds the convenience that when expression (3.6) is evaluated at nodal points, the value of the dependent function at that point is obtained.
4. The type of interpolation chosen defines the "shape" that the dependent variable can take within an element, i.e., linear, quadratic, etc. Therefore, the name "shape function" is used to denote the functions N_i that define the element. It should also be noted that higher-order functions always reduce exactly to the lower-order ones; i.e., quadratic elements exactly represent linear and constant functions, cubic elements represent quadratic, linear, and constant functions, etc. This fact guarantees that a better approximation can be obtained if higher-order elements are used. It is left to Exercise 3.4 to show this property of shape functions.

3.3 STEADY CONDUCTION EQUATION

3.3.1 Galerkin Formulation

Consider the problem of finding the temperature distribution $T = T(x)$ on the interval $0 \le x \le L$ that satisfies the steady one-dimensional equation with internal heat source:

$$-K\frac{d^2T}{dx^2} = Q \qquad 0 < x < L \tag{3.27}$$

with

$$-K\frac{dT}{dx} = q \qquad at\ x = 0 \tag{3.28}$$

and

$$T = T_L \qquad at\ x = L \tag{3.29}$$

Equations (3.27) to (3.29) are identical to Eqs. (2.1) to (2.3) used to describe heat conduction in a rod in Chapter 2.

We introduce the basic concepts using a discretization consisting of two linear elements of equal length, thus leading to a set of simultaneous equations containing three nodal values T_1, T_2, and T_3. A weighted residuals expression for (3.27) gives

$$\int_0^L W\left(-K\frac{d^2T}{dx^2} - Q\right)dx = 0 \tag{3.30}$$

Using the additive property of the integral and two elements, expression (3.30) can be written as

$$\sum_{i=1}^{2}\int_{x_i}^{x_{i+1}} W\left(-K\frac{d^2T}{dx^2} - Q\right)dx = \int_0^{L/2} W\left(-K\frac{d^2T}{dx^2} - Q\right)dx + \int_{L/2}^{L} W\left(-K\frac{d^2T}{dx^2} - Q\right)dx = 0 \tag{3.31}$$

The function $T(x)$ in the global system will be approximated by

$$T(x) = \sum_{j=1}^{3} N_j(x)T_j = N_1(x)T_1 + N_2(x)T_2 + N_3(x)T_3 \tag{3.32}$$

where the shape functions $N_i(x), i = 1,2,3,$ are given by expressions (3.2), (3.3), and (3.4), respectively, with n = 2 and $x_1 = 0, x_2 = L/2,$ and $x_3 = L.$

The Galerkin form of (3.31) becomes, with $W_i(x) = N_i(x)$,

$$\int_0^{L/2}\left[K\frac{dN_i}{dx}\left(\sum_{j=1}^{3}\frac{dN_j}{dx}T_j\right)-N_iQ\right]dx+\left[N_i\left(-K\frac{dT}{dx}\right)\right]_0^{L/2}$$

$$+\int_{L/2}^{L}\left[K\frac{dN_i}{dx}\left(\sum_{j=1}^{3}\frac{dN_j}{dx}T_j\right)-N_iQ\right]dx+\left[N_i\left(-K\frac{dT}{dx}\right)\right]_{L/2}^{L}=0 \quad i=1,2,3 \tag{3.33}$$

The first two terms in Eq. (3.33) correspond to element e_1 and the last two to e_2. Let us now consider each of these, one at a time, using local element coordinates. For element e_1, $N_3(x) = 0$; hence we can write the equations arising from $i = 1,2$ in matrix form as

$$\int_0^{L/2}\left\{K\left(\begin{bmatrix}\dfrac{dN_1^{(e_1)}}{dx}\\[2ex]\dfrac{dN_2^{(e_1)}}{dx}\end{bmatrix}\begin{bmatrix}\dfrac{dN_1^{(e_1)}}{dx} & \dfrac{dN_2^{(e_1)}}{dx}\end{bmatrix}\right)\begin{bmatrix}T_1^{(e_1)}\\ T_2^{(e_1)}\end{bmatrix}-\begin{bmatrix}N_1^{(e_1)}\\ N_2^{(e_1)}\end{bmatrix}Q\right\}dx \tag{3.34}$$

$$+\left\{\begin{bmatrix}N_1^{(e_1)}\\ N_2^{(e_1)}\end{bmatrix}\left(-K\frac{dT}{dx}\right)\right\}_0^{L/2}=\begin{bmatrix}0\\0\end{bmatrix}$$

Using Eqs. (3.14) through (3.16) and the fact that $h^{(e_1)} = L/2$, Eq. (3.34) can be written as

$$\int_0^{L/2}\left\{K\begin{bmatrix}\dfrac{4}{L^2} & -\dfrac{4}{L^2}\\[2ex] -\dfrac{4}{L^2} & \dfrac{4}{L^2}\end{bmatrix}\begin{bmatrix}T_1^{(e_1)}\\ T_2^{(e_1)}\end{bmatrix}-\begin{bmatrix}1-\dfrac{2x}{L}\\[2ex]\dfrac{2x}{L}\end{bmatrix}Q\right\}dx+\begin{bmatrix}-\left(-K\dfrac{dT}{dx}\right)_{x=0}\\[2ex]\left(-K\dfrac{dT}{dx}\right)_{x=L/2}\end{bmatrix}=\begin{bmatrix}0\\0\end{bmatrix}$$

Using (3.28) and integrating, we finally obtain the element equations for e_1 in the form

$$\frac{2K}{L}\begin{bmatrix}1 & -1\\ -1 & 1\end{bmatrix}\begin{bmatrix}T_1\\ T_2\end{bmatrix}=\frac{QL}{4}\begin{bmatrix}1\\1\end{bmatrix}+\begin{bmatrix}q\\[1ex] -\left(-K\dfrac{dT}{dx}\right)_{x=L/2}\end{bmatrix} \tag{3.35}$$

where we have used the fact that the global numbers of the degrees of freedom in element e_1 and T_1 and T_2. In a similar way (Exercise 3.5), we obtain the equations for element e_2 as

$$\frac{2K}{L}\begin{bmatrix} 1 & -1 \\ -1 & 1 \end{bmatrix}\begin{bmatrix} T_2 \\ T_3 \end{bmatrix} = \frac{QL}{4}\begin{bmatrix} 1 \\ 1 \end{bmatrix} + \begin{bmatrix} \left(-K\dfrac{dT}{dx}\right)_{x=L/2} \\ -\left(-K\dfrac{dT}{dx}\right)_{x=L} \end{bmatrix} \tag{3.36}$$

The element equations must now be assembled into global 3 × 3 matrix. This is done using global numbering of the equations and degrees of freedom; the process is illustrated in Figure 3.7 and is equivalent to adding equations corresponding to the same weighting function in different elements. In this case, the contributions over each element corresponding to $W_2 = N_2$ are added up as shown in the figure. In practice this is achieved by using the global degree of freedom numbers in the element matrices and adding their contributions to the corresponding locations in the global matrix. Thus, for example, in element 2, the location (1,2) corresponds to (2,3) in the global system and the entry $-2k/L$ is added to the position (2,3) in the assembled matrix. For another interpretation of the assembly process see Section 3.7.

It can be clearly seen in Figure 3.7 that the expressions involving the fluxes at the interior nodes cancel out; in fact, this just states that fluxes must be continuous in the interior. For this reason, these terms are always omitted when constructing the element equations. However, one must be careful that if the terms are omitted, the equality does not hold in the element equations. It is, nevertheless, customary to set the term involving the fluxes, e.g., Eq. (3.6), equal to zero. The global system of equations thus becomes

$$\frac{2K}{L}\begin{bmatrix} 1 & -1 & 0 \\ -1 & 2 & -1 \\ 0 & -1 & 1 \end{bmatrix}\begin{bmatrix} T_1 \\ T_2 \\ T_3 \end{bmatrix} = \frac{Ql}{4}\begin{bmatrix} 1 \\ 2 \\ 1 \end{bmatrix} + \begin{bmatrix} q \\ 0 \\ -\left(-K\dfrac{dT}{dx}\right)_{x=L} \end{bmatrix} \tag{3.37}$$

In the last equation, the heat flux at $x = L$ appears. However, if we think of this as an equation for T_3, since T_3 is known, this equation can be discarded and the system rewritten as

$$\frac{2K}{L}\begin{bmatrix} 1 & -1 \\ -1 & 2 \end{bmatrix}\begin{bmatrix} T_1 \\ T_2 \end{bmatrix} = \frac{QL}{4}\begin{bmatrix} 1 \\ 2 \end{bmatrix} + \begin{bmatrix} q \\ 0 \end{bmatrix} + \frac{2K}{L}T_L\begin{bmatrix} 0 \\ 1 \end{bmatrix} \tag{3.38}$$

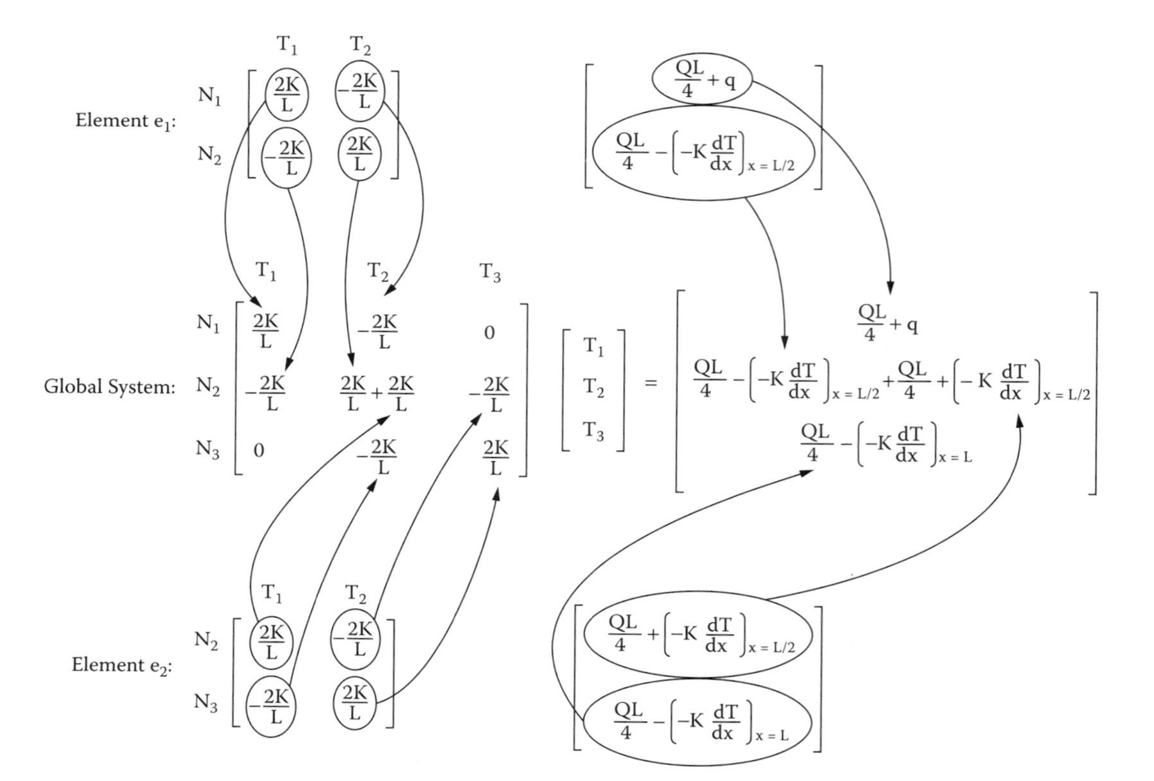

Figure 3.7 Schematic of assembly process for two-element system

and (3.38) can now be solved for T_1 and T_2. Once this has been done, the third equation can be used to calculate the heat transfer at the right boundary, given by

$$\left(-K\frac{dT}{dx}\right)_{x=L} = \frac{2K}{L}(T_2 - T_L) + \frac{QL}{4} \tag{3.39}$$

Let us now approximate Eqs. (3.27) through (3.29) using one quadratic element to discretize the domain. We are using the same number of nodal points as in the previous case, but the next higher approximation. The Galerkin statement is the same as before. Since the local and global systems are the same for the case of one element, we can write

$$\int_0^L \left[K\frac{dN_i}{dx}\left(\sum_{j=1}^{3}\frac{dN_j}{dx}T_j\right) - N_iQ\right]dx + \left[N_i\left(-K\frac{dT}{dx}\right)\right]_0^L = 0 \tag{3.40}$$

Using Eqs. (3.19), (3.22), and (3.23), we have in matrix form:

$$\int_0^L \left\{ K\left(\begin{bmatrix} \frac{1}{L}\left(\frac{4x}{L}-3\right) \\ \frac{4}{L}\left(1-\frac{2x}{L}\right) \\ \frac{1}{L}\left(\frac{4x}{L}-1\right) \end{bmatrix} \begin{bmatrix} \frac{1}{L}\left(\frac{4x}{L}-3\right) & \frac{4}{L}\left(1-\frac{2x}{L}\right) & \frac{1}{L}\left(\frac{4x}{L}-1\right) \end{bmatrix}\right) \begin{bmatrix} T_1 \\ T_2 \\ T_3 \end{bmatrix} - \begin{bmatrix} 1-\frac{3x}{L}+2\left(\frac{x}{L}\right)^2 \\ \frac{4x}{L}\left(1-\frac{x}{L}\right) \\ \frac{x}{L}\left(\frac{2x}{L}-1\right) \end{bmatrix} Q \right\} dx = \begin{bmatrix} q \\ 0 \\ -\left(-K\frac{dT}{dx}\right)_{x=L} \end{bmatrix} \tag{3.41}$$

Integrating, we get

$$\frac{K}{6L}\begin{bmatrix} 14 & -16 & 2 \\ -16 & 32 & -16 \\ 2 & -16 & 14 \end{bmatrix}\begin{bmatrix} T_1 \\ T_2 \\ T_3 \end{bmatrix} = \frac{QL}{6}\begin{bmatrix} 1 \\ 4 \\ 1 \end{bmatrix} + \begin{bmatrix} q \\ 0 \\ -\left(-K\frac{dT}{dx}\right)_{x=L} \end{bmatrix} \tag{3.42}$$

Again, the last equation can be eliminated using $T_3 = T_L$; the final system is

$$\frac{K}{6L}\begin{bmatrix} 14 & -16 \\ -16 & 32 \end{bmatrix}\begin{bmatrix} T_1 \\ T_2 \end{bmatrix} = \frac{QL}{6}\begin{bmatrix} 1 \\ 4 \end{bmatrix} + \begin{bmatrix} q \\ 0 \end{bmatrix} + \frac{K}{6L}T_L\begin{bmatrix} -2 \\ 16 \end{bmatrix} \tag{3.43}$$

with the heat flux at the right-hand boundary given by

$$\left(-K\frac{dT}{dx}\right)_{x=L} = \frac{K}{6L}\left(16T_2 - 2T_1 - 14T_L\right) + \frac{QL}{6} \tag{3.44}$$

3.3.2 Variable Conduction and Boundary Convection

We now extend the finite element algorithm to approximately solve for the steady-state temperature distribution in a thin rod of length L, subject to a convective heat load at the left end, $x = 0$, and held at a fixed temperature, T_L, at the right end, $x = L$. We also assume that there are no internal heat sources (i.e., $Q = 0$), but that the rod thermal conductivity varies with x (as a result of changes in material composition or variations in the rod's cross section), i.e., $K = K(x)$. The differential equation describing this problem is

$$-\frac{d}{dx}\left(K(x)\frac{dT}{dx}\right) = 0 \tag{3.45}$$

with

$$-K\frac{dT}{dx} + h(T - T_\infty) = 0 \qquad at\ x = 0 \tag{3.46}$$

where T_∞ is the external reference temperature and h the convective heat transfer coefficient, and

$$T = T_L \qquad at\ x = L \tag{3.47}$$

The weighted residual formulation is obtained as before:

$$\int_0^L K(x)\frac{dW}{dx}\frac{dT}{dx}dx - \left[W\left(-K(x)\frac{dT}{dx}\right)\right]_{x=0} + \left[W\left(-K(x)\frac{dT}{dx}\right)\right]_{x=L} = 0 \tag{3.48}$$

From now on, we drop the flux term corresponding to a boundary where the temperature is prescribed, in this case at $x = L$, as is customary in finite element modeling. If we also replace (3.46) for the flux at the left-hand boundary, we have

$$\int_0^L K(x)\frac{dW}{dx}\frac{dT}{dx}dx + Wh\left(T - T_\infty\right)\Big|_{x=0} = 0 \tag{3.49}$$

Notice that nothing has been lost in going from (3.45) to (3.49); however, we do not seek the analytical solution to Eq. (3.49), but instead a computable finite element approximation utilizing our defined shape functions and mesh discretization.

As before, we define a mesh over $0 \le x \le L$ and we approximate $T(x)$ by

$$T(x) = \sum_{i=1}^{n+1} N_i(x)T_i \tag{3.50}$$

where n is the number of elements in the mesh. If $N_i, i = 1, \ldots, n+1$, are linear shape functions, the Galerkin form of Eq. (3.49) is then

$$\int_0^L K(x)\frac{dN_i}{dx}\left(\sum_{j=1}^{n+1}\frac{dN_j}{dx}T_j\right)dx + N_i h(T - T_\infty)\Big|_{x=0} = 0 \tag{3.51}$$

Rather than plugging in a specific form for $K(x)$, we interpolate $K(x)$ in terms of its nodal values using the same shape functions as chosen for $T(x)$, in this case, linear. The element equations for the first element, which includes the convective boundary condition, then yield

$$\int_{x_1}^{x_2}\begin{bmatrix} N_1^{(e_1)} & N_2^{(e_1)}\end{bmatrix}\begin{bmatrix} K_1^{(e_1)} \\ K_2^{(e_1)}\end{bmatrix}\left(\begin{bmatrix}\frac{dN_1^{(e_1)}}{dx} \\ \frac{dN_2^{(e_1)}}{dx}\end{bmatrix}\begin{bmatrix}\frac{dN_1^{(e_1)}}{dx} & \frac{dN_2^{(e_1)}}{dx}\end{bmatrix}\right)dx\begin{bmatrix} T_1^{(e_1)} \\ T_2^{(e_1)}\end{bmatrix}$$

$$+\begin{bmatrix} h & 0 \\ 0 & 0\end{bmatrix}\begin{bmatrix} T_1^{(e_1)} \\ T_2^{(e_1)}\end{bmatrix} = hT_\infty\begin{bmatrix} 1 \\ 0\end{bmatrix}$$

We have used the fact that $N_1^{(e_1)}(0) = 1$ and $N_2^{(e_1)}(0) = 0$ in the convective boundary term, that is,

$$N_i h(T-T_\infty)|_{x=0} = \begin{bmatrix} N_1 \\ N_2 \end{bmatrix} h[N_1\ N_2] \begin{bmatrix} T_1 - T_\infty \\ T_2 - T_\infty \end{bmatrix}|_{x=0}$$

$$= \left(h \begin{bmatrix} N_1 \\ N_2 \end{bmatrix} [N_1\ N_2] \begin{bmatrix} T_1 \\ T_2 \end{bmatrix} - \begin{bmatrix} N_1 \\ N_2 \end{bmatrix} hT_\infty \right)|_{x=0}$$

$$= \left(h \begin{bmatrix} N_1N_1 & N_1N_2 \\ N_2N_1 & N_2N_2 \end{bmatrix} \begin{bmatrix} T_1 \\ T_2 \end{bmatrix} - \begin{bmatrix} N_1 \\ N_2 \end{bmatrix} hT_\infty \right)|_{x=0}$$

$$= h \begin{bmatrix} 1 & 0 \\ 0 & 0 \end{bmatrix} \begin{bmatrix} T_1 \\ T_2 \end{bmatrix} - hT_\infty \begin{bmatrix} 1 \\ 0 \end{bmatrix}$$

Performing the integration, we get

$$\frac{1}{2h^{(e_1)}}(K_1^{(e_1)} + K_2^{(e_1)}) \begin{bmatrix} 1 & -1 \\ -1 & 1 \end{bmatrix} \begin{bmatrix} T_1^{(e_1)} \\ T_2^{(e_1)} \end{bmatrix} + \begin{bmatrix} h & 0 \\ 0 & 0 \end{bmatrix} \begin{bmatrix} T_1^{(e_1)} \\ T_2^{(e_1)} \end{bmatrix} = h\ T_\infty \begin{bmatrix} 1 \\ 0 \end{bmatrix} \tag{3.52}$$

For all other elements, $e_1 \neq e_1$, the equations are

$$\frac{1}{2h^{(e_1)}}(K_1^{(e_1)} + K_2^{(e_1)}) \begin{bmatrix} 1 & -1 \\ -1 & 1 \end{bmatrix} \begin{bmatrix} T_1^{(e_1)} \\ T_2^{(e_1)} \end{bmatrix} = \begin{bmatrix} 0 \\ 0 \end{bmatrix} \tag{3.53}$$

If the region $0 \le x \le L$ has been discretized using two elements of equal length $L/2$, the resulting assembled system of equations can be written as

$$\frac{1}{L} \begin{bmatrix} K_1 + K_2 + Lh & -(K_1 + K_2) & 0 \\ -(K_1 + K_2) & K_1 + 2K_2 + K_3 & -(K_2 + K_3) \\ 0 & -(K_2 + K_3) & K_2 + K_3 \end{bmatrix} \begin{bmatrix} T_1 \\ T_2 \\ T_3 \end{bmatrix} = \begin{bmatrix} hT_\infty \\ 0 \\ 0 \end{bmatrix} \tag{3.54}$$

If a convective boundary condition of the form (3.46) is given at $x = L$, the element equations for the last element follow from the weighted residuals form (3.48). Neglecting the first boundary term and replacing the convective condition in the second term leads to

$$\frac{1}{2h^{(e_n)}}(K_1^{(e_n)} + K_2^{(e_n)}) \begin{bmatrix} 1 & -1 \\ -1 & 1 \end{bmatrix} \begin{bmatrix} T_1^{(e_n)} \\ T_2^{(e_n)} \end{bmatrix} + \begin{bmatrix} 0 & 0 \\ 0 & -h \end{bmatrix} \begin{bmatrix} T_1^{(e_n)} \\ T_2^{(e_n)} \end{bmatrix} = hT_\infty \begin{bmatrix} 0 \\ -1 \end{bmatrix} \tag{3.55}$$

A convective boundary condition can be applied at either or both boundaries in the mesh; Eqs. (3.52), (3.53), and (3.55) are valid for any mesh (uniform or nonuniform), for variable conductivity, and for convection at either or both boundaries. A specific solution can be obtained upon defining the nodal coordinates of the mesh and the various problem data h, T_∞, and $K(x)$.

Example 3.1*

Let us solve Eq. (3.45) for L = 10 cm assuming that $K(x)$ varies linearly between 40 and 60 W/mC, h = 100 W/m²C, T_∞, and T_L = 39.18°C at $x = L$. For these data, the exact solution to the problem produces a convection surface temperature of 100°C. First, we solve the problem using the coarsest possible grid, i.e., one linear element. Replacing the problem data into Eq. (3.52), we have

$$500\begin{bmatrix} 1 & -1 \\ -1 & 1 \end{bmatrix}\begin{bmatrix} T_1 \\ T_2 \end{bmatrix} + \begin{bmatrix} 100 & 0 \\ 0 & 0 \end{bmatrix}\begin{bmatrix} T_1 \\ T_2 \end{bmatrix} = \begin{bmatrix} 40{,}000 \\ 0 \end{bmatrix}$$

which reduces to

$$\begin{bmatrix} 6 & -5 \\ -5 & 5 \end{bmatrix}\begin{bmatrix} T_1 \\ T_2 \end{bmatrix} = \begin{bmatrix} 400 \\ 0 \end{bmatrix} \tag{3.56}$$

The equations can be solved at this point without considering the right-hand-side boundary condition. The solution gives $T_1 = T_2$ = 400°C, i.e., $T(x)$ = constant = T_∞. This is exactly the solution one should obtain for an adiabatic right-end boundary condition. If we recall Eq. (2.18) and the discussion following Eq. (3.37), the second element on the right-hand side of (3.56) corresponds to

$$-K\frac{dT}{dx}\Big|_{x=L} = 0$$

which is an adiabatic condition at that end. Hence, in the absence of a fixed temperature boundary condition, also called a *Dirichlet* condition, if no action is taken, the finite element method automatically enforces a vanishing derivative (for free).

If we now enforce the Dirichlet condition $T(L) = T_2$ = 39.18°C, Eq. (3.56) is modified to

$$\begin{bmatrix} 6 & -5 \\ 0 & 1 \end{bmatrix}\begin{bmatrix} T_1 \\ T_2 \end{bmatrix} = \begin{bmatrix} 400 \\ 39.18 \end{bmatrix} \tag{3.57}$$

* FEM-1D and COMSOL files available.

and the solution yields $T_1 = 99.317°C$ Recalling that the exact solution is $T_1 = 100°C$, the one-element solution is less that 0.7% in error. This shows us that the finite element method effectively incorporates flux boundary conditions of the general form

$$aT + b\frac{dT}{dx} + c = 0 \tag{3.58}$$

for arbitrary values of the coefficients a, b, and c. In the present problem, we have $a = h$, $b = -K(0)$, and $c = -hT_\infty$. Conditions of the form (3.58) are known as *Neumann* conditions when $a = 0$, and as *mixed* conditions if $a \neq 0$ and $b \neq 0$; if $b = 0$, we have the already-introduced *Dirichlet* condition.

We now generate the finite element solution for a uniform two-element mesh using linear shape functions. The element and problem data are now

$$\text{for element } e_1 : h^{(e_1)} = \frac{L}{2} = 0.05$$

$$\begin{bmatrix} K_1^{(e_1)} \\ K_2^{(e_1)} \end{bmatrix} = \begin{bmatrix} 40 \\ 60 \end{bmatrix}$$

$$\text{for element } e_2 : h^{(e_2)} = \frac{L}{2} = 0.05$$

$$\begin{bmatrix} K_1^{(e_2)} \\ K_2^{(e_2)} \end{bmatrix} = \begin{bmatrix} 50 \\ 60 \end{bmatrix}$$

Using Eq. (3.52), element e_1 yields

$$900\begin{bmatrix} 1 & -1 \\ -1 & 1 \end{bmatrix}\begin{bmatrix} T_1 \\ T_2 \end{bmatrix} + \begin{bmatrix} 100 & 0 \\ 0 & 0 \end{bmatrix}\begin{bmatrix} T_1 \\ T_2 \end{bmatrix} = \begin{bmatrix} 40{,}000 \\ 0 \end{bmatrix}$$

and element e_2 yields

$$1100\begin{bmatrix} 1 & -1 \\ -1 & 1 \end{bmatrix}\begin{bmatrix} T_2 \\ T_3 \end{bmatrix} = \begin{bmatrix} 0 \\ 0 \end{bmatrix}$$

Assembling the equations, we have

$$\begin{bmatrix} 10 & -9 & 0 \\ -9 & 20 & -11 \\ 0 & -11 & 11 \end{bmatrix} \begin{bmatrix} T_1 \\ T_2 \\ T_3 \end{bmatrix} = \begin{bmatrix} 400 \\ 0 \\ 0 \end{bmatrix}$$

and enforcing the Dirichlet condition $T_3 = 39.18\,C$, we finally obtain

$$\begin{bmatrix} 10 & -9 & 0 \\ -9 & 20 & -11 \\ 0 & 0 & 0 \end{bmatrix} \begin{bmatrix} T_1 \\ T_2 \\ T_3 \end{bmatrix} = \begin{bmatrix} 400 \\ 0 \\ 39.18 \end{bmatrix}$$

with solution

$$\begin{bmatrix} T_1 \\ T_2 \\ T_3 \end{bmatrix} = \begin{bmatrix} 99.822 \\ 66.469 \\ 39.18 \end{bmatrix}$$

Comparing with the solution of (3.57), the error in the convection surface temperature T_1 has decreased from less than 0.7% to less than 0.2%. Linear shape functions applied to boundary value problems yield second-order-accurate results, i.e., the error in the nodal temperatures decrease as $(h^{(e)})^2$ for uniform refinement of the mesh. In the above example, note that $h^{(e)}$ was halved and the resulting error decreased by a factor of approximately four, i.e., $(1/2)^2$.

Problems 3.13 and 3.14 offer a comparison between evaluating the boundary flux using the finite element interpolant to the temperature and the boundary residual Eq. (3.39). This and other issues related to the calculation of fluxes are beyond the scope of this book. The interested reader is referred to Heinrich and Pepper (1999) for a more complete discussion.

3.4 AXISYMMETRIC HEAT CONDUCTION

Many problems in steady heat conduction with boundary convection involve fluids flowing in pipes. To develop the corresponding finite element algorithm, the axisymmetric form of Eqs. (3.45) to (3.47) is

$$-\frac{1}{r}\frac{d}{dr}\left(rK\frac{dT}{dr}\right) = 0 \qquad r_1 < r < r_2 \tag{3.59}$$

$$-K\frac{dT}{dr}+h(T-T_\infty)=0 \qquad at\ r=r_1 \tag{3.60}$$

$$T=T_L \qquad at\ r=r_2 \tag{3.61}$$

The weighted residuals form of (3.59) is

$$\int_0^{2\pi}\int_{r_1}^{r_2} W\left[-\frac{d}{dr}\left(rK\frac{dT}{dr}\right)\right]drd\theta=0 \tag{3.62}$$

Integrating by parts with respect to r and performing the integral with respect to θ, we obtain

$$2\pi\int_{r_1}^{r_2} rK\frac{dW}{dr}\frac{dT}{dr}dr+(2\pi rWh(T-T_\infty))_{r=r_1}=0 \tag{3.63}$$

Comparing Eq. (3.63) with (3.48) the significant distinction for the axisymmetric case is that $K(x)$ is replaced by rK. A smoothly varying thermal conductivity is not too important physically, while an abrupt change in (constant) conductivity is, e.g., a change in material or thermal insulation wrapped around a pipe. Hence, Eq. (3.63) can be simplified by bringing K outside the integral, with the assumption that the integral will be evaluated by segments according to step changes in K. The axisymmetric Galerkin integral can be written (deleting the 2π factor) for an element $e_1=\{r\,|\,r_1^{(e_1)}\le r\le r_2^{(e_1)}\}$ as

$$\left\{\frac{K^{(e_i)}}{2h^{(e_i)}}(r_1^{(e_i)}+r_2^{(e_i)})\begin{bmatrix}1 & -1\\ -1 & 1\end{bmatrix}+r_1h\begin{bmatrix}1 & 0\\ 0 & 0\end{bmatrix}\delta_{e_1}-r_2h\begin{bmatrix}0 & 0\\ 0 & 1\end{bmatrix}\delta_{e_n}\right\}\begin{bmatrix}T_1^{(e_i)}\\ T_2^{(e_i)}\end{bmatrix}$$

$$=\left\{r_1hT_\infty\begin{bmatrix}1\\ 0\end{bmatrix}\delta_{e_1}-r_2hT_\infty\begin{bmatrix}0\\ 1\end{bmatrix}\delta_{e_n}\right\} \tag{3.64}$$

where δ_e is an "element" Kronecker delta function, i.e., $\delta_e = 1$ in element e_i and zero in all elements e_j with $j \neq i$; n denotes the number of elements so e_n is the element adjacent to the right-hand boundary.

Expression (3.64) differs from Eqs. (3.52), (3.53), and (3.55) only in the handling of the element data. We are thus led to the next example problem.

EXAMPLE 3.2*

Solve Eq. (3.59) for a uniform two-element discretization of a thick-walled circular pipe containing a flowing hot fluid. The pipe has a thickness of 10 cm and an inner radius of 20 cm;

* FEM-1D and COMSOL files are available.

the thermal conductivity is $K = 20$ W/mC. The fluid heat exchange temperature is 400°C, $h =$ 50 W/m²C, and the pipe outer wall is $T_L = 39.18$°C. Thus,

$$\text{element } e_1 : \; h^{(e_1)} = r_2^{(e_1)} - r_1^{(e_1)}$$

$$= 0.25 - 0.2 = 0.05$$

$$K^{(e_1)} = 20$$

$$h = 50$$

$$T_\infty = 400$$

$$r_1^{(e_1)} + r_2^{(e_1)} = 0.2 + 0.25 = 0.45$$

$$\text{element } e_2 : \; h^{(e_2)} = r_2^{(e_2)} - r_1^{(e_2)}$$

$$= 0.3 - 0.25 = 0.05$$

$$K^{(e_2)} = 20$$

$$r_1^{(e_2)} + r_2^{(e_2)} = 0.25 + 0.3 = 0.55$$

Using Eq. (3.64), with $\delta_{e_1} = 1$ and $\delta_{e_n} = 0$, element e_1 yields

$$\left(90\begin{bmatrix} 1 & -1 \\ -1 & 1 \end{bmatrix} + 10\begin{bmatrix} 1 & 0 \\ 0 & 0 \end{bmatrix}\right)\begin{bmatrix} T_1 \\ T_2 \end{bmatrix} = 4{,}000\begin{bmatrix} 1 \\ 0 \end{bmatrix}$$

and element e_2 gives

$$110\begin{bmatrix} 1 & -1 \\ -1 & 1 \end{bmatrix}\begin{bmatrix} T_2 \\ T_3 \end{bmatrix} = \begin{bmatrix} 0 \\ 0 \end{bmatrix}$$

Comparing these results to the element equations obtained in Example 3.1 confirms that the finite element equations are identical for the data selected. Assembling the equations and imposing the right-hand boundary condition gives the nodal solution, and the convection temperature error is again less than 0.2%. Notice that these examples lend credence to the assertion of flexibility in the finite element method.

3.5 NATURAL COORDINATE SYSTEM

We now introduce a natural coordinate system for specifying the local operations on an element. We define the elements in the interval $-1 \le \xi \le 1$, together with a transformation of the form:

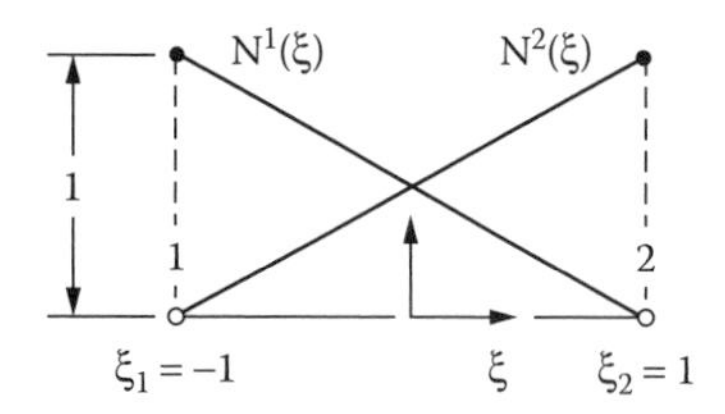

Figure 3.8 Notation and linear shape functions in natural local coordinate system

$$x = \frac{1}{2}(1-\xi)x_i + \frac{1}{2}(1+\xi)x_{i+1} \tag{3.65}$$

that maps the interval $-1 \le \xi \le 1$ into the element interval $x_i \le x \le x_{i+1}$. The reasons and advantages of doing so will become evident later when we deal with numerical integration and the more complicated two- and three-dimensional geometries.

The linear element shape functions are shown in Figure 3.8 and are given by

$$\left.\begin{aligned} N_1(\xi) &= \frac{1}{2}(1-\xi) \\ N_2(\xi) &= \frac{1}{2}(1+\xi) \end{aligned}\right\} \tag{3.66}$$

Utilizing this type of coordinate system allows the products of shape functions and derivatives of the shape functions to be easily evaluated numerically by means of the transformation (3.65) if we note that

$$\frac{d}{dx} = \frac{d}{d\xi} \cdot \frac{d\xi}{dx} = \frac{2}{h^{(e_i)}} \frac{d}{d\xi} \tag{3.67}$$

and

$$dx = \frac{h^{(e_1)}}{2} d\xi \tag{3.68}$$

Notice also that Eq. (3.65) can be written as

$$x = N_1(\xi)x_1 + N_2(\xi)x_2$$

Hence, the transformations can be written strictly in terms of the shape functions and the nodal coordinates. Integrals and derivatives of the shape functions take the form

$$\int_{x_i}^{x_{i+1}} N_j(x) N_k(x)\, dx = \frac{h^{(e_1)}}{2} \int_{-1}^{1} N_j(\xi) N_k(\xi)\, d\xi \tag{3.69a}$$

$$\int_{x_i}^{x_{i+1}} \frac{dN_j}{dx}\frac{dN_k}{dx}dx = \frac{2}{h^{(e_i)}}\int_{-1}^{1}\frac{dN_j}{d\xi}\frac{dN_k}{d\xi}d\xi \tag{3.69b}$$

$$\int_{x_i}^{x_{i+1}} \frac{dN_j}{dx}N_k\left(x\right)dx = \int_{-1}^{1}\frac{dN_j}{d\xi}N_k\left(\xi\right)d\xi \tag{3.69c}$$

and so on. Furthermore, the products of functions in Eq. (3.69a) can be easily evaluated analytically in terms of exponents. This procedure is well known in structural mechanics and can be readily used for any one-dimensional element mesh; see Zienkiewicz (1977). The integral relation is expressed as

$$\int_{-1}^{1}(N_1(\xi))^a(N_2(\xi))^b d\xi = 2\frac{a!b!}{(1+a+b)!} \tag{3.70}$$

where a and b are non-negative integers. Use of Eq. (3.70) is relatively easy. For example,

$$\begin{aligned}
\left[\int_{x_i}^{x_{i+1}} N_i(x)N_j(x)dx\right] &= \frac{h^{(e_i)}}{2}\int_{-1}^{1}\begin{bmatrix} N_1(\xi) \\ N_2(\xi)\end{bmatrix}[N_1(\xi)\ \ N_2(\xi)]d\xi \\
&= \frac{h^{(e_i)}}{2}\int_{-1}^{1}\begin{bmatrix} (N_1(\xi))^2 & N_1(\xi)N_2(\xi) \\ N_2(\xi)N_1(\xi) & (N_2(\xi))^2\end{bmatrix}d\xi \\
&= \frac{h^{(e_i)}}{2}\begin{bmatrix} 2\dfrac{2!0!}{(1+1+1)!} & 2\dfrac{1!\ 1!}{(1+1+1)!} \\ 2\dfrac{1!\ 1!}{(1+1+1)!} & 2\dfrac{0!\ 2!}{(1+0+2)!}\end{bmatrix} \\
&= \frac{h^{(e_1)}}{6}\begin{bmatrix} 2 & 1 \\ 1 & 2\end{bmatrix}
\end{aligned} \tag{3.71}$$

Notice that this is a general relation for any element involving the product $[N]^T[N]$. Also, from (3.69b)

$$\left[\int_{x_i}^{x_{i+1}} \frac{dN_j}{dx}\frac{dN_k}{dx}dx\right] = \frac{2}{h^{(e_i)}}\int_{-1}^{1}\begin{bmatrix}\frac{dN_1}{d\xi}\\ \frac{dN_2}{d\xi}\end{bmatrix}\begin{bmatrix}\frac{dN_1}{d\xi} & \frac{dN_2}{d\xi}\end{bmatrix}d\xi$$

$$= \frac{2}{h^{(e_i)}}\int_{-1}^{1}\frac{1}{4}\begin{bmatrix}-1\\ 1\end{bmatrix}[-1\ 1]d\xi \tag{3.72}$$

$$= \frac{1}{h^{(e_i)}}\begin{bmatrix}1 & -1\\ -1 & 1\end{bmatrix}$$

which is the familiar diffusion integral. We can easily utilize the natural coordinate system for higher-order elements, in particular, the quadratic element. The shape functions in natural coordinates are

$$N_1(\xi) = \frac{1}{2}\xi(\xi - 1) \tag{3.73a}$$

$$N_2(\xi) = 1 - \xi^2 \tag{3.73b}$$

$$N_3(\xi) = \frac{1}{2}\xi(\xi + 1) \tag{3.73c}$$

The transformation from the interval $-1 \le \xi \le 1$ to the element domain $x_{2i-1} \le x \le x_{2i+1}$ in e_i is given by

$$x = N_1(\xi)x_{2i-1} + N_2(\xi)x_{2i} + N_3(\xi)x_{2i+1} \tag{3.74}$$

Notice that this expression reduces to Eq. (3.65) when $x_{2i} = (x_{2i-1} + x_{2i+1})/2$. The details on this are left to Exercise 3.9. This representation is more general than the one of Section 3.2, as it allows for the interior node to be located at points other than the element midpoint.

The derivative of $N_i(\xi)$ with respect to x is obtained from

$$\frac{dN_i}{d\xi} = \frac{dN_i}{dx}\cdot\frac{dx}{d\xi}$$

and is given by

$$\frac{dN_i}{dx} = \frac{1}{dx/d\xi}\frac{dN_i}{d\xi} \tag{3.75}$$

where now

$$\frac{dx}{d\xi} = \frac{dN_1}{d\xi} x_{2i-1} + \frac{dN_2}{d\xi} x_{2i} + \frac{dN_3}{d\xi} x_{2i+1} \tag{3.76}$$

The global values of the element coordinates $x_1^{(e_i)}, x_2^{(e_i)}$, and $x_3^{(e_i)}$ are used in (3.76). The term $dx/d\xi$ is known as the *Jacobian* of the coordinate transformation, and is usually denoted as **J**. This term will become particularly important when dealing with multidimensional elements.

Example 3.3

Using the natural coordinate system for a quadratic element, we wish to evaluate the term

$$-K\int_{e_k} W \frac{d^2T}{dx^2} dx$$

which is the diffusion term with constant diffusivity. Setting $W_i = N_i$, the Galerkin procedure yields

$$\left[K\int_{x_{2k-1}}^{x_{2k+1}} \frac{dN_i^{(e_k)}}{dx} \frac{dN_j^{(e_k)}}{dx} dx\right] T_j + \left(-K\frac{dT}{dx} N_i\right)_{x=x_{2k-1}}^{x=x_{2k+1}} = 0 \tag{3.77}$$

and the Jacobian of the transformation (3.74), if we take $x_{2k-1} = 0$, $x_{2k} = L/2$, and $x_{2k+1} = L$, is

$$J = \frac{dN_1(\xi)}{d\xi} \cdot 0 + \frac{dN_2(\xi)}{d\xi} \cdot \frac{L}{2} + \frac{dN_3(\xi)}{d\xi} \cdot L = (-2\xi) \cdot \frac{L}{2} + \left[\xi + \frac{1}{2}\right] \cdot L = \frac{L}{2}$$

which is to be expected since, in this case, the transformation is the same as the linear transformation (3.65). Hence, the derivative values are

$$\frac{dN_1^{(e_k)}}{dx} = J^{-1} \frac{dN_1}{d\xi} = \frac{2}{L} \frac{dN_i}{d\xi} \tag{3.78}$$

which can be written in vector form as

$$\begin{bmatrix} \frac{dN_1^{(e_k)}}{dx} \\ \frac{dN_2^{(e_k)}}{dx} \\ \frac{dN_3^{(e_k)}}{dx} \end{bmatrix} = \frac{1}{L} \begin{bmatrix} -1+2\xi \\ -4\xi \\ 1+2\xi \end{bmatrix} \tag{3.79}$$

In this natural coordinate system, we need an expression for the integration differential. For one-dimensional space, this is given by $dx = |\mathbf{J}| d\xi$, where $|\mathbf{J}|$ denotes the magnitude of the Jacobian. Later on, in higher dimensions, $\mathbf{J}$ will be a matrix and $|\mathbf{J}|$ its determinant.

We are now ready to evaluate the diffusion integral. Assuming, for simplicity, that the boundary terms in Eq. (3.77) vanish, we get

$$
\begin{aligned}
&K\int_{x_{2k-1}}^{x_{2k+1}} \begin{bmatrix} \dfrac{dN_1^{(e_k)}}{dx} \\ \dfrac{dN_2^{(e_k)}}{dx} \\ \dfrac{dN_3^{(e_k)}}{dx} \end{bmatrix} \begin{bmatrix} \dfrac{dN_1^{(e_k)}}{dx} & \dfrac{dN_2^{(e_k)}}{dx} & \dfrac{dN_3^{(e_k)}}{dx} \end{bmatrix} dx \\
&= \frac{K}{2L}\int_{-1}^{1} \begin{bmatrix} -1+2\xi \\ -4\xi \\ 1+2\xi \end{bmatrix} \begin{bmatrix} -1+2\xi & -4\xi & 1+2\xi \end{bmatrix} d\xi \qquad (3.80) \\
&= \frac{K}{2L}\int_{-1}^{1} \begin{bmatrix} (-1+2\xi)^2 & -4\xi(-1+2\xi) & (-1+2\xi)(1+2\xi) \\ -4\xi(-1+2\xi) & 16\xi^2 & -4\xi)(1+2\xi) \\ (1+2\xi)(1+2\xi) & -4\xi(1+2\xi) & (1+2\xi)^2 \end{bmatrix} d\xi \\
&= \frac{K}{6} \begin{bmatrix} 14 & -16 & 2 \\ -16 & 32 & -16 \\ 2 & -16 & 14 \end{bmatrix}
\end{aligned}
$$

which is the same matrix for conduction as given in Eq. (3.42). In the same fashion, if we assume that a constant heat flux q occurs only at node 1 of a quadratic element, then at $\xi = -1$ we have $N_1 = 1$ and $N_2 = N_3 = 0$. For a prescribed heat flux q at $x = 0$, the boundary term in Eq. (3.77) now renders a contribution at $x = x_{2k-1}$, which is

$$
K\frac{dT}{dx}\Big|_{x=0} = -q
$$

and the total contribution of the element becomes

$$
\frac{K}{6} \begin{bmatrix} 14 & -16 & 2 \\ -16 & 32 & -16 \\ 2 & -16 & 14 \end{bmatrix} \begin{bmatrix} T_1 \\ T_2 \\ T_3 \end{bmatrix} - q \begin{bmatrix} 1 \\ 0 \\ 0 \end{bmatrix} = 0 \qquad (3.81)
$$

as we expected.

Example 3.4

Let us now evaluate the integral

$$\int_{e_i} WQdx$$

where Q is a constant internal heat source, for both linear and quadratic elements, using natural coordinates.

For linear elements, we have, in Galerkin form

$$Q\int_{x_k}^{x_{k+1}} \begin{bmatrix} N_1^{(e_k)} \\ N_2^{(e_k)} \end{bmatrix} dx = \frac{QL}{2}\int_{-1}^{1} \begin{bmatrix} \frac{1}{2}(1-\xi) \\ \frac{1}{2}(1+\xi) \end{bmatrix} d\xi = \frac{QL}{2}\begin{bmatrix} 1 \\ 1 \end{bmatrix} \tag{3.82}$$

where we used Eqs. (3.66) and (3.68).

For quadratic elements, from Eqs. (3.73) and the expression for the Jacobian $J = (L/2)$, we have

$$Q\int_{x_{2k-1}}^{x_{2k+1}} \begin{bmatrix} N_1^{(e_k)} \\ N_2^{(e_k)} \\ N_3^{(e_k)} \end{bmatrix} dx = \frac{QL}{2}\int_{-1}^{1} \begin{bmatrix} \frac{1}{2}\xi(\xi-1) \\ 1-\xi^2 \\ \frac{1}{2}\xi(\xi+1) \end{bmatrix} d\xi = \frac{QL}{6}\begin{bmatrix} 1 \\ 4 \\ 1 \end{bmatrix} \tag{3.83}$$

Equations (3.82) and (3.83) are the same expressions obtained before the source term in Eqs. (3.35) and (3.42), respectively, if we replace L by $L/2$ in Eq. (3.82).

Example 3.5*

Let us go back once more to the simple steady-state problem defined in Example 3.1. We wish to find the temperature distribution over one quadratic element. Recall the $h^{(e)} = 0.1$ and $[K^{(e)}]^T = [40 \quad 50 \quad 60]$. Since there is no need for assembly over one element, the Galerkin finite element expression can be written as

$$\sum_{j=1}^{3}\left(\int_0^1 \frac{dN_i^{(e)}}{dx} K(x) \frac{dN_j^{(e)}}{dx} dx\right) T_j + N_i^{(e)} h(T - T_\infty)|_{x=0} = 0 \quad i = 1,2,3 \tag{3.84}$$

Transforming to natural coordinates, we can write now in matrix form

* FEM-1D and COMSOL files are available.

$$\left\{\frac{1}{2}\int_{-1}^{1}\begin{bmatrix}\frac{dN_1}{d\xi}\\ \frac{dN_2}{d\xi}\\ \frac{dN_3}{d\xi}\end{bmatrix}\left([N_1\ N_2\ N_3]\begin{bmatrix}40\\50\\60\end{bmatrix}\right)\begin{bmatrix}\frac{dN_1}{d\xi} & \frac{dN_2}{d\xi} & \frac{dN_3}{d\xi}\end{bmatrix}d\xi+100\begin{bmatrix}1&0&0\\0&0&0\\0&0&0\end{bmatrix}\right\}\begin{bmatrix}T_1\\T_2\\T_3\end{bmatrix}$$

$$=\begin{bmatrix}40{,}000\\0\\0\end{bmatrix}$$

Utilizing Eqs. (3.73) for the shape functions, we have

$$\left\{5\int_{-1}^{1}(5+\xi)\begin{bmatrix}(2\xi-1)^2 & 4\xi(1-2\xi) & (2\xi-1)(2\xi+1)\\ 4\xi(1-2\xi) & 16\xi^2 & -4\xi(1+2\xi)\\ (2\xi-1)(2\xi+1) & -4\xi(1+2\xi) & (2\xi+1)^2\end{bmatrix}d\xi+\begin{bmatrix}100&0&0\\0&0&0\\0&0&0\end{bmatrix}\right\}\begin{bmatrix}T_1\\T_2\\T_3\end{bmatrix}=\begin{bmatrix}40{,}000\\0\\0\end{bmatrix}$$

Performing the integrations, and setting $T_3 = 39.18°C$ produces the final set of equations:

$$\begin{bmatrix}68 & -72 & 10\\ -72 & 160 & -88\\ 0 & 0 & 1\end{bmatrix}\begin{bmatrix}T_1\\T_2\\T_3\end{bmatrix}=\begin{bmatrix}2400\\0\\39.18\end{bmatrix} \tag{3.85}$$

Equation (3.85) yields the solution $T_1 = 99.992°C$ and $T_2 = 66.546°C$. Comparing with the solution obtained using two linear elements, a remarkable improvement in accuracy is achieved in the convection surface temperature (recall that the exact solution is 100°C). This is a consequence of the fact that the error $e(x)$ in a finite element approximation behaves like

$$|e(x)| \le ch^{n+1} \tag{3.86}$$

where c is a constant, n is the order of interpolation polynomial, and h is the distance between nodes. Hence, for linear elements, the error decreases as $O(h^2)$ and for quadratics as $O(h^3)$. Since the distance between nodes is the same using two linear elements and one quadratic element, we can expect the solution obtained using the quadratic approximation to be more accurate, as is indeed the case.

3.6 TIME DEPENDENCE

We now extend our finite element algorithm to the unsteady heat diffusion equation. Let us first assume for simplicity that $Q = 0$; i.e., there are no sources or sinks. The governing equation for heat conduction with constant diffusivity is usually as

$$\frac{\partial T}{\partial t} - \alpha \frac{\partial^2 T}{\partial x^2} = 0 \tag{3.87}$$

where $\partial T/\partial t$ is the time rate of change of temperature and $\alpha(\alpha = \mathbf{K}/\rho c_p)$ is the thermal diffusivity; ρ, is the density and c_p is the heat capacity (specific heat) of the material. Previously, T was a function of x only; now $T = T(x,t)$ a function of space (x) and time (t). Consequently, in addition to boundary conditions, we need to specify an initial condition, i.e.,

$$-K\frac{\partial T}{\partial t} + h\left(T - T_\infty\right)|_{(0,t)} = 0 \tag{3.88}$$

$$T\left(L,t\right) = T_L \tag{3.89}$$

and

$$T\left(x,0\right) = T_0\left(x\right) \tag{3.90}$$

with the additional generalization that $T_\infty = T_\infty(t)$ and $T_L = T_L(t)$; $T_0(x)$ is the temperature distribution in the rod at time $t = 0$.

3.6.1 Spatial Discretization

The weighted residual form of Eq. (3.87) is

$$\int W(x)\left[\frac{\partial T}{\partial t} - \alpha \frac{\partial^2 T}{\partial x^2}\right]dx = 0 \tag{3.91}$$

Note that the weighting functions are chosen to be a function of x only, i.e., related strictly to the discretization in space. Moreover, we assume that the temperature can be approximated using the same shape functions as before, i.e.,

$$T(x,t) = \sum_{i=1}^{n+1} N_i(x)\, T_i(t) \tag{3.92}$$

where n is the number of elements in the grid.

The time dependence *does not* affect the shape functions and is kept in the dependent variable. The derivatives of the temperature are then given by

$$\frac{\partial T}{\partial t} = \sum_{i=1}^{n+1} \frac{\partial N_i}{\partial x} T_i \equiv \left[\frac{\partial N_i}{\partial x}\right]\left[T_i\right] \tag{3.93}$$

which is the same as before. We use partial derivative symbols to denote derivatives of the shape function and discretized variable, even though these are only functions of one independent variable and the derivatives are the total derivatives. We also have

$$\frac{\partial T}{\partial t} = \sum_{i=1}^{n+1} N_i \frac{\partial T}{\partial t} \equiv \sum_{i=1}^{n+1} N_i \dot{T}_i \equiv \left[N_i\right]\left[\dot{T}_i\right] \tag{3.94}$$

where the dot $(\cdot)$ above T_i denotes time differentiation. Recall from Eq. (3.22) that [**N**] is a row matrix and [**T**] is a column matrix.

The Galerkin formulation of Eq. (3.91) is now

$$\int_0^L \left\{ N_i \left(\sum_{j=1}^{n+1} N_j \dot{T}_j \right) + \alpha \frac{\partial N_i}{\partial x} \left(\sum_{j=1}^{n+1} \frac{\partial N_j}{\partial x} T_j \right) \right\} dx + \left[N_i \left(-\alpha \frac{\partial T}{\partial x} \right) \right]_{x=0}^{x=L} = 0 \tag{3.95}$$

At this stage, we introduce indicial notation to replace the summation signs in the expressions, as in Eq. (3.95). The following definitions are now used to simplify notation:

$$\sum_{i=1}^{n+1} a_i b_i \equiv a_i b_i \tag{3.96}$$

For example, the gradient expression for temperature is

$$\frac{\partial N_j}{\partial x} T_j \equiv \frac{\partial N_1}{\partial x} T_1 + \frac{\partial N_2}{\partial x} T_2 + \ldots + \frac{\partial N_{n+1}}{\partial x} T_{n+1}$$

When indicial notation is used, we must remember that the range of the summation index is clear from the context, since it will not appear explicitly in the expressions.

Equation (3.95) is conveniently written in a more compact form as

$$\left[\int_0^L N_i N_j dx\right] \dot{T}_j + \left[\alpha \int_0^L \frac{\partial N_i}{\partial x} \frac{\partial N_j}{\partial x} dx\right] T_j + \left[N_i \left(-\alpha \frac{\partial T}{\partial x} \right) \right]_{x=0}^{x=L} = 0 \tag{3.97}$$

where the time-dependence expression in the first integral is rewritten as

$$\int_0^L N_i \frac{\partial T}{\partial t} dx = \left(\int_0^L N_i N_j dx \right) \dot{T}_j$$

The coefficient $\dot{T}_j$ in the above equation is a matrix commonly referred to as the *mass matrix*, since the integral of the product $N_i\,N_j$ represents the area of elements common to node i associated to all connected nodes j. The formulation given by Eq. (3.97) is known as a *semi-discrete Galerkin* formulation, since only the spatial variable has been discretized; nothing has been said yet about the time derivative term. The presence of the mass matrix arising from the Galerkin discretization of the time derivative term is a major difference between a finite element and a finite difference discretization of Eq. (3.87) (see Pepper and Baker, 1979).

3.6.2 Time Discretization

There are a number of ways to deal with the time integration of Eq. (3.97) in finite element methodology; see Zienkiewicz (1977). Here, we will introduce the so-called θ-*method*, which leads to the most commonly used algorithms for time integration. In the θ-method, the time derivative is replaced by a simple difference as

$$\frac{\partial T}{\partial t} = \frac{T^{n+1} - T^n}{\Delta t} \tag{3.98}$$

where $T^n = T(x, t_n)$ denotes the variable's value at time $t = t_n$, Δt is the time increment and $t_{n+1} = t_n + \Delta t$. In general, we assume that $T(x, t_n)$ is already known and is used as an initial condition to advance the solution to time level t_{n+1}, where it is not yet known. We now introduce a relaxation parameter θ and write the solution T in the form

$$T = \theta\, T^{n+1} + (1-\theta)\, T^n, \qquad t_n \le t \le_{n+1} \tag{3.99}$$

The parameter θ is usually specified with the range $0 \le \theta \le 1$ and is used to control the accuracy and stability of the algorithm. The most commonly used values of θ are 0, $1/2$, and 1. It is well known that when $\theta < 1/2$, only conditional stability is attained (Richtmeyer and Morton, 1967).

$\theta = 1$ leads to the backwards implicit method.
$\theta = 1/2$ gives a second-order, centered implicit method (*Crank-Nicolson-Galerkin*).
$\theta = 0$ gives the explicit Euler forward scheme.*

Substituting Eqs. (3.98) and (3.99) into (3.97), we obtain

* Notice that setting $\theta = 0$ in Eq. (3.99) does not yield a fully explicit method in Eq. (3.97) because of the presence of the mass matrix, which must always be inverted. To produce fully explicit algorithms, we must use the idea of *mass lumping*, which is introduced in Chapter 4.

$$\left(\int_0^L N_i N_j dx\right)\left(\frac{T_j^{n+1}-T_j^n}{\Delta t}\right)+\left(\alpha\int_0^L \frac{\partial N_i}{\partial x}\frac{\partial N_j}{\partial x}dx\right)(\theta T_j^{n+1}+(1-\theta)T_j^n)$$
$$+\left[N_i\left(-\alpha\left\{\theta\frac{\partial T^{n+1}}{\partial x}+(1-\theta)\frac{\partial T^n}{\partial x}\right\}\right)\right]_{x=0}^{x=L}=0$$

which can be rewritten as

$$\left[\int_0^L N_i N_j dx+\alpha\Delta t\theta\int_0^L \frac{\partial N_i}{\partial x}\frac{\partial N_j}{\partial x}dx\right]T_j^{n+1}+\theta\Delta t\left[N_i\left(-\alpha\frac{\partial T^n}{\partial x}\right)\right]_{x=0}^{x=L}$$
$$=\left[\int_0^L N_i N_j dx+\alpha\Delta t(\theta-1)\int_0^L \frac{\partial N_i}{\partial x}\frac{\partial N_j}{\partial x}dx\right]T_j^{n}+(\theta-1)\Delta t\left[N_i\left(-\alpha\frac{\partial T^n}{\partial x}\right)\right]_{x=0}^{x=L} \tag{3.100}$$

In the above expression, it can be clearly seen that only the mass matrix term survives on the left-hand side if $\theta = 0$. This term results in a greater degree of "connectivity" among adjacent elements than in conventional finite difference methods.

Example 3.6*

We wish to solve for the time-dependent conduction of heat in a thin rod. The thermal diffusivity is constant over each element and the right-hand-side wall of the rod is held at a fixed temperature. The left wall has a convective flux of heat out of the surface. Initial temperature conditions are constant everywhere. We establish the finite element equations using two linear elements to span the slab.

From Eq. (3.100), we obtain, for the first element,

$$\left\{\frac{L}{12}\begin{bmatrix}2&1\\1&2\end{bmatrix}+\frac{2\alpha\Delta t\theta}{L}\begin{bmatrix}1&-1\\-1&1\end{bmatrix}+\bar{h}\theta\Delta t\begin{bmatrix}1&0\\0&0\end{bmatrix}\right\}\begin{bmatrix}T_1^{n+1}\\T_2^{n+1}\end{bmatrix}$$
$$=\left\{\frac{L}{12}\begin{bmatrix}2&1\\1&2\end{bmatrix}+\frac{2\alpha\Delta t(\theta-1)}{L}\begin{bmatrix}1&-1\\-1&1\end{bmatrix}\right.$$
$$\left.+\bar{h}\Delta t(\theta-1)\begin{bmatrix}1&0\\0&0\end{bmatrix}\right\}\begin{bmatrix}T_1^{n}\\T_2^{n}\end{bmatrix}+\bar{h}\Delta t T_\infty\begin{bmatrix}1\\0\end{bmatrix} \tag{3.101}$$

where $\bar{h}=h/\rho c_p$ and the boundary condition (3.88) has been replaced in the boundary terms. For the second element, we obtain similarly

* FEM-1D and COMSOL files are available.

$$\left\{\frac{L}{12}\begin{bmatrix}2 & 1\\ 1 & 2\end{bmatrix}+\frac{2\alpha\Delta t\theta}{L}\begin{bmatrix}1 & -1\\ -1 & 1\end{bmatrix}\right\}\begin{bmatrix}T_2^{n+1}\\ T_3^{n+1}\end{bmatrix} \tag{3.102}$$

$$=\left\{\frac{L}{12}\begin{bmatrix}2 & 1\\ 1 & 2\end{bmatrix}\right.$$

$$\left.+\frac{2\alpha\Delta t\,(\theta-1)}{L}\right\}\begin{bmatrix}1 & -1\\ -1 & 1\end{bmatrix}\begin{bmatrix}T_2^{n}\\ T_3^{n}\end{bmatrix}$$

Let us now utilize the data of Example 3.1, assuming that $\rho c_p = 4\times10^6\ W\,s/m^3C$, the average element diffusivities are $\alpha^{(e_1)} = 1.125\times10^{-5}\ m^2/s, \alpha^{(e_2)} =$ and $\bar{h} = 2.5\times10^{-5} m/s$. We also have that $L = 10\,cm$, $T_L = 39.18\,C$, and $T_\infty = 400\,C$.

If we choose θ = 1 for a fully implicit backward scheme, after multiplying by 120 on both sides, the element equations become

element e_1 :

$$\begin{bmatrix}2+0.03\ \Delta t & 1-0.027\ \Delta t\\ 1-0.027\ \Delta t & 2+0.027\ \Delta t\end{bmatrix}\begin{bmatrix}T_1^{n+1}\\ T_2^{n+1}\end{bmatrix}$$

$$=\begin{bmatrix}2 & 1\\ 1 & 2\end{bmatrix}\begin{bmatrix}T_1^{n}\\ T_2^{n}\end{bmatrix}+\begin{bmatrix}1.2\ \Delta t\\ 0\end{bmatrix}$$

element e_2 :

$$\begin{bmatrix}2+0.033\ \Delta t & 1-0.027\ \Delta t\\ 1-0.033\ \Delta t & 2+0.033\ \Delta t\end{bmatrix}\begin{bmatrix}T_2^{n+1}\\ T_3^{n+1}\end{bmatrix}=\begin{bmatrix}2 & 1\\ 1 & 2\end{bmatrix}\begin{bmatrix}T_2^{n}\\ T_3^{n}\end{bmatrix}$$

Assembling yields the matrix expression

$$\begin{bmatrix}2+0.030\Delta t & 1-0.027\Delta t & 0\\ 1-0.027\Delta t & 4+0.060\Delta t & 1-0.033\Delta t\\ 0 & 1-0.033\Delta t & 2+0.033\Delta t\end{bmatrix}\begin{bmatrix}T_1^{n+1}\\ T_2^{n+1}\\ T_3^{n+1}\end{bmatrix} \tag{3.103a}$$

$$=\begin{bmatrix}2 & 1 & 0\\ 1 & 4 & 1\\ 0 & 1 & 2\end{bmatrix}\begin{bmatrix}T_1^{n}\\ T_2^{n}\\ T_3^{n}\end{bmatrix}+\begin{bmatrix}1.2\Delta t\\ 0\\ 0\end{bmatrix}$$

Imposing the right-hand-side boundary condition, Eq. (3.103a) becomes

$$\begin{bmatrix} 2+0.030\Delta t & 1-0.027\Delta t & 0 \\ 1-0.027\Delta t & 4+0.060\Delta t & 1-0.033\Delta t \\ 0 & 0 & 1 \end{bmatrix} \begin{bmatrix} T_1^{n+1} \\ T_2^{n+1} \\ T_3^{n+1} \end{bmatrix} = \begin{bmatrix} 2 & 1 & 0 \\ 1 & 4 & 1 \\ 0 & 0 & 0 \end{bmatrix} \begin{bmatrix} T_1^{n} \\ T_2^{n} \\ T_3^{n} \end{bmatrix} + \begin{bmatrix} 1.2\Delta t \\ 0 \\ 39.18 \end{bmatrix} \tag{3.103b}$$

which can be solved for T_1^{n+1} and T_2^{n+1} at each time step. We see the influence of the mass matrix on the time derivative term. In a finite difference method, this 3×3 matrix reduces to a diagonal matrix, i.e., the off-diagonal terms are added or "lumped" into the diagonal terms.

We now set Δt = 100 s, for convenience in the calculations, and assume that $T_0(x)$ = 39.18°C; we are starting from a slab at a uniform temperature equal to the fixed temperature on the right-hand side. The temperature distribution $T_1(x)$ at $t = \Delta t$ is thus obtained by simultaneously solving the 3 × 3 system of linear equations:

$$\begin{bmatrix} 5 & -1.7 & 0 \\ -1.7 & 10 & -2.3 \\ 0 & 0 & 1 \end{bmatrix} \begin{bmatrix} T_1 \\ T_2 \\ T_3 \end{bmatrix} = \begin{bmatrix} 237.54 \\ 235.08 \\ 39.18 \end{bmatrix}$$

which yields the solution T_1 = 62.157°C and T_2 = 43.086°C. The results show that the temperature rises with time at nodes one and two, as expected. The calculation can be continued for further time using $T^1(x)$ as the initial condition.

At this point, it is time to introduce matrix representation for the finite element equations and the integral terms. This representation enables us to express the finite element solution procedure more clearly and concisely in solving both steady-state and time-dependent problems; it will also become mandatory when dealing with multidimensional elements.

3.7 MATRIX FORMULATION

The finite element method is based on the numerical approximation of the dependent variable at specific nodal locations; a set of simultaneous linear algebraic equations is produced that must be solved either directly or iteratively. For the example problems previously discussed, the number of unknowns, and hence equations, were few and could be solved easily by hand. However, for most problems, the number of nodes, and therefore unknowns, will become much larger and a computer will be needed to perform the solution.

The coefficients matrices that we have used previously are formulated from evaluations performed locally over each element and assembled into global arrays. In other words, the local coefficients matrices obtained from each element are "stuffed" into a large matrix, which contains all the local element contributions. This procedure is easily performed via "do loops" within a computer program. Once we formulate the finite element algorithm for one element, the general nature of the method allows us to use the same procedure for all elements. Thus, we can construct the set of global matrices, which are based on the local element matrix evaluations, and then solve the resulting matrix equation however we wish.

It is convenient to define the operations in matrix notation. The local mass matrix for the time derivative term is evaluated as

$$\mathbf{M}^{(e_k)} = \left[m_{ij}^{e_k} \right] = \int_0^{h^{(e_k)}} \left[N_i^{(e_k)} \; N_j^{(e_k)} \right] dx \tag{3.104}$$

The global matrix $\mathbf{M}$ is formed as

$$\mathbf{M} = \left[m_{ij} \right] = \sum_{k=1}^{n} \mathbf{M}^{(e_k)} = \sum_{k=1}^{n} \int_0^{h^{(e_k)}} \left[N_i^{(e_k)} \; N_j^{(e_k)} \right] dx \tag{3.105}$$

where n is the number of elements. The summation implies that each element matrix is an $(n+1)\times(n+1)$ matrix obtained from expanding the element matrices with zeros in all other locations.

As an example, consider the discretization of the interval $0 \le x \le 1$ using three linear elements of equal length. Evaluation of the local mass matrices using Eq. (3.104) yields

$$\mathbf{M}^{(e_1)} = \mathbf{M}^{(e_2)} = \mathbf{M}^{(e_3)} = \frac{1}{18}\begin{bmatrix} 2 & 1 \\ 1 & 2 \end{bmatrix}$$

and the assembled matrix $\mathbf{M}$ is

$$\mathbf{M} = \begin{bmatrix} 1/9 & 1/18 & 0 & 0 \\ 1/18 & 1/9 & 0 & 0 \\ 0 & 0 & 0 & 0 \\ 0 & 0 & 0 & 0 \end{bmatrix} + \begin{bmatrix} 0 & 0 & 0 & 0 \\ 0 & 1/9 & 1/18 & 0 \\ 0 & 1/18 & 1/9 & 0 \\ 0 & 0 & 0 & 0 \end{bmatrix}$$

$$+ \begin{bmatrix} 0 & 0 & 0 & 0 \\ 0 & 0 & 0 & 0 \\ 0 & 0 & 1/9 & 1/18 \\ 0 & 0 & 1/18 & 1/9 \end{bmatrix} = \begin{bmatrix} 1/9 & 1/18 & 0 & 0 \\ 1/18 & 2/9 & 1/18 & 0 \\ 0 & 1/18 & 2/9 & 1/18 \\ 0 & 0 & 1/18 & 1/9 \end{bmatrix}$$

Because a number of entries in the matrix **M** are zero, it is called a *sparse matrix.* In fact, all global matrices arising from finite element discretizations are sparse. In the special cases of one-dimensional linear and quadratic elements, the matrices will be tri-diagonal and penta-diagonal, respectively.

In a similar fashion, the diffusion term gives rise to

$$\mathbf{K} = \sum_{k=1}^{n} \mathbf{K}^{(e_k)} = \sum_{k=1}^{n} \int_0^{h^{(e_k)}} K^{(e_k)} \left[\frac{dN_i^{(e_k)}}{dx} \frac{dN_j^{(e_k)}}{dx} \right] dx \tag{3.106}$$

The matrix **K** is normally referred to as the "stiffness" or "conductance" matrix.

The integral containing the contributions of known functions such as the source term is formulated as a column matrix, with length equal to the number of nodes. We write it as

$$\mathbf{F} = \sum_{k=1}^{n} F^{(e_k)} = \sum_{k=1}^{n} \int_0^{h^{(e_k)}} \left[N_i^{(e_k)} \right]^T Q^{(e_k)} dx \tag{3.107}$$

where $Q^{(e_k)}$ denotes the restriction of Q to element e_k and $[N_i^{(e_k)}]^T$ is a column matrix.

Contributions from a heat flux boundary condition are added to the stiffness matrix (3.106) if the dependent variable appears, or to the vector **F** if only known data are involved. The vector **F** is usually called the *load* vector.

The time-dependent conduction equation can then be expressed as

$$\mathbf{M}\dot{\mathbf{T}} + \mathbf{K}\mathbf{T} = \mathbf{F} \tag{3.108}$$

where **T** is the vector of nodal unknowns and $\dot{\mathbf{T}}$ the time derivatives. Using Eqs. (3.98) and (3.99) to replace the time derivative $\dot{\mathbf{T}}$ in Eq. (3.108), the fully discretized system of linear algebraic equations that results can be written as

$$(\mathbf{M} + \theta \Delta t \mathbf{K})\mathbf{T}^{n+1} = (\mathbf{M} + (\theta - 1)\Delta t \mathbf{K})\mathbf{T}^n + \Delta t(\theta \mathbf{F}^{n+1} + (1-\theta)\mathbf{F}^n) \tag{3.109}$$

Equation (3.109) represents the solution algorithm for the time-dependent conduction equation using the θ-method. The advantage of this kind of representation is that now we can work symbolically with Eq. (3.109) using the theory of matrices. For example, the solution $\mathbf{T}^{n+1}$ is given by

$$\mathbf{T}^{n+1} = (\mathbf{M} + \theta \Delta t \mathbf{K})^{-1} \left[(\mathbf{M} + (\theta - 1)\Delta t \mathbf{K})\mathbf{T}^n + \Delta t(\theta\ \mathbf{F}^{n+1} + (1-\theta)\mathbf{F}^n) \right] \tag{3.110}$$

if $(\mathbf{M} + \theta \Delta t \mathbf{K})$ is an invertible, nonsingular matrix.

As we proceed throughout the rest of this book, we utilize matrix notation to simplify our algebraic expressions. However, we illustrate the local, elemental evaluations of the matrices upon which the computer programs are based.

3.8 SOLUTION METHODS

In all the preceding sections, application of the finite element method to the governing equations led to a set of linear equations consisting of matrices associated with various terms of the original differential equations, such as Eq. (3.109). We can further express this equation in the form

$$\mathbf{A}\phi = \mathbf{B} \tag{3.111}$$

where

$$\mathbf{A} = \mathbf{M} + \theta\Delta t\mathbf{K} \tag{3.112a}$$

$$\phi = \mathbf{T}^{n+1} \tag{3.112b}$$

and

$$\mathbf{B} = \left(\mathbf{M} + \left(\theta - 1\right)\Delta t\mathbf{K}\right)\mathbf{T}^{n} + \Delta t\left(\theta\mathbf{F}^{n+1} + \left(1 - \theta\right)\mathbf{F}^{n}\right) \tag{3.112c}$$

All the terms appearing in **A** and **B** are known, so we can readily solve for the unknown values ϕ in Eq. (3.111).

However, as indicated earlier, solving large systems of equations can become very time-consuming and expensive, even with mainframe computers, and more so using a personal computer. This situation introduces the need of special methods for the solution of large systems of linear equations such as Eq. (3.111). The mathematical discipline that deals with these methods is called *numerical linear algebra*, and the reader with some background will recognize methods by the name of Gaussian elimination, Jacobi and Gauss-Seidel iterations, LU decompositions, successive overrelaxation, conjugate gradient, etc. Development of matrix solution techniques in numerical linear algebra is a field in itself (see Varga, 1962; Isaacson and Keller, 1966; Hageman and Young, 1981); our purpose here is only to familiarize the reader with the basic ideas in order to understand their implementation into computer programs.

The solution of Eq. (3.111) is obtained by multiplying both sides by A^{-1} (note that $A^{-1}A \equiv I$, which is the identity matrix). Thus,

$$\phi = \mathbf{A}^{-1}\mathbf{B} \tag{3.113}$$

which is difficult to execute if **A** is large. Fortunately, efficient numerical methods exist that allow us to find ϕ without the need of finding A^{-1}. Matrix algebra, including finding the inverse of a matrix, is discussed in more detail in Appendix A.

One of the simplest and most efficient schemes for solving equations of the form of Eq. (3.113) is the family of iterative methods. An iterative method is an approximate method; i.e., it uses initial guesses to start the solution procedure, then iterates to obtain refined estimates of the solution. Equation (3.113) consists of a set of n equations, either linear or nonlinear, with n unknowns (ϕ). Matrix **A** contains $n \times n$

coefficients (although in practice many of the coefficients are zero). If we assume the diagonal coefficients $a_{ij} (i = 1, ..., n)$ are not zero (which is valid in the finite element method), it is relatively easy to rearrange the equations in a way that leads to a method of solving for the unknown values ϕ_i. Consider

$$\begin{aligned}
\phi_1^{(k+1)} &= -\frac{1}{a_{11}}(a_{12}\phi_2^{(k)} + a_{13}\phi_3^{(k)} + ... + a_{1n}\phi_{n+1}^{(k)}) + \frac{b_1}{a_{11}} \\
\phi_2^{(k+1)} &= -\frac{1}{a_{22}}(a_{21}\phi_1^{(k)} + a_{23}\phi_3^{(k)} + ... + a_{2n}\phi_{n+1}^{(k)}) + \frac{b_2}{a_{22}} \\
&\vdots \\
\phi_{n+1}^{(k+1)} &= -\frac{1}{a_{nn}}(a_{n1}\phi_1^{(k)} + a_{n2}\phi_2^{(k)} + ... + a_{n,n-1}\phi_n^{(k)}) + \frac{b_n}{a_{nn}}
\end{aligned} \tag{3.114}$$

where k denotes the iteration index.

To solve Eq. (3.111), an initial guess is made, $\phi^{(0)}$, and the guessed value ($\phi_i^{(0)} = 0$, for example) is substituted into the right-hand side of Eq. (3.114). This yields a new estimate $\phi^{(0)}$, to the solution ϕ, which is hoped to be better than the previous estimate, $\phi^{(0)}$. We continue this procedure to obtain $\phi^{(2)}$, $\phi^{(3)}$, ... $\phi^{(k)}$, until the solution converges. Convergence is usually determined by calculating either the relative or absolute error (differences) between iterates (k), i.e.,

$$\max \frac{\left|\phi_i^{(k+1)} - \phi_i^{(k)}\right|}{\left|\phi_i^{(k+1)}\right|} < \varepsilon \tag{3.115}$$

or

$$\max \left|\phi_i^{(k+1)} - \phi_i^{(k)}\right| < \varepsilon \tag{3.116}$$

This type of iteration is known as linear iteration. Although simple, the method may require many iterations before convergence is achieved in large problems. Acceleration of the convergence is necessary to produce an efficient, practical scheme for solution on small computers.

The matrix **A** must also satisfy certain conditions in order for convergence to occur. It is not within the scope of this text, however, to discuss these conditions except in their simplest form. Except for a few comments, we will be satisfied to state that for the type of finite element discretizations considered here, these conditions are indeed satisfied and convergence will generally occur. For more detailed discussions on convergence of iterative methods for the solution of linear algebraic systems of equations, the reader should consult the books by Varga (1962), Isaacson and Keller (1966), and Hageman and Young (1981).

The Gauss-Seidel method is particularly well suited for solving large equation sets. The method is simple to implement, is computationally efficient, and less prone to round-off error (as dictated by the number of iterations) than are direct elimination methods. However, a cautionary note must be made regarding such schemes—there are situations in which the Gauss-Seidel method will not converge. These conditions generally occur when the matrix is ill-conditioned.* For the interested reader, the text by Chapra and Canale (1988) describes such situations in more detail.

Because matrix inversion is computationally slow and can require excessive storage, matrix multiplication and scalar diversion are used to obtain the solution of Eq. (3.113). Although somewhat redundant in iterative methods, such operations are quick. The algorithm for Gauss-Seidel iterations can be expressed as

$$\phi_i^{(k+1)} = \frac{b_i}{a_{ij}} - \sum_{j=1}^{i-1} \frac{a_{ij}}{a_{ii}} \phi_j^{(k)} \quad k = 1, 2, \ldots \tag{3.117}$$

Notice that the most recent update for $\phi_i^{(k+1)}$ is used in the iterative process, which makes the scheme faster and more efficient. A condition for convergence requires that the matrix be *diagonally dominant*, that is,

$$|a_{ii}| > \sum_{\substack{j=1 \\ j \neq 1}}^{n} |a_{ij}| \qquad i = 1, 2, \ldots, n \tag{3.118}$$

If this condition is satisfied, the solution will converge no matter what value is used for the initial vector (which is why so many use zero as an initial estimate).

We can express Eq. (3.113) in terms of upper and lower triangular matrices:

$$(\mathbf{L} + \mathbf{D})\phi = -\mathbf{U}\phi + b \tag{3.119}$$

where **L, D,** and **U** are square matrices defined as follows:

$$\mathbf{L} = [\ell_{ij}] = \begin{cases} a_{ij} & if\ i > j \\ 0 & if\ i \le j \end{cases} \tag{3.120}$$

which is called a lower triangular matrix;

$$\mathbf{D} = [d_{ij}] = \begin{cases} a_{ij} & if\ i = j \\ 0 & if\ i \neq j \end{cases} \tag{3.121}$$

* Ill-conditioning of a matrix occurs when small changes in the coefficients result in large changes in the solution.

which is called a diagonal matrix; and

$$\mathbf{U} = \left[u_{ij} \right] = \begin{cases} a_{ij} & \text{if } i < j \\ 0 & \text{if } i \geq j \end{cases} \tag{3.122}$$

which is an upper triangular matrix. We can rewrite Eq. (3.117) in matrix form as

$$\phi^{(k+1)} = -\mathbf{D}^{-1}\mathbf{L}\phi^{(k+1)} - \mathbf{D}^{-1}\mathbf{U}\phi^{(k)} + \mathbf{D}^{-1}\mathbf{b} \tag{3.123}$$

or, multiplying through $\mathbf{D}$ and solving for $\phi^{(k+1)}$,

$$\phi^{(k+1)} = (\mathbf{D} + \mathbf{L})^{-1}(\mathbf{b} - \mathbf{U}\phi^{(k)}) \tag{3.124}$$

The speed with which the iterations converge to the solution depends on the magnitude of the diagonal terms as given in Eq. (3.118). More strongly diagonally dominant matrices will yield faster convergence. Obviously, when the initial (guessed) values are close to the true solution, only a few iterations are required. Successive overrelaxation (SOR) is a popular variant of the Gauss-Seidel method in which acceleration parameters, or relaxation factors, are used to accelerate convergence (Conte, 1965; Chapra and Canale, 1988).

Actual implementation of the algorithm defined by Eq. (3.124) is fairly easy with a computer code. Formation of the upper and lower triangular matrices are conveniently handled through a simple "do loop" instruction. Examples of source code listings in FORTRAN, C/C++, and JAVA can be found in the series of *Numerical Recipes* books published by Cambridge University Press (1999).

Iterative methods are advantageous when the mesh is structured so that the banded character of the matrix is not altered (e.g., in one-dimensional problems or in two- and three-dimensional problems defined over rectangular or parallelepiped domains) and when only a few solutions are needed, as in the case of linear steady-state problems. If the computational mesh is irregular, or if solution of the same linear system of equations must be performed over and over again (as would be the case in a linear time-dependent problem or in an optimization problem where the same system must be solved for many different right-hand sides), iterative methods become less attractive. The main advantage of iterative methods lies in the fact that only the nonzero elements of the matrix must be stored in memory. This can mean enormous storage savings when compared to direct elimination procedures. The disadvantage is that the implementation of the algorithm procedure must be repeated every time the right-hand-side **B** in Eq. (3.111) is changed.

Elimination methods in which the coefficients matrix **A** is decomposed into the product of a lower and an upper triangular matrix, known as LU decomposition,*

* These matrices are not upper and lower triangular in the same sense as the definitions given by Eqs. (3.120) and (3.122). In this case, the matrices **L** and **U** both contain nonzero diagonal terms. Furthermore, the nonzero coefficients are, in general, different from the corresponding coefficients a_{ij} of the matrix **A**.

are attractive alternatives to iterative methods. The time-consuming elimination step can be written to take advantage of the sparse nature of the coefficients matrix **A**. Elimination procedures which employ LU decomposition are the most popular techniques for the direct solution of systems of equations (Atkinson, 1985; Chapra and Canale, 1988).

LU decomposition methods are a variant of Gauss elimination; by "decomposing" the **A** matrix into the product of an **L** and **U** matrix, an algorithm can be obtained which is more efficient than the original elimination. Let us rewrite Eq. (3.111) in the form:

$$\mathbf{A}\phi - \mathbf{B} = 0 \tag{3.125}$$

We can express Eq. (3.125) in the form of an upper triangular system, that is,

$$\mathbf{U}\phi - \mathbf{D} = 0 \tag{3.126}$$

If we now premultiply Eq. (3.126) by a lower triangular matrix **L** and require the resulting system to be equal to the original system of equations given by Eq. (3.125), we have

$$\mathbf{L}\left(\mathbf{U}\phi - \mathbf{D}\right) = \mathbf{A}\phi - \mathbf{B} \tag{3.127}$$

The conditions under which (3.127) is valid are

$$\mathbf{LU} = \mathbf{A} \tag{3.128}$$

and

$$\mathbf{LD} = \mathbf{B} \tag{3.129}$$

Equation (3.128) is referred to as an LU decomposition of **A**. A decomposition of this type can be carried out for any nonsingular matrix **A**; however, there is no unique way to perform it. Different methods to obtain such decompositions are found in the literature. For example, if all the diagonal elements of the matrix **L** are set to equal to one, a unique decomposition follows, which is known as a Doolittle reduction. If the matrix **A** is symmetric, the decomposition can be carried out in such a way that

$$\mathbf{L} = \mathbf{U}^{\mathrm{T}} \tag{3.130}$$

This is called Cholesky's method, or the method of square roots, because a square root operation is needed to obtain the diagonal elements u_{ii} (this is discussed in more detail later).

Equation (3.129) is known as a forward substitution and yields the vector **D**. To obtain the solution vector, ϕ, one more step is needed. This is provided by the solution of Eq. (3.126) once **D** has been obtained and involves a backward substitution of the same kind as required in a direct Gaussian elimination. Algorithms and subroutines listings employing these procedures are presented in most texts on numerical methods (see Conte, 1965; Isaacson and Keller, 1966; Atkinson, 1985; and Chapra and Canale, 1988).

A symmetric banded matrix is a square matrix, which has all coefficients equal to zero except for a band centered on the main diagonal. Such systems frequently occur in the solution of differential equations, particularly in engineering and scientific problems. In finite elements, such matrices occur in linear structural and diffusion-related problems. When the equations are nonlinear, as in fluid flow, nonsymmetric banded matrices may occur. In this introductory text, we are concerned only with linear symmetric banded systems equations; however, the matrix solver included with the FORTRAN software will handle either symmetric or nonsymmetric banded matrices. Direct methods for the solution of systems of linear algebraic equations, such as Gauss elimination or LU decomposition, offer great advantages over iterative methods when the geometry of the problem to be solved is irregular and the methods yield matrices with a poor banded structure. They are also more convenient when the same system must be solved for many different right-hand sides since the decomposition only needs to be done once. Solutions are obtained by simple forward and backward substitutions that can be performed very efficiently. For the solution of strongly banded systems of equations, such methods are typically less efficient and introduce unnecessary computational effort in storage and manipulation of zero values.

For a symmetric matrix with a half- or semi-bandwidth of ℓ, as depicted in Figure 3.9, only the upper triangular portion of order $n \times \ell$ must be stored in core at all times. The total storage requirement is $\ell\left[n-\left(\ell-1\right)/2\right]$.

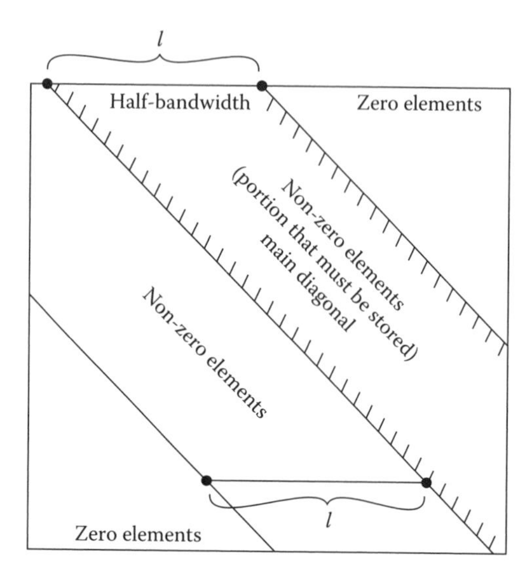

Figure 3.9 Banded $n \times n$ matrix **A** with half-bandwidth ℓ and bandwidth $2\ell-1$, indicating storage requirement in **A** in symmetric

The banded nature of the matrix, and hence the size of ℓ, depends directly on the way the nodes are numbered. The bandwidth is determined by the maximum difference in node numbers within an element over all elements. This fact becomes very important when we advance to two- and three-dimensional elements, and points out the necessity for "optimizing" nodal numbering. For one-dimensional elements, the node numbering is always sequential and the resulting global matrix tri-diagonal for linear elements and penta-diagonal for quadratic elements. However, the reader should note that if the node numbering is not sequential, this will not be so. A further example is given in Exercise 3.34, where the use of linear elements does not lead to tri-diagonal matrices.

One of the most efficient, and most popular, approaches to solving banded systems of equations is the Cholesky decomposition mentioned earlier. This algorithm relies on the fact that a symmetric matrix can be decomposed into lower triangular matrices, which are transposes of each other, i.e.,

$$\mathbf{A} = \mathbf{U}^{\mathrm{T}}\mathbf{U} \tag{3.131}$$

where superscript **T** denotes the transpose (see Appendix A). Equation (3.131) can be written in recurrence form as

$$u_{ii} = \left(a_{ii} - \sum_{k=1}^{i-1} u_{ki}^2 \right)^{1/2}, \qquad i = 1, 2 \ldots, n \tag{3.132}$$

with

$$u_{ij} = \frac{1}{u_{ii}} \left(a_{ij} - \sum_{k=1}^{i-1} u_{ki} u_{kj} \right), \qquad j < i \tag{3.133}$$

providing the matrix is positive definite.* This procedure significantly speeds the LU decomposition step. Forward and backward substitution can then be performed to yield the solution vector. In this method, pivoting is not required to avoid division by zero. The positive definiteness of the matrix **A** guarantees that $u_{ii} > 0$ for all values of i.

Sophisticated algorithms exist in the open literature. The reader interested in matrix solvers can find such methods described in several technical journals, e.g., *Journal of Computational Physics* or the series of journals published by the Society of Industrial and Applied Mathematics (*SIAM*); an especially fast matrix solver is described by Young (1989).

* A positive definite matrix is one in which the product $\phi^{\mathrm{T}}\mathbf{A}\phi > 0$ for all nonzero values of ϕ. In most cases, engineering problems yield positive definite matrices.

3.9 CLOSURE

The purpose of this chapter has been to illustrate and establish application of the finite element method to a linear partial differential equation. Beginning with the simple steady-state diffusion equation for a scalar quantity, the finite element method was applied to progressively more difficult equations, ultimately leading to the one-dimensional, time-dependent diffusion transport equation commonly found in engineering and scientific problems. Following derivation and use of various degrees of shape functions in several example problems, a generalized finite element procedure was seen to evolve which could be applied to a wide range of problems. The process of assemblage over the entire set of elements, or mesh, was shown to produce a set of global matrices. Reduction of the global matrix system to a simple linear (matrix) equation of the form $\mathbf{A}\phi = \mathbf{B}$, with $\phi \equiv$ column vector containing the unknown variables, permits standard matrix solution routines to be utilized in solving for ϕ.

Most of the fundamental "tools" that pertain to the finite element method can be developed from analysis of one-dimensional problems. The basic ingredients of the finite element method can be summarized as follows:

1. Apply the weighted residuals method to the governing equations and associated boundary conditions.
2. Choose appropriate shape functions and weighting factors (linear, quadratic, cubic, etc.).
3. Discretize the x-axis and evaluate the Galerkin approximation for the general finite element domain.
4. Assemble the element contributions into a set of global matrices.
5. Apply the necessary boundary condition data (consisting of known values) to the load vector.
6. Solve for the column vector of unknown values, ϕ, using a suitable matrix solution routine, starting with initial condition data, $\phi_{t=0} = \phi_0$, and proceeding through time for $t^{n+1} = t^n + \Delta t$, or to convergence to steady-state conditions.

At this point in the text, the reader should begin to gain insight into the mathematical basis of finite elements stemming from these simple one-dimensional examples. For those less familiar with finite element methodology, we suggest at this point they review these least few sections once more before continuing on to the more advanced material in the remaining chapters. In the next chapter, we extend the algorithm to two-dimensional problems.

EXERCISES

3.1 Repeat the development of expressions (3.7) to (3.12) using two linear elements to discretize the interval $0 \leq x \leq L$. Then combine the element shape functions to construct the global shape functions and compare your result to that obtained using expressions (3.2) to (3.4) directly, with $n = 2$.

3.2 Show that if in Eq. (3.6) we replace $T_{n+1} = T_L$, the boundary condition (2.3) is automatically satisfied.

3.3 Derive expression (3.25) from (3.24) for cubic elements. Then find a relation between global and local reference systems such as the ones given in Figures 3.3 and 3.6 for linear and quadratic elements.

3.4 Write the matrix expression for the derivatives of $T^{(e)}(x)$ using cubic elements corresponding to (3.16) and (3.23) for the linear and quadratic cases.

3.5 Interpolate the function $T(x) = \sin x$ in the interval $0 \le x \le \pi$ using two equal-length quadratic elements. Calculate the derivative of $T(x)$ at $x = (\pi/2)$ from each element and compare.

(a) Show that if $T(x)$ is linear function over the interval $0 \le x \le L$, then its one quadratic element interpolant over the interval gives

$$\sum_{i=1}^{3} N_i^{(e)}(x)T_i \equiv T(x)$$

(b) Repeat part (a) for the case where $T(x)$ is quadratic using a cubic element interpolant.

(c) Show how the above leads to the conclusion that for linear, quadratic, and cubic elements we have

$$\sum_{i=1}^{k} N_i^{(e)}(x) = 1$$

for all x and

$$\sum_{i=1}^{k} \frac{dN_i^{(e)}(x)}{dx} \equiv 0$$

where $k = 2, 3, 4$ for linear, quadratic, and cubic elements, respectively.

3.7 Using integration by parts, obtain Eq. (3.33) from Eq. (3.31).

3.8 Show that after substituting Eq. (3.32) into the first line of (3.33) we obtain (3.34).

3.9 Fill in the details leading to Eq. (3.35) from (3.34).

3.10 Derive Eq. (3.36) from the second line of Eq. (3.33).

3.11 (a) Solve Eq. (3.37) for T_1 and T_2. Then compare with the exact solution for Eqs. (3.27) to (3.29).

(b) Find the solution using one quadratic element, i.e., solve Eq. (3.43).

(c) Let T_L = 100°C, q = 0.15 cal/cm², $\mathbf{K}$ = 0.15 cal/cm²s C, L =10 cm, and Q = 1.5 cal/cm³s. Sketch the solutions obtained using two linear elements and one quadratic element and compare the results.

3.12 Set $T_3 = T_L$ in Eq. (3.37) and show that the resulting system is equivalent to the uncoupled systems (3.38) and (3.39).

3.13 Calculate the heat flux at the left-hand boundary point from the solution obtained using linear elements in (3.38) differentiating (3.32), and compare with q. What happens? Now rewrite the solution in terms of $h^{(e_1)}$ Instead of L, and show that as $h^{(e_1)} \to 0$, we do obtain q. This is because the boundary condition (3.28) at $x = 0$ has been imposed "weakly" through the integral statement, and not forced to be satisfied by the solution, as in the case of (3.29) at $x = L$.

3.14 Now calculate the flux in Exercise 3.13 from Eq. (3.39). Discuss the differences between this solution and the one in the previous problem.

3.15 Fill in the details leading from Eq. (3.30) to (3.41).

3.16 Discretize Eq. (3.27) using five linear elements of equal length and find the assembled global system of equations assuming constant K and q.

3.17 Starting from Eq. (3.51) fill in the details leading to Eqs. (3.52) and (3.53).

3.18 Verify the solution obtained for Example 3.1.

3.19 Solve Example 3.1 using three equal size elements.

3.20 Show that Eq. (3.64) yields the element equations for the axisymmetric problem with convective boundary conditions applied at either or both ends.

3.21 The spherical coordinates of the energy equation take the form:

$$\frac{1}{r^2}\frac{d}{dr}\left(Kr^2\frac{dT}{dr}\right) = Q$$

Find the weak weighted residuals formulation for general convective boundary conditions at both ends.

3.22 Solve the energy equation in spherical coordinates using three linear elements for an insulating layer of a cryogenic storage tank with $r_1 = 0.3$ m and $r_2 = 0.35$ m. The inner wall is in perfect contact with a metal sheet at 800 K, the outer wall is exposed to convection with $T = 300$ K and $h = 25$ W/m^{2}K. The heat conduction coefficient is $K = 0.0017$ W/mK.

3.23 Consider a solid tube of length 20 cm with uniform heat generation $Q = 20{,}000$ W/m^3. The linear surface is kept cool at $T_1 = 20$°C and the outer surface is insulated, $K = 0.6$ W/mK, $r_1 = 10$ cm.
(a) Solve using three linear elements.
(b) Solve using one quadratic element.

3.24 Solve Exercise 3.23 with a convective boundary condition at the right-hand side, $h = 25$ W/m^{2}K and $T = 100$°C.

3.25 In a composite slab material 1 has a thickness of 2 cm, $K_1 = 0.3$ W/mK, $h_1 = 10$ W/m^{2}K and the left end is in contact with fluid at $T_1 = 35$°C. Material 2 is 3 cm thick and in perfect contact at the interface with material 1. $K_2 = 1$ W/mK, $h_2 = 5$ W/m^{2}K, and $T_2 = 20$°C. Solve using two linear elements.

3.26 The error in finite element approximations behaves like $e = Ch^p$, where C is a constant, h the uniform mesh size, and p a power that determines the order of the approximation. We look at this expression as giving us a function of h. To determine p, we perform two or more calculations using different uniform meshes. Then, taking logarithms on both sides of the equation, we get $\ell n\, e = p$ $\ln h + \ell n\, c$. Since the constant only provides a shift of the line, we can assume $\ell n\, C = 0$ and plot $\ell n\, e$ vs. $\ell n\, h$. The slope of this line gives us p, the rate of convergence. In Example 3.1, use this method to determine the effective order of convergence at the convective boundary.

3.27 Solve the problem of Example 3.2 using three equally spaced finite elements. Then obtain a solution using three linear elements of sizes $h^{(e_1)} = 0.2$, $h^{(e_2)} =$ 0.3, and $h^{(e_3)} = 0.5$, and compare the solutions. The analytical solution for this problem is given by

$$T(r) = \frac{h r_1 (T_L - T_\infty)}{K - h r_1 \ell n\left(\frac{r_1}{r_2}\right)} \ell n\left(\frac{r_1}{r_2}\right) + T_L$$

3.28 (a) Show that if $x_{2i} = (x_{2i-1} + x_{2i+1})/2$ is replaced into Eq. (3.74) together with Eqs. (3.73) for the shape functions, it reduces to Eq. (3.65) in terms of the element end points x_{2i-1} and x_{2i+1}.

(b) Find the element shape functions for a quadratic element with nodes at $x_1 = 0$, $x_2 = 1/3$, and $x_3 = 1$.

3.29 Write the expressions for the shape functions of a cubic element in natural coordinates and the associated coordinate transformation. Show that the transformation reduces to Eq. (3.65) when $x_2 = (2x_1 + x_4)/3$ and $x_3 = (x_1 + 2x_4)/3$.

3.30 Adding time dependence to Exercise 3.22 and = 5 × 10^{-7} m^2/s, solve for the first two time steps using two equal length elements with $T(x,0) = 80$ and $t = 10$ s.

3.31 Consider the time-dependent heat conduction defined in Exercise 3.23 with = 8 × 10^{-5} m^2/s and $T(x,0)$ = 20°C. Use a computer code to find the time evaluation to steady-state at the outer surface of the cylinder.

3.32 Repeat for Exercise 3.24.

3.33 Solve Exercise 3.25 as a time-dependent problem with $\alpha_1 = 4 \times 10^{-6}$ m^2/s, $\alpha_2 = 6 \times 10^{-7}$ m^2/s, and $T(x,0)$ = 35°C.

3.34 Consider the problem of heat conduction in a circular ring discretized using n linear elements, as shown in (a). The x-coordinate is measured in the circular direction, as indicated in the figure, and, in the x-coordinate the problem can be considered one-dimensional. The boundary conditions are that the temperature and heat flux must both be continuous at $x = 0$. Describe the structure of the coefficient matrix obtained from a finite element discretization using

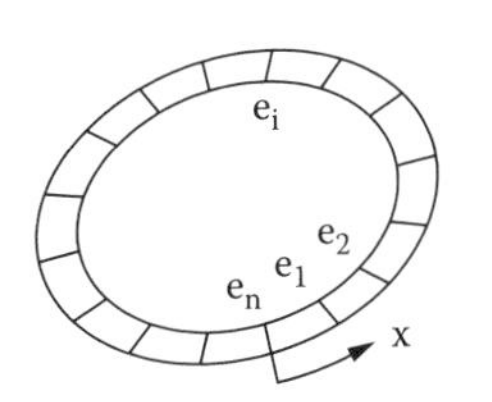

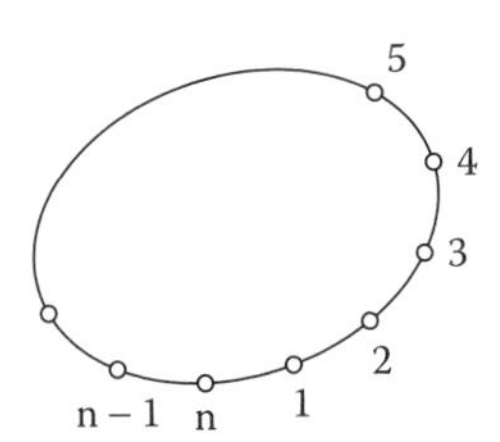

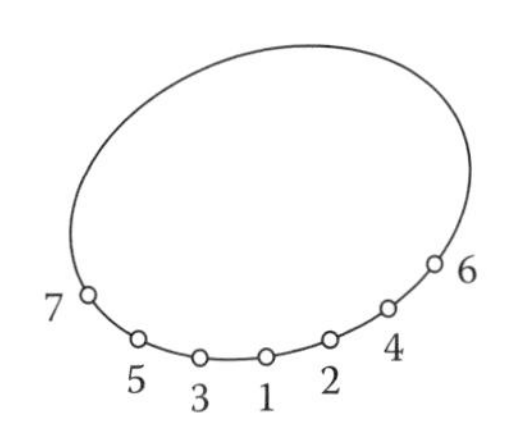

linear elements when the nodes are numbered in the two different ways indicated in (b) and (c). In both cases, find the half-bandwidth ℓ when $n = 20$.

3.35 Determine the temperature at a point 1.5 cm from the left end of the linear element. Show, in the figure, if the end temperatures are 100°C and 33°C, respectively. Also, determine the temperature gradient.

$x_1 = 1$ cm   2.5 cm $x_2 = 4$ cm   x

1   2

100 C   33 C

3.36 Find the nodal temperatures in a two-element model used to discretize a rod attached to a wall, as shown in the figure, where $K = 75$ W/m²C, $h = 15$ W/m²C, $L = 10$ cm, and $T = 60$°C. The temperature at the wall is $T_w = 150$°C.

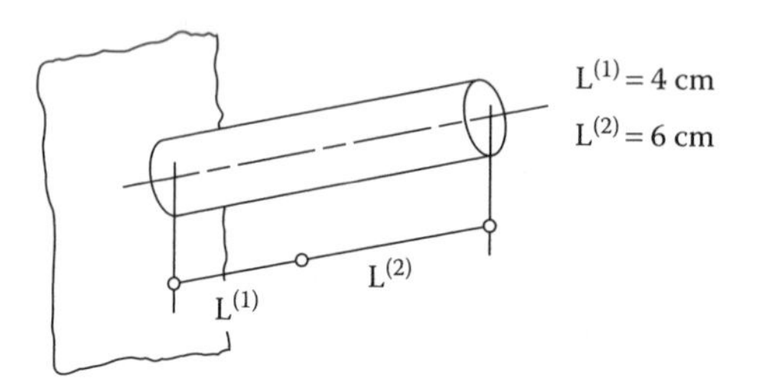

3.37 The deflection of a simply supported beam is governed by the following equation and boundary conditions:

$$E_1 \frac{d^2 y}{dx^2} + M = 0 \text{ and } y(0) = y(L) = 0$$

where E_1 is a constant and M is a distributed bending moment. Derive the weak statement form of the governing equation.

3.38 Solve Example 3.1 using six linear elements of equal length, and compare results with Example 3.1 and Example 3.5.

3.39 Consider Example 3.6 with an added source term $\overline{Q} = 10C/s$ (which is Q/c_p) and eight linear elements.

(a) Find the matrices **M** and **K** and the vector **F** in expression (3.108).

(b) Write Eq. (3.109) for $\phi = 1/2$ and $t = 0.01$ s.

(c) Repeat parts (a) and (b) using four quadratic elements.

3.40 Solve the transient one-dimensional heat conduction problem of Example 3.6 using the backwards implicit marching scheme. What is the temperature after 50 time steps if $t = 0.001$ s.

3.41 Fill in the details leading to Eq. (3.79).

3.42 Verify Eq. (3.85) by filling in the missing details.

3.43 Starting from Eq. (3.97), use (3.98) and (3.99) to obtain Eq. (3.100).

REFERENCES

Atkinson, K. (1985). *Elementary Numerical Analysis*. New York: John Wiley & Sons.

Chapra, S. C. and Canale, R. P. (1988). *Numerical Methods for Engineers*. New York: McGraw-Hill.

Conte, S. D. (1965). *Elementary Numerical Analysis*. New York: McGraw-Hill.

Hageman, L. A. and Young, D. M. (1981). *Applied Iterative Methods*. New York: Academic Press.

Heinrich, J. C. and Pepper, D. W. (1999). *Intermediate Finite Element Method: Fluid Flow and Heat Transfer Applications*. Philadelphia: Taylor & Francis.

Isaacson, E. and Keller, H. B. (1966). *Analysis of Numerical Methods*. New York: John Wiley & Sons.

Numerical Recipes in C/C++ (1999). Cambridge: Cambridge University Press.

Numerical Recipes in FORTRAN (1999). Cambridge: Cambridge University Press.

Numerical Recipes in JAVA (1999). Cambridge: Cambridge University Press.

Pepper, D. W. and Baker, A. J. (1979). "A Simple One-Dimensional Finite Element Algorithm with Multidimensional Capabilities." *Num. Heat Transfer* 2:81–95.

Richtmeyer, R. D. and Morton, K. W. (1967). *Difference Methods for Initial Value Problems*. New York: John Wiley & Sons.

Varga, R. S. (1962). *Matrix Iterative Analysis*. Englewood Cliffs, N.J.: Prentice-Hall.

Young, R. C. (1989). *Fast Matrix Solver*, version 4.0. Marlborough, Mass.: Multipath Corp.

Zienkiewicz, O. C. (1977). *The Finite Element Method*. London: McGraw-Hill.

CHAPTER

4

THE TWO-DIMENSIONAL TRIANGULAR ELEMENT

4.1 BACKGROUND

In the engineering world, we quickly discover that most problems are not amendable to solution using one-dimensional analysis. On the other hand, many problems can be "analyzed" and realistic solutions obtained using two-dimensional concepts. It is in the two-dimensional (and three-dimensional) problem domains that the strength of the finite element method becomes evident. The concept of a mesh and the choices available for element discretization become significantly more important than in the one-dimensional examples discussed in Chapter 3.

Close examination of the one-dimensional finite element concepts described in Chapter 3 will show that the only constraint placed on the solution procedure was one dimensionality in the basis functions. Otherwise, the methodology is generic and is applicable to multidimensional problems as well. This is of paramount importance—we can use the basic formulation supplied in Chapter 3 and only have to extend our formulation of the basis functions to two dimensions.

In two-dimensional problems, the physical (or problem) domain is subdivided into subregions, or elements. Many polygonal shapes can be used to define the elements. For example, rectangles are commonly used when the problem domain itself is rectangular (which is a requirement in simple finite difference methods). However, rectangles do not easily fit well when the domain is irregular (an exception to this is the general quadrilateral, which we discuss in Chapter 6). The simplest geometric structure that easily accommodates irregular surfaces is the triangle, and it is one of the most popular element shapes used today.

Two additional constraints now surface when dealing with two-dimensional problems. The first constraint is that considerably more nodes will become involved in the solution process. Recalling the general matrix equation

$$\mathbf{A}\phi = \mathbf{B}$$

one can quickly see that the global matrix, **A**, becomes much larger ($n \times n$ nodes) and sparser; hence, straightforward, direct solution is usually limited to small problems. Second, local mesh refinement is generally required in those regions where the variables of interest vary rapidly or where discontinuities exist. The importance of using efficient matrix solvers becomes evident. Our approach in this chapter, and in the succeeding chapters, is to use either Gauss–Siedel iteration or Cholesky decomposition, which is a modified form of Gauss elimination commonly used in many finite element programs.

In this chapter, we introduce the first two general element forms for discretizing a two-dimensional domain—the triangle. Historically, the finite element method first employed triangular elements to model structural problems. Today, many commercially available finite element codes for structural analysis use a mixture of triangles and quadrilaterals.

4.2 THE MESH

Our selected element of choice is a triangle consisting of three vertex nodes. Knowing where to place optimally and size the elements is more an art than a science. In general, one places more elements in those regions of the physical domain where functions are *expected* to change rapidly. As one can quickly surmise, mesh generation may take several "tries" before a good mesh is achieved. Solution of diffusion-type equations is rather forgiving—we generally obtain solutions (although not highly accurate), even on the coarsest meshes. On the other hand, solutions of nonlinear equations, particularly those which are advection-dominated, are almost guaranteed to require several remeshings before a suitable solution is obtained. Most of the effort expended in using finite elements deals with generating appropriate meshes: hence, practice and careful attention to creating a good mesh pay for itself in the long run. Most practitioners of the finite element method routinely ask their colleagues for opinions regarding their meshes when reviewing problems.

When creating a triangular element, it is recommended that the sides run in the direction of the largest gradient. Elements that are equilateral in shape are more accurate than long, narrow triangular elements. When dealing with curved or irregular boundaries, sides of the element should closely approximate the boundary—using linear, three-noded elements (straight sides) requires many elements to reduce the error in the solution; the use of quadratic, curve-sided triangular elements naturally accommodated curved boundaries but requires more computational effort in the solution process. The choice of mesh generation is conveniently performed using computer programs. The mesh-generation computer code, MESH-2D (available on the Web at femcodes.nscee.edu), permits either 2-D linear or quadratic triangular elements to be generated. The included COMSOL example files were created using the mesh generation preprocessor module that accompanies COMSOL. Persson and Strang (2004) describe a simple 2-D mesh generator that can be used in MATLAB.

The first task in developing a mesh is to subdivide the problem domain into regular geometric shapes, i.e., macroregions consisting of rectangles, circles, etc. Nearly every irregular shape can be reduced to simple geometric primitives. By dealing with primitive simple shapes, it is relatively easy to grid large domains and interlace the elements at the boundaries. Note that we can make use of the general quadrilateral primitive to generate our triangles—a quadrilateral consists of at least two triangular elements. Figure 4.1 shows an irregular shape discretized using linear triangular elements embedded within two macro-quadrilateral elements. Use of quadratic (curve-sided) elements would more accurately approximate the curved boundaries—this concept is utilized in MESH-2D to model irregular shapes accurately.

Because the finite element method is based on the use of unstructured grids, i.e., calculations are performed on an element-by-element basis, we are free to place our elements where we wish and to connect them to other elements without regard for "orthogonality" or sequential node numbering (as in the one-dimensional element). We can, in fact, create a *universal*, or generic element, which will then serve as a model for all our elements in the problem domain. In keeping with the procedure commonly practiced in most finite element schemes, the node numbers associated with a triangular element will be designated 1, 2, and 3 in a counterclockwise order. A typical linear triangular element containing three corner nodes is shown in Figure 4.2.

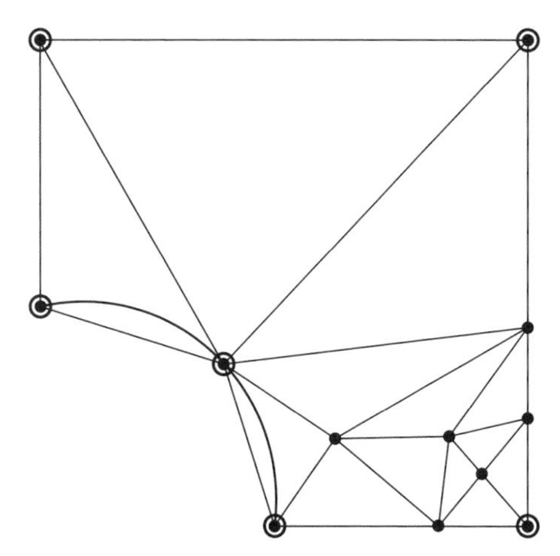

Figure 4.1 Irregular region divided into linear triangular elements; ⊙ denote corners of quadrilateral macro elements

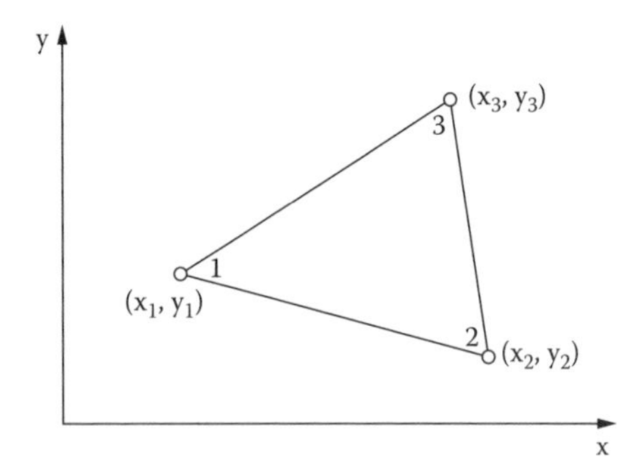

Figure 4.2 Two-dimensional linear triangular element

The generation of a simple mesh using linear triangular elements is relatively easy for simple geometries, and in some instances may not require sophisticated mesh-generation techniques. In this book, we have deliberately kept the problem geometries simple and amendable to coarse-mesh solution. However, the time will quickly come when generation of elements by hand becomes intractable—as in most practical engineering problems—and a detailed mesh will be required. The necessity for automating this procedure via a computer code will be fully appreciated when dealing with two- and three-dimensional geometries. In finite difference methods, it is relatively easy to create rows and columns of lines to create a mesh—providing the lines are orthogonal; i.e., the problem domain permits a simple, *structured* mesh to be generated. However, with the advent of the boundary-fitted coordinates in use today, even finite difference methods require the use of mesh-generating preprocessor codes; these codes are typically more complex and demanding than finite element mesh generators. The EAGLE code, developed by Thompson (1987) and 3DGRAPE, written by Sorenson (1989), are finite difference mesh generators based on boundary-fitted coordinate transformations. GRIDGEN is a commercial mesh generator that spun off from the EAGLE code. pcGRIDGEN is an inexpensive boundary-fitted coordinate mesh generator developed some years ago for the PC market.

PATRAN is a well-known commercial finite element mesh generator that has been around for many years. More recent mesh generators include FEMAP, True-Grid, GAMBIT, and COMSOL. Nearly all commercial finite element codes contain a mesh-generating preprocessor module.

Some of the example problems beginning with this chapter will require the use of a computer code to solve the problems. These problems must first be preprocessed using a mesh generator that creates a mesh consisting of linear or quadratic two-dimensional triangles. Implementation of the resulting element algorithm over all elements follows in exactly the same manner as for the one-dimensional finite element procedure.

There are two mesh generators available on the Web. MESH-1D is a simple one-dimensional mesh generator that is used as a preprocessor for FEM-1D (as discussed in Chapter 3). MESH-2D generates a two-dimensional mesh consisting of either triangles or quadrilaterals, and serves as the preprocessor for FEM-2D. MESH-2D automatically generates node numbers, element numbers, and x, y coordinates after the user has input required boundary data for the geometry. This preprocessor is not overly sophisticated and is written with the intention of illustrating a fundamental mesh generation capability that permits irregular domains to be discretized. A simple 2-D mesh generator developed by Persson and Strang (2004) for MATLAB is also available on the web.

4.3 SHAPE FUNCTIONS (LINEAR, QUADRATIC)

Finite elements can normally be classified into three groups according to their polynomial order: simplex, complex, and multiplex (Oden, 1972). In a simplex element, the number of coefficients in the polynomial is equal to the problem's dimensional space, plus one. The polynomial consists of a constant term, plus linear terms. The two-dimensional triangular simplex element is represented by the polynomial.

$$\phi = \alpha_1 + \alpha_2 x + \alpha_3 y \tag{4.1}$$

The polynomial is linear in x and y and contains three coefficients since the triangle has three nodes, i.e., vertices.

Complex elements utilize a polynomial that contains constant and linear terms, plus higher-order terms. While the shape may be identical to a simplex element, the complex element has additional boundary nodes and can have internal nodes. The interpolating polynomial for a quadratic complex element is

$$\phi = \alpha_1 + \alpha_2 x + \alpha_3 y + \alpha_4 x^2 + \alpha_5 xy + \alpha_6 y^2 \tag{4.2}$$

The six coefficients indicate that the element must have six nodes.

The multiplex elements also use polynomials with higher-order terms; however, the boundaries of the element must be parallel to the coordinate axes. An example of a multiplex element is a rectangular, or quadrilateral, element, which is discussed later in Chapter 5.

4.3.1 Linear Shape Functions

The interpolating polynomial throughout a linear triangular element is defined by relation (4.1), where ϕ is used to represent any unknown variable. At each vertex node,

$$\left.\begin{aligned} \phi &= \phi_1 \qquad at\ x = x_1,\ y = y_1 \\ \phi &= \phi_2 \qquad at\ x = x_2,\ y = y_2 \\ \phi &= \phi_3 \qquad at\ x = x_3,\ y = y_3 \end{aligned}\right\} \tag{4.3}$$

Thus,

$$\left.\begin{aligned} \phi_1 &= \alpha_1 + \alpha_2 x_1 + \alpha_3 y_1 \\ \phi_2 &= \alpha_1 + \alpha_2 x_2 + \alpha_3 y_2 \\ \phi_3 &= \alpha_1 + \alpha_2 x_3 + \alpha_3 y_3 \end{aligned}\right\} \tag{4.4}$$

It is a simple matter now to find the values for α_1, α_2, and α_3, which are

$$\left.\begin{aligned} \alpha_1 &= \frac{1}{2A^{(e)}}\left[\left(x_2y_3 - x_3y_2\right)\phi_1 + \left(x_3y_1 - x_1y_3\right)\phi_2 + \left(x_1y_2 - x_2y_1\right)\phi_3\right] \\ \alpha_2 &= \frac{1}{2A^{(e)}}\left[\left(y_2 - y_3\right)\phi_1 + \left(y_3 - y_1\right)\phi_2 + \left(y_1 - y_2\right)\phi_3\right] \\ \alpha_3 &= \frac{1}{2A^{(e)}}\left[\left(x_3 - x_2\right)\phi_1 + \left(x_1 - x_3\right)\phi_2 + \left(x_2 - x_1\right)\phi_3\right] \end{aligned}\right\} \tag{4.5}$$

where the area, $A^{(e)}$, is defined by

$$2A^{(e)} = \left(x_1y_2 - x_2y_1\right) + \left(x_3y_1 - x_1y_3\right) + \left(x_2y_3 - x_3y_2\right) \tag{4.6}$$

Notice that the area of the element is defined in terms of the nodal coordinates. To obtain the shape functions, we approximate ϕ as a function of the three nodal values such that

$$\phi = N_1\phi_1 + N_2\phi_2 + N_3\phi_3 \tag{4.7}$$

just as in the one-dimensional element. The shape functions are formally defined as

$$N_1^{(e)}(x,y) = \frac{1}{2A^{(e)}}\left[\left(x_2y_3 - x_3y_2\right) + \left(y_2 - y_3\right)x + \left(x_3 - x_2\right)y\right] \tag{4.8}$$

$$N_2^{(e)}(x,y) = \frac{1}{2A^{(e)}}\left[\left(x_3y_1 - x_1y_3\right) + \left(y_3 - y_1\right)x + \left(x_1 - x_3\right)y\right] \tag{4.9}$$

$$N_3^{(e)}\left(x,y\right) = \frac{1}{2A^{(e)}}\left[\left(x_1y_2 - x_2y_1\right) + \left(y_1 - y_2\right)x + \left(x_2 - x_1\right)y\right] \tag{4.10}$$

which conforms to the general relation expressed by Eq. (4.1). An example, the value of N_1 at node 1 can be obtained by substituting $x = x_1$ and $y = y_1$ into Eq. (4.8) (assuming the functional relationships (x, y) and superscript (e) are understood); i.e.,

$$N_1 = \frac{1}{2A}\left(x_2y_3 - x_3y_2 + y_2x_1 - y_3x_1 + x_3y_1 - x_2y_1\right) \tag{4.11}$$

N_1 is zero at nodes 2 and 3 and all points on a line passing through these nodes. We also note that the value in parentheses in Eq. (4.11) is also equal to $2A$; hence,

$$N_1 = \frac{2A}{2A} = 1 \quad x = x_1,\ y = y_1$$

$$N_1 = 0 \qquad x = x_2,\ y = y_2 \text{ and } x = x_3,\ y = y_3$$

The gradient of the variable ϕ is given by the expression

$$\left.\begin{aligned} \frac{\partial \phi}{\partial x} &= \frac{\partial N}{\partial x}\phi_1 + \frac{\partial N_2}{\partial x}\phi_2 + \frac{\partial N_3}{\partial x}\phi_3 \\ \frac{\partial \phi}{\partial y} &= \frac{\partial N_1}{\partial y}\phi_1 + \frac{\partial N_2}{\partial y}\phi_2 + \frac{\partial N_3}{\partial y}\phi_3 \end{aligned}\right\} \tag{4.12}$$

The value of $\partial N_1 / \partial x$, for example, is easily determined from Eq. (4.8):

$$\frac{\partial N_1}{\partial x} = \frac{y_2 - y_3}{2A} \tag{4.13}$$

The remaining derivatives for the shape functions are obtained in a straightforward manner. The value of $\partial \phi / \partial x$ becomes

$$\frac{\partial \phi}{\partial x} = \frac{1}{2A}[(y_2 - y_3)\phi_1 + (y_3 - y_1)\phi_2 + (y_1 - y_2)\phi_3] \tag{4.14}$$

EXAMPLE 4.1

As an illustrative example, we wish to calculate the temperature T, at a specific point within an element. We define the nodal values for T as $T_1 = 10°C$, $T_2 = 20°C$, and $T_3 = 30°C$. The coordinates are $x_1 = 0,\ y_1 = 0;\ x_2 = 2,\ y_2 = 1;\ x_3 = 1,\ y_3 = 2$ (cm). The specific point within the element is located at $x = 0.5,\ y = 0.5$. The element and coordinate system are shown in Figure 4.3.

We let

$$T = N_1 T_1 + N_2 T_2 + N_3 T_3 \tag{4.15}$$

where

$$\left.\begin{aligned} N_1 &= \frac{1}{2A}\left[(2\cdot 2 - 1\cdot 1) + (1-2)x + (1-2)y\right] \\ N_2 &= \frac{1}{2A}\left[(1\cdot 0 - 0\cdot 2) + (2-0)x + (0-1)y\right] \\ N_3 &= \frac{1}{2A}\left[(0\cdot 1 - 2\cdot 0) + (0-1)x + (2-0)y\right] \end{aligned}\right\} \tag{4.16}$$

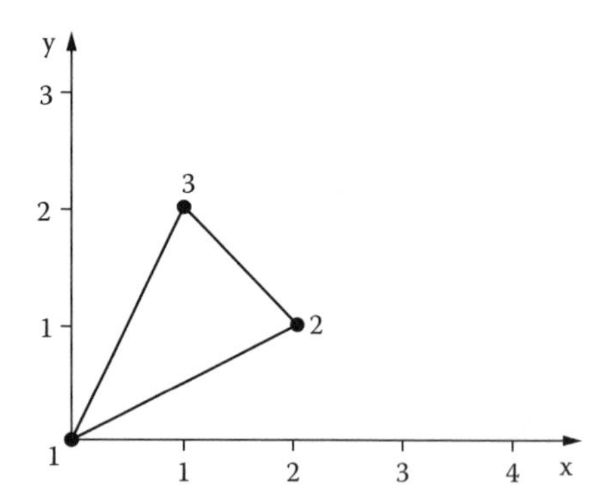

Figure 4.3 Calculation of temperature within a triangular element

and

$$2A = (0 \cdot 1 - 2 \cdot 0) + (1 \cdot 0 - 0 \cdot 2) + (2 \cdot 2 - 1 \cdot 1) = 3 \tag{4.17}$$

Thus, at $x = 0.5$, $y = 0.5$,

$$\left.\begin{aligned} N_1 &= \frac{1}{3}\left[3 - 0.5 - 0.5\right] = \frac{2.0}{3} \\ N_2 &= \frac{1}{3}\left[1 - 0.5\right] = \frac{0.5}{3} \\ N_3 &= \frac{1}{3}\left[-0.5 + 1\right] = \frac{0.5}{3} \end{aligned}\right\} \tag{4.18}$$

The value of T at the point in question is obtained as

$$T(0.5, 0.5) = \frac{2.0}{3}(10) + \frac{0.5}{3}(20) + \frac{0.5}{3}(30) = 15.0\,C \tag{4.19}$$

4.3.2 Quadratic Shape Functions

As in the one-dimensional element, we can likewise write a quadratic approximation over the triangular element. The quadratic triangular element has six nodes, as shown in Figure 4.4. The reasons for adopting this particular node-numbering system will become evident later when we consider area coordinates.

The interpolating polynomial is written as a series expansion consisting of six terms:

$$\phi = \alpha_1 + \alpha_2 x + \alpha_3 y + \alpha_4 x^2 + \alpha_5 xy + \alpha_6 y^2 \tag{4.20}$$

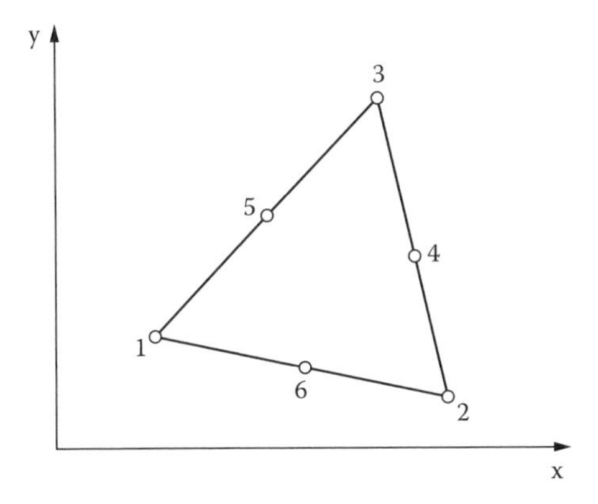

Figure 4.4 Quadratic triangular element

The α_i terms are determined in exactly the same manner as for the linear element. Defining ϕ at each node location,

$$\left.\begin{aligned}\phi_1 &= \alpha_1 + \alpha_2 x_1 + \alpha_3 y_1 + \alpha_4 x_1^2 + \alpha_5 x_1 y_1 + \alpha_6 y_1^2\\ \phi_2 &= \alpha_1 + \alpha_2 x_2 + \alpha_3 y_2 + \alpha_4 x_2^2 + \alpha_5 x_2 y_2 + \alpha_6 y_2^2\\ &\;\;\vdots\\ \phi_6 &= \alpha_1 + \alpha_2 x_6 + \alpha_3 y_6 + \alpha_4 x_6^2 + \alpha_5 x_6 y_6 + \alpha_6 y_6^2\end{aligned}\right\} \tag{4.21}$$

The α_i terms can be solved from Eq. (4.21). However, the amount of effort to obtain the α_i terms and, subsequently, the shape functions N_1, N_2, N_3, N_4, N_5, and N_6 becomes rather burdensome. Equation (4.21) represents the 6 × 6 square coefficients matrix associated with the six equations for ϕ_i, $i = 1, \ldots 6$. Solving Eq. (4.21) by hand is tedious and unnecessary—a much simpler method for establishing the shape functions exists that is based on an area coordinate system, which is introduced in the next section.

If we wished to use a cubic triangular element, ten nodes would be required to define the element—four along each side of the triangle and one at the centroid. Thus, Eq. (4.21) would consist of ten equations, i.e.,

$$\phi = \alpha_1 + \alpha_2 x + \alpha_3 y + \alpha_4 x^2 + \alpha_5 xy + \alpha_6 y^2 + \alpha_7 x^3 + \alpha_8 x^2 y + \alpha_9 xy^2 + \alpha_{10} y^3 \tag{4.22}$$

In this instance, the algebra becomes considerable in determining the α_i terms and, subsequently, the shape functions, N_1 through N_{10}.

4.4 AREA COORDINATES

To simplify the solution process when using triangular elements, an area, or natural, coordinate system is introduced that is analogous to the natural coordinate system devised for the one-dimensional element. This two-dimensional coordinate system permits easy evaluation of the integral relations using the expression represented by Eq. (3.70).

We define three coordinate variables, L_1, L_2, and L_3, by

$$\left.\begin{aligned} L_1 &= \frac{A_1}{A} \\ L_2 &= \frac{A_2}{A} \\ L_3 &= \frac{A_3}{A} \end{aligned}\right\} \tag{4.23}$$

where $A_1, A_2,$ and A_3 are partial areas defined by joining a point P in the triangle with the corner nodes (shown in Figure 4.5) in such a way that A_i is the area opposite node i.

Lines of constant L_i will run parallel to the triangle's side opposite node i. Notice that $L_1 + L_2 + L_3 = 1$ at any point within the triangle. Figure 4.6 shows the increasing direction for each coordinate. It is clear that each L_i varies between 0 along the side opposite node i and 1 at node i.

For a linear triangle, we can define the shape functions as the coordinate variables

$$\left.\begin{aligned} N_1 &= L_1 \\ N_2 &= L_2 \\ N_3 &= L_3 \end{aligned}\right\} \tag{4.24}$$

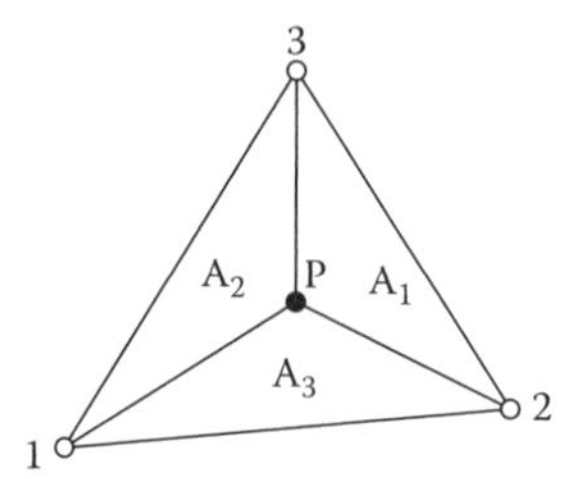

Figure 4.5 Area coordinate system

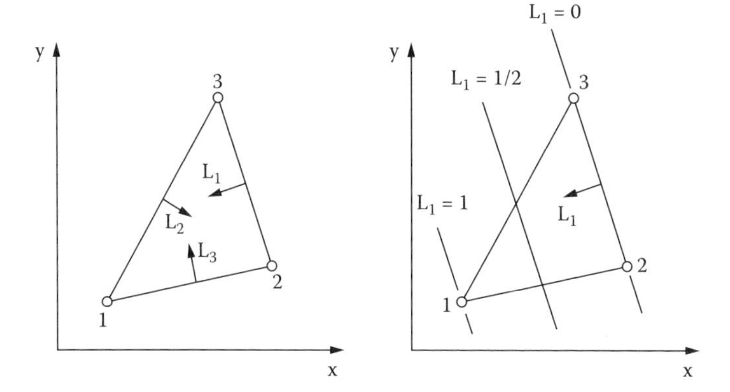

Figure 4.6 Area coordinates in a linear triangle

Hence, the coordinates (x, y) are related to the coordinates (L_1, L_2, L_3) through the expressions:

$$\left.\begin{aligned} x &= L_1x_1 + L_2x_2 + L_3x_3 \\ y &= L_1y_1 + L_2y_2 + L_3y_3 \\ 1 &= L_1 + L_2 + L_3 \end{aligned}\right\} \tag{4.25}$$

The interested reader can find more extensive descriptions on the area coordinate system in Segerlind (1984), Zienkiewicz (1977), and Huebner and Thornton (1982).

Example 4.1 Revisited

Let us return to Example 4.1, only this time we shall solve for the temperature at a point within the triangular element using area coordinates. We retain the assumption that the temperature varies linearly, but this time the point is located to the distance from corner 1 to the 2–3 (side 1) and ½ the distance from corner 3 to the line 1–2 (side 3).

The natural coordinates of the point in the triangle are (see Figure 4.6)

$$L_1 = \frac{1}{3}, \quad L_3 = \frac{1}{2}, \quad \mathrm{L}_2 = 1 - L_1 - L_3 = \frac{1}{6}$$

Thus,

$$\begin{aligned} T_{point} &= L_1T_1 + L_2T_2 + L_3T_3 \\ &= \frac{1}{3}T_1 + \frac{1}{6}T_2 + \frac{1}{2}T_3 \\ &= \frac{1}{3}(10) + \frac{1}{6}(20) + \frac{1}{2}(3) \\ &= 25.0\,C \end{aligned}$$

We see that the natural area coordinates greatly simplify the calculation. Consider now the use of area coordinates in higher-order elements.

For the quadratic triangle of Figure 4.4, the shape functions are also given as functions of L_1, L_2, and L_3 following Figure 4.7:

$$\left.\begin{aligned} N_1 &= L_1(2L_1 - 1) \\ N_2 &= L_2(2L_2 - 1) \\ N_3 &= L_3(2L_3 - 1) \\ N_4 &= 4L_2L_3 \\ N_5 &= 4L_1L_3 \\ N_6 &= 4L_1L_2 \end{aligned}\right\} \tag{4.26}$$

The calculation of the derivatives of the shape functions require some modifications because the coordinates are not independent, i.e., $L_1 + L_2 + L_3 = 1$. The usual practice is to assume L_1 and L_2 are independent coordinates. Thus, using the chain rule,

$$\left.\begin{aligned} \frac{\partial N_i}{\partial L_1} &= \frac{\partial N_i}{\partial x}\frac{\partial x}{\partial L_1} + \frac{\partial N_i}{\partial y}\frac{\partial y}{\partial L_1} \\ \frac{\partial N_i}{\partial L_2} &= \frac{\partial N_i}{\partial x}\frac{\partial x}{\partial L_2} + \frac{\partial N_i}{\partial y}\frac{\partial y}{\partial L_2} \end{aligned}\right\} \tag{4.27}$$

which can be rewritten as

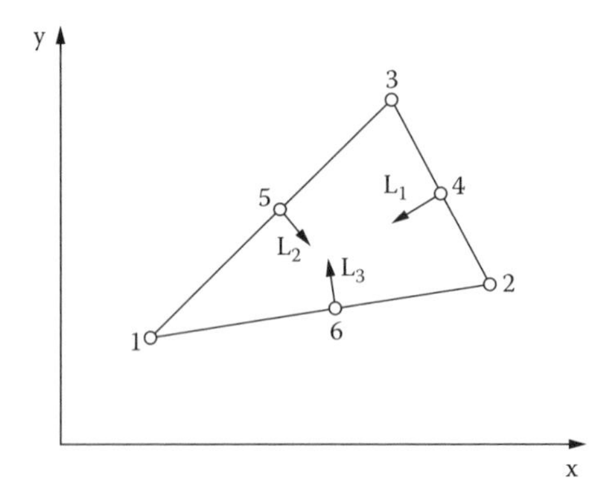

Figure 4.7 Area coordinate system for quadratic triangular element

$$\begin{bmatrix} \dfrac{\partial N_i}{\partial L_1} \\ \dfrac{\partial N_i}{\partial L_2} \end{bmatrix} = \begin{bmatrix} \dfrac{\partial x}{\partial L_1} & \dfrac{\partial y}{\partial L_1} \\ \dfrac{\partial x}{\partial L_2} & \dfrac{\partial y}{\partial L_2} \end{bmatrix} \begin{bmatrix} \dfrac{\partial N_i}{\partial x} \\ \dfrac{\partial N_i}{\partial y} \end{bmatrix} = \mathbf{J} \begin{bmatrix} \dfrac{\partial N_i}{\partial x} \\ \dfrac{\partial N_i}{\partial y} \end{bmatrix} \tag{4.28}$$

The 2 × 2 matrix, **J**, is called the Jacobian matrix and was introduced earlier in the one-dimensional element. To account for L_3 , we see that for any function of ϕ,

$$\frac{\partial \phi}{\partial L_1} = \frac{\partial \phi}{\partial L_1}\frac{\partial L_1}{\partial L_1} + \frac{\partial \phi}{\partial L_2}\frac{\partial L_2}{\partial L_1} + \frac{\partial \phi}{\partial L_3}\frac{\partial L_3}{\partial L_1} \tag{4.29}$$

However, L_1 and L_2 are assumed to be independent, and $L_3 = 1 - L_1 - L_2$. Thus,

$$\frac{\partial L_3}{\partial L_1} = -1 \tag{4.30}$$

Consequently, the derivatives for ϕ with respect the to the natural coordinates become*

$$\left.\begin{aligned} \frac{\partial \phi}{\partial L_1} &\equiv \frac{\partial \phi}{\partial L_1} - \frac{\partial \phi}{\partial L_3} \\ \frac{\partial \phi}{\partial L_2} &\equiv \frac{\partial \phi}{\partial L_2} - \frac{\partial \phi}{\partial L_3} \end{aligned}\right\} \tag{4.31}$$

Example 4.2

To illustrate the use of area coordinates, find the value of $\partial N_5 / \partial x$ and $\partial N_5 / \partial y$ at the point (1, 1) in the triangle shown in Figure 4.8. The values of x and y are obtained from the linear shape functions and corner nodes:

$$\left.\begin{aligned} x &= L_1 x_1 + L_2 x_2 + L_3 x_3 \\ y &= L_1 y_1 + L_2 y_2 + L_3 y_3 \end{aligned}\right\} \tag{4.32}$$

If we substitute for the nodal values,

$$\left.\begin{aligned} x &= 3L_2 + L_3 \\ y &= L_2 + 3L_3 \end{aligned}\right\} \tag{4.33}$$

* The notation here is a little bit ambiguous. For this reason, we have used Eq. (4.31) as a definition, where the partial derivatives on the left-hand side do not have the same meaning as those on the right-hand side. The difference should always be clear from the context.

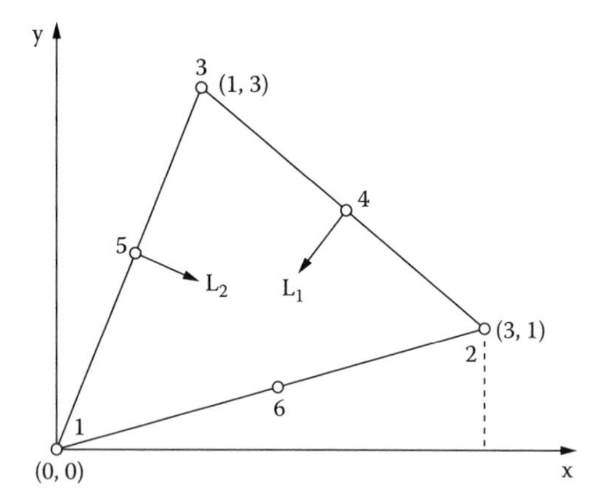

Figure 4.8 Example problem used to calculate $\partial N_5/\partial x$ and $\partial N_5/\partial y$

The derivatives in Eq. (4.31), which are used in the Jacobian matrix, become

$$\left.\begin{aligned}
\frac{\partial x}{\partial L_1} &\equiv \frac{\partial x}{\partial L_1} - \frac{\partial x}{\partial L_3} = -1 \\
\frac{\partial x}{\partial L_2} &\equiv \frac{\partial x}{\partial L_2} - \frac{\partial x}{\partial L_3} = 2 \\
\frac{\partial y}{\partial L_1} &\equiv \frac{\partial y}{\partial L_1} - \frac{\partial y}{\partial L_3} = -3 \\
\frac{\partial y}{\partial L_2} &\equiv \frac{\partial y}{\partial L_2} - \frac{\partial y}{\partial L_3} = -2
\end{aligned}\right\} \tag{4.34}$$

The Jacobian in Eq. (4.28) is

$$\mathbf{J} = \begin{bmatrix} -1 & -3 \\ 2 & -2 \end{bmatrix} \tag{4.35}$$

Because we want to find the $\partial N_i/\partial x$ and $\partial N_i/\partial y$ terms, we rearrange Eq. (4.28) into the form

$$\begin{bmatrix} \dfrac{\partial N_i}{\partial x} \\ \dfrac{\partial N_i}{\partial y} \end{bmatrix} = \mathbf{J}^{-1} \begin{bmatrix} \dfrac{\partial N_i}{\partial L_1} \\ \dfrac{\partial N_i}{\partial L_2} \end{bmatrix} \tag{4.36}$$

Now we determine the $\partial N_i/\partial L_1$ and $\partial N_i/\partial L_2$ terms for node 5 using Eqs. (4.26) and (4.31):

$$\left.\begin{aligned}\frac{\partial N_5}{\partial L_1} &\equiv \frac{\partial N_5}{\partial L_1} - \frac{\partial N_5}{\partial L_3} = 4L_3 - 4L_1 \\ \frac{\partial N_5}{\partial L_2} &\equiv \frac{\partial N_5}{\partial L_2} - \frac{\partial N_5}{\partial L_3} = -4L_1\end{aligned}\right\} \tag{4.37}$$

Substituting the derivative evaluations into Eq. (4.36),

$$\begin{bmatrix}\dfrac{\partial N_5}{\partial x} \\ \dfrac{\partial N_5}{\partial y}\end{bmatrix} = \mathbf{J}^{-1}\begin{bmatrix}4(L_3 - L_1) \\ -4L_1\end{bmatrix} \tag{4.38}$$

where the Jacobian is

$$\mathbf{J}^{-1} = \frac{1}{8}\begin{bmatrix}-2 & 3 \\ -2 & -1\end{bmatrix} \tag{4.39}$$

Equation (4.38) thus becomes

$$\begin{aligned}\begin{bmatrix}\dfrac{\partial N_5}{\partial x} \\ \dfrac{\partial N_5}{\partial y}\end{bmatrix} &= \frac{1}{8}\begin{bmatrix}-2 & 3 \\ -2 & -1\end{bmatrix}\begin{bmatrix}4(L_3 - L_1) \\ -4L_1\end{bmatrix} \\ &= \frac{1}{8}\begin{bmatrix}-8L_3 - 4L_1 \\ -8L_3 + 12L_1\end{bmatrix} \\ &= \begin{bmatrix}-L_3 - \dfrac{1}{2}L_1 \\ -L_3 + \dfrac{3}{2}L_1\end{bmatrix}\end{aligned} \tag{4.40}$$

These two equations are applicable over the entire element. To find the actual values for the two derivatives, we determine L_1, L_2, and L_3 at point (1, 1) using Eq. (4.32) and the fact that $L_1 + L_2 + L_3 = 1$:

$$\left.\begin{aligned}1 &= 0 \cdot L_1 + 3L_2 + L_3 \\ 1 &= 0 \cdot L_1 + L_2 + 3 \cdot L_3 \\ 1 &= L_1 + L_2 + L_3\end{aligned}\right\} \tag{4.41}$$

Solving for the three unknown coordinate values gives

$$\left.\begin{aligned} L_1 &= \frac{1}{2} \\ L_2 &= \frac{1}{4} \\ L_3 &= \frac{1}{4} \end{aligned}\right\} \tag{4.42}$$

The final values for the two derivatives are

$$\left.\begin{aligned} \frac{\partial N_5}{\partial x} &= -L_3 - \frac{1}{2}L_1 = -\frac{1}{2} \\ \frac{\partial N_5}{\partial y} &= -L_3 + \frac{3}{2}L_1 = \frac{1}{2} \end{aligned}\right\} \tag{4.43}$$

4.5 NUMERICAL INTEGRATION

The advantage of using area coordinates lies in the ability to evaluate the integral equations from integration formulae, as in the one-dimensional case. For the two-dimensional element with two coordinates (L_1, L_2), Eq. (3.70) takes the form

$$\int_0^L L_1^0 L_2^b dx = \frac{a!\ b!}{(1+a+b)!} \cdot L \tag{4.44a}$$

which holds for a line integral of length L defined between two nodes and non-negative integers a, b. This relation will also be valid when defining integrals that are only a function of the length along an edge of an element. For our two-dimensional element with L_1, L_2, and L_3,

$$\int_A L_1^0 L_2^b L_3^c dA = \frac{a!\ b!\ c!}{(2+a+b+c)!} 2A \tag{4.44b}$$

where A denotes area. For example, in a linear triangle, the area integral of the product $N_1 N_2$ is given as

$$\int_A N_1 N_2 dA = \int_A L_1^1 L_2^1 L_3^0 dA = \frac{1!\ 1!\ 0!}{(2+1+1+0)!} 2A = \frac{A}{12} \tag{4.45}$$

The area integrals that arise in an element take the general form:

$$\int_0^1 \int_0^{1-L_2} f(L_1, L_2, L_3) \left|\mathbf{J}\right| dL_1 dL_2 \tag{4.46}$$

When the higher-order interpolation functions (quadratic, cubic, etc.) are written in terms of the area coordinates, the matrix integrals must be evaluated numerically—this is due to the inverse of the Jacobian matrix, which is a rational function of the area coordinates themselves, and appears in Eq. (4.36) when derivatives of the shape functions are involved. The formulae employed in Eqs. (4.44) are valid for simple integral relations. However, for complex integrals it is best to let the computer do the integration to avoid human error.

One of the most common methods for evaluating integrals numerically in triangular elements was developed by Hammer et al. (1956) and can be found in many introductory texts on computational methods. In the finite element method we assume

$$\int_0^1 \int_0^{1-L_2} f(L_1, L_2, L_3)\left|\mathbf{J}\right| dL_1 dL_2 = \frac{1}{2}\sum_{m=1}^{M} w_m g\left[(L_1)_m, (L_2)_m, (L_3)_m\right] \tag{4.47}$$

where $g(\cdot)$ includes $\left|\mathbf{J}\right|$, the $1/2$ factor accounts for the area in the local coordinate system, $\left(L_i\right)_m$ denote specific points in the triangle, and w_m are the weights associated with the procedure. The order of the quadrature formula must be at least one integer larger than the sum of the powers of the coordinates L_1, L_2, and L_3. For example, to integrate the product $L_1L_2L_3$ the sum of the exponentials is three, and a quartic integration scheme would be used. Table 4.1 lists the weights and coordinate values for numerical integration formulae of various orders. Determination of the order of the integration formulae, as well as more comprehensive list, is found in Cowper (1973).

Example 4.3

To illustrate the procedure, find the integral value of the product

$$\frac{\partial N_5}{\partial x}\frac{\partial N_5}{\partial y}$$

for the element of Figure 4.8, which was used previously. Recall that $\left|\mathbf{J}\right| = 8$ from Eq. (4.35). From Eq. (4.43),

$$\left.\begin{aligned} \frac{\partial N_5}{\partial x} &= \left(-\frac{1}{2}L_1 - L_3\right) \\ \frac{\partial N_5}{\partial y} &= \left(\frac{3}{2}L_1 - L_3\right) \end{aligned}\right\} \tag{4.48}$$

Thus,

$$\frac{\partial N_5}{\partial x}\cdot\frac{\partial N_5}{\partial y} = -\frac{3}{4}L_1^2 - L_1L_3 + L_3^2 \tag{4.49}$$

Table 4.1 Numerical Integration Formulae for Triangles

Gauss Point	m	L_1	L_2	L_3	Weight	Order
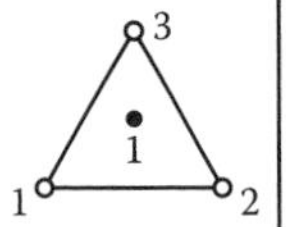	1	$\frac{1}{3}$	$\frac{1}{3}$	$\frac{1}{3}$	1	2
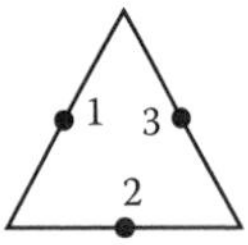	1	$\frac{1}{2}$	0	$\frac{1}{2}$	$\frac{1}{3}$	3
	2	$\frac{1}{2}$	$\frac{1}{2}$	0	$\frac{1}{3}$	
	3	0	$\frac{1}{2}$	$\frac{1}{2}$	$\frac{1}{3}$	
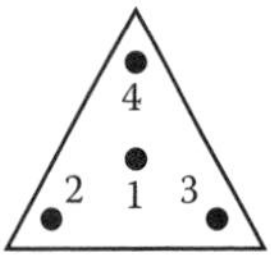	1	$\frac{1}{3}$	$\frac{1}{3}$	$\frac{1}{3}$	$-\frac{27}{48}$	4
	2	$\frac{11}{15}$	$\frac{2}{15}$	$\frac{2}{15}$	$\frac{25}{48}$	
	3	$\frac{2}{15}$	$\frac{2}{15}$	$\frac{11}{15}$	$\frac{25}{48}$	
	4	$\frac{2}{15}$	$\frac{11}{15}$	$\frac{2}{15}$	$\frac{25}{48}$	
	1	$\frac{1}{3}$	$\frac{1}{3}$	$\frac{1}{3}$	0.225	6
	2	β	α	β		
	3	β	β	α	0.13239415	
	4	α	β	β		
	5	γ	δ	γ		
	6	γ	γ	δ	0.12593918	
	7	δ	γ	γ		

α = 0.05971587 γ = 0.10128651

β = 0.47014206 δ = 0.79742699

*Cowper (1973)

Source: Cowper, G. R. (1973). "Gaussian Quadrature Formulas for Triangles." *Int. J. Num. Mech. Eng.* 7: 405–408. With permission.

We shall use a cubic-order integration formula since each term is second order. In this case, using Table 4.1, the sampling points are located at

$$\left.\begin{array}{ll}(1) & L_1 = L_3 = \dfrac{1}{2},\ L_2 = 0 \\ (2) & L_1 = L_2 = \dfrac{1}{2},\ L_3 = 0 \\ (3) & L_2 = L_3 = \dfrac{1}{2},\ L_1 = 0\end{array}\right\} \tag{4.50}$$

and each sampling point has weight 1/3. Thus,

$$\int_0^1 \int_0^{1-L_2} \left(-\frac{3}{4}L_1^2 - L_1L_3 + L_3^2\right) 8dL_1dL_2 = \frac{1}{2}\sum_{m=1}^{3} w_m g_m(L_1, L_2, L_3) \tag{4.51}$$

Now,

$$g_m(L_1, L_2, L_3) = \left(-\frac{3}{4}L_1^2 - L_1L_3 + L_3^2\right)8 \tag{4.52}$$

or

$$g_m(L_1, L_2, L_3) = -6L_1^2 - 8L_1L_3 + 8L_3^2 \tag{4.53}$$

At the first sampling point, $L_1 = L_3 = 1/2, L_2 = 0$,

$$g_1 = -6L_1^2 - 8L_1L_3 + 8L_3^2 = -\frac{6}{4} - 2 + 2 = -\frac{3}{2} \tag{4.54}$$

At $L_1 = L_2 = 1/2, L_3 = 0$,

$$g_2 = -6L_1^2 = -\frac{3}{2} \tag{4.55}$$

At $L_2 = L_3 = 1/2, L_1 = 0$,

$$g_3 = 8L_3^2 = 2 \tag{4.56}$$

If we now substitute these values into Eq. (4.51), we obtain

$$\frac{1}{2}\sum_{m=1}^{3} w_m g_m(L_1, L_2, L_3) = \frac{1}{6}g_1 + \frac{1}{6}g_2 + \frac{1}{6}g_3 = \frac{1}{6}\left(-\frac{3}{2} - \frac{3}{2} + 2\right) = -\frac{1}{6} \tag{4.57}$$

In this instance, the product integral could also have been determined from Eq. (4.44) and would be found to yield exactly the same result. This is left to Exercise 4.7.

4.6 CONDUCTION IN A TRIANGULAR ELEMENT

Consider the problem of determining the steady-state temperature distribution due to conduction within an isotropic two-dimensional domain Ω with boundary Γ, as depicted in Figure 4.9. The governing equation for $T(x, y)$ is written as

$$-K\left(\frac{\partial^2 T}{\partial x^2}+\frac{\partial^2 T}{\partial y^2}\right)=Q \tag{4.58}$$

where K is the thermal conductivity, assumed constant, and Q is a source/sink term. We wish to formulate a set of finite element expressions unique to the linear triangular element. The boundary conditions associated with Eq. (4.58) are written as

$$-K\frac{\partial^2 T}{\partial x^2}=q \quad \text{along } \Gamma_B \tag{4.59a}$$

$$T=T_A \quad \text{along } \Gamma_A \tag{4.59b}$$

where Γ_A and Γ_B are portions of the boundary Γ, as depicted in Figure 4.9, and the normal derivative is defined as

$$\frac{\partial T}{\partial n}\equiv\frac{\partial T}{\partial x}n_x+\frac{\partial T}{\partial y}n_y \tag{4.60}$$

where n_x and n_y are the components or direction cosines of the unit outward vector normal to the boundary Γ.

Using the method of weighted residuals,

$$\int_\Omega WRd\Omega=0 \tag{4.61}$$

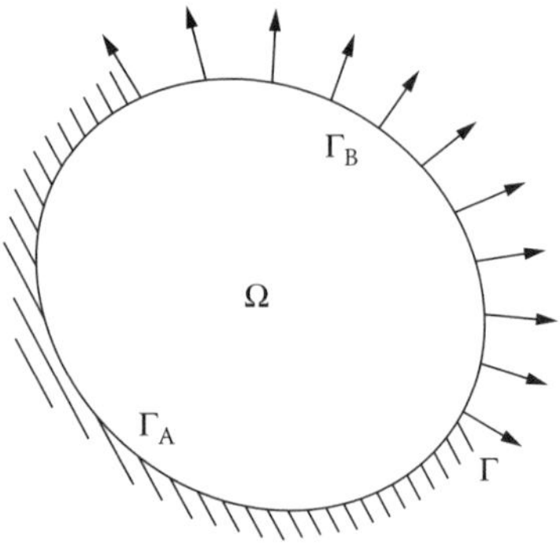

Figure 4.9 Two-dimensional domain for heat conduction

where

$$R = -K\left(\frac{\partial^2 T}{\partial x^2} + \frac{\partial^2 T}{\partial y^2}\right) - Q \tag{4.62}$$

and W is the weighting function. Employing the (generic) formula for differentiation

$$\frac{\partial}{\partial \alpha}\left(F\frac{\partial G}{\partial \alpha}\right) = \frac{\partial F}{\partial \alpha}\frac{\partial G}{\partial \alpha} + F\frac{\partial^2 G}{\partial \alpha^2} \tag{4.63}$$

and Green's theorem in the plane,

$$\int_\Omega \left(\frac{\partial F}{\partial x} + \frac{\partial G}{\partial y}\right) dx\, dy = \int_\Gamma (Fdy - Gdx) \tag{4.64}$$

Eq. (4.61) can be written in the form

$$\int_\Omega \left[K\left(\frac{\partial W}{\partial x}\frac{\partial T}{\partial x} + \frac{\partial W}{\partial y}\frac{\partial T}{\partial y}\right) - WQ\right] d\Omega + \int_\Gamma W\left(-K\frac{\partial T}{\partial n}\right) d\Gamma = 0 \tag{4.65}$$

We now approximate the temperature field using the shape functions

$$T(x, y) = \sum_{i=1}^{M} N_i(x, y) T_i \tag{4.66}$$

as was done before for one-dimensional elements, where M is the number of nodes in the domain Ω. Using Eq. (4.59a) in the line integral term, and an argument similar to that invoked in one dimension to set the boundary flux terms to zero over the portion Γ_A, we set $W_i = N_i$ in a Galerkin formulation. Equation (4.65) becomes

$$\sum_{j=1}^{M}\left[\int_\Omega K\left(\frac{\partial N_i}{\partial x}\frac{\partial N_j}{\partial x} + \frac{\partial N_i}{\partial y}\frac{\partial N_j}{\partial y}\right) d\Omega\right] T_j = \int_\Omega N_i Q d\Omega - \int_{\Gamma_B} N_i q d\Gamma \tag{4.67}$$

where $i = 1, \ldots, M$. Equation (4.67) is valid for any region Ω and any type of element that we may want to use, as long as it is geometrically compatible, whether it be a linear or quadratic triangular element, or quadrilateral elements, or a combination of both. It is clear that Eq. (4.67) can be written as

$$KT = F \tag{4.68}$$

where

$$\mathbf{K} = \left[k_{ij}\right] = \left[\int_{\Omega} K\left(\frac{\partial N_i}{\partial x}\frac{\partial N_j}{\partial x} + \frac{\partial N_i}{\partial y}\frac{\partial N_j}{\partial y}\right) d\Omega\right] \tag{4.69}$$

and

$$\mathbf{F} = \left[f_i\right] = \left\{\int_{\Omega} N_i Q \, d\Omega - \int_{\Gamma_B} N_i q \, d\Gamma\right\} \tag{4.70}$$

As in Chapter 3, the **K** term is called the “conduction” or “stiffness” matrix and **F** is the “load vector.” To simplify the algebra involved in establishing the matrix equations, we will rewrite the shape functions in a more general form. For node points 1, 2, and 3 in a linear triangular element.

$$\left.\begin{aligned} N_1^{(e)} &= \frac{1}{2A}\left(a_1 + b_1 + c_1 y\right) \\ N_2^{(e)} &= \frac{1}{2A}\left(a_2 + b_2 x + c_2 y\right) \\ N_3^{(e)} &= \frac{1}{2A}\left(a_3 + b_3 x + c_3 y\right) \end{aligned}\right\} \tag{4.71}$$

where the values for a_i, b_i, and c_i, $i = 1,2,3,$ are given in Table 4.2.

Notice that $\mathbf{K}^{(e)}$ is symmetric and that all terms under the integral are constant:

$$\mathbf{k}_{ij}^{(e)} = \int_{A^{(e)}} \frac{K}{4(A^{(e)})^2}(b_i b_j + c_i c_j) dA \quad i = 1,2,3, \; j = 1,2,3$$

$$= \frac{K}{4A^{(e)}}\begin{bmatrix} b_1 b_1 + c_1 c_1 & b_2 b_1 + c_2 c_1 & b_3 b_1 + c_3 c_1 \\ b_2 b_1 + c_2 c_1 & b_2 b_2 + c_2 c_2 & b_2 b_3 + c_2 c_3 \\ b_3 b_1 + c_3 c_1 & b_3 b_2 + c_3 c_2 & b_3 b_3 + c_3 c_3 \end{bmatrix} \tag{4.72}$$

Table 4.2 Coefficient Values for Linear Triangular Element Shape Functions

i	a_i	b_i	c_i
1	$x_2 y_3 - x_3 y_2$	$y_2 - y_3$	$x_3 - x_2$
2	$x_3 y_1 - x_1 y_3$	$y_3 - y_1$	$x_1 - x_3$
3	$x_1 y_2 - x_2 y_1$	$y_1 - y_2$	$x_2 - x_1$

The components of the load vector **F** consist of two integrals. Since $N_i = L_i$, the first integral for the heat source can be explicitly written as

$$\mathbf{F}_Q^{(e)} = Q\int_{\mathbf{A}^{(e)}} \begin{bmatrix} N_1^{(e)} \\ N_2^{(e)} \\ N_3^{(e)} \end{bmatrix} dA = Q\int_A \begin{bmatrix} L_1^1 L_2^0 L_3^0 \\ L_1^0 L_2^1 L_3^0 \\ L_0^0 L_2^0 L_3^1 \end{bmatrix} dA = \frac{Q\,A^{(e)}}{3}\begin{bmatrix} 1 \\ 1 \\ 1 \end{bmatrix} \tag{4.73}$$

The second integral term accounts for the boundary condition on surface Γ_B, which is composed of element sides. The results depend on which side of the element is subjected to the heat flux, q. If we assume that the heat flux is constant over the surface,

$$\left.\begin{aligned} F_{q_{1-2}}^{(e)} &= -q\int_{\Gamma_B^{(e)}} \begin{bmatrix} N_1^{(e)} \\ N_2^{(e)} \\ N_3^{(e)} \end{bmatrix} dS = -q\int_{\Gamma_B^{(e)}} \begin{bmatrix} L_1^1 L_2^0 \\ L_1^1 L_2^1 \\ 0 \end{bmatrix} dS = -\frac{q\ell_{1-2}^{(e)}}{2}\begin{bmatrix} 1 \\ 1 \\ 0 \end{bmatrix} \\ F_{q_{2-3}}^{(e)} &= -\frac{q\ell_{2-3}^{(e)}}{2}\begin{bmatrix} 0 \\ 1 \\ 1 \end{bmatrix} \\ F_{q_{3-1}}^{(e)} &= -\frac{q\ell_{2-3}^{(e)}}{2}\begin{bmatrix} 1 \\ 0 \\ 1 \end{bmatrix} \end{aligned}\right\} \tag{4.74}$$

where $\ell_{1-2}^{(e)}$, $\ell_{2-3}^{(e)}$, and $\ell_{3-1}^{(e)}$ are the lengths of sides 1–2, 2–3, and 3–1, respectively, in the triangular element. Thus,

$$\mathbf{F}^{(e)} = \mathbf{F}_Q^{(e)} + \mathbf{F}_{q_{1-2}}^{(e)} + \mathbf{F}_{q_{2-3}}^{(e)} + \mathbf{F}_{q_{3-1}}^{(e)} \tag{4.75}$$

where the flux boundary values in the last three terms depend on the specified element side. Equations (4.72), (4.73), and (4.74) serve as the general finite element relations for triangular elements. We have only to input actual values to obtain the nodal values for temperature as specified by the element mesh. In the next section, we apply the general relations for a one-element problem with convection along one face.

4.7 STEADY-STATE CONDUCTION WITH BOUNDARY CONVECTION

Assume that we have a discretization that contains an element as shown in Figure 4.10, with specified convection along the face joining nodes 2 and 3. We want to solve the steady-state conduction Eq. (4.58) with the boundary condition

$$-K\frac{\partial T}{\partial n} = h(T - T_\infty) \qquad \text{along } \ell_{2-3}^{(e)} \tag{4.76}$$

The conduction matrix is given by

$$\mathbf{K}^{(e)} = \left[k_{ij}^{(e)}\right] = \left[\int_{A^{(e)}} K\left(\frac{\partial N_i^{(e)}}{\partial x}\frac{\partial N_j^{(e)}}{\partial x} + \frac{\partial N_i^{(e)}}{\partial y}\frac{\partial N_j^{(e)}}{\partial y}\right) dA + \int_{\ell_{2-3}^{(e)}} hN_i^{(e)}N_j^{(e)}\,d\Gamma\right] \tag{4.77}$$

where the additional term comes from the boundary condition for convection, Eq. (4.76), which involves hT. Evaluating expression (4.77), we get

$$\mathbf{K}^{(e)} = \frac{K}{4A^e}\begin{bmatrix} b_1b_1 + c_1c_1 & b_1b_2 + c_1c_2 & b_1b_3 + c_1c_3 \\ b_2b_1 + c_2c_1 & b_2b_2 + c_2c_2 & b_2b_3 + c_2c_3 \\ b_3b_1 + c_3c_1 & b_3b_2 + c_3c_2 & b_3b_3 + c_3c_3 \end{bmatrix} + \frac{h\ell_{2-3}^{(e)}}{6}\begin{bmatrix} 0 & 0 & 0 \\ 0 & 2 & 1 \\ 0 & 1 & 2 \end{bmatrix} \tag{4.78}$$

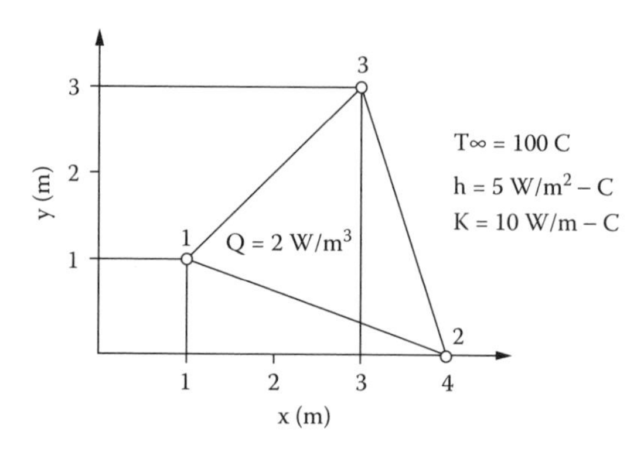

Figure 4.10 Conduction of heat in an element with convection on one face

where

$$\int_{\ell_{2-3}^{(e)}} hN_i^{(e)}N_j^{(e)}dS = h\int_{\ell_{2-3}^{(e)}} \begin{bmatrix} 0 & 0 & 0 \\ 0 & L_2L_2 & L_2L_3 \\ 0 & L_2L_3 & L_3L_3 \end{bmatrix} dS \tag{4.79}$$

and

$$\int_{\ell_{2-3}^{(e)}} L_2^2 dS = \int_{\ell_{2-3}^{(e)}} L_2^2 L_3^0 dS = \frac{2!\,0!}{(2+0+1)!}\ell_{2-3}^{(e)} = \frac{\ell_{2-3}^{(e)}}{3} \tag{4.80}$$

The integral

$$\int_{\ell_{2-3}^{(e)}} L_2^2 dS = \int_{\ell_{2-3}^{(e)}} L_3^2 dS$$

Likewise,

$$\int_{\ell_{2-3}^{(e)}} L_2L_3 dS = \frac{\ell_{2-3}^{(e)}}{6}$$

The b_i and c_i values are

$$\left.\begin{aligned} b_1 &= y_2 - y_3 = -3 \qquad c_1 = x_3 - x_2 = -1 \\ b_2 &= y_3 - y_1 = 2 \qquad c_2 = x_1 - x_3 = -2 \\ b_3 &= y_1 - y_2 = 1 \qquad c_3 = x_2 - x_1 = 3 \end{aligned}\right\} \tag{4.81}$$

The area of the element is

$$A^{(e)} = \frac{1}{2}\begin{vmatrix} 1 & x_1 & y_1 \\ 1 & x_2 & y_2 \\ 1 & x_3 & y_3 \end{vmatrix} = \frac{1}{2}\begin{vmatrix} 1 & 1 & 1 \\ 1 & 4 & 0 \\ 1 & 3 & 3 \end{vmatrix} = \frac{8}{2} = 4 \tag{4.82}$$

which is written using Eq. (4.6). The length of side $\ell_{2-3}^{(e)}$ is calculated from

$$\ell_{2-3}^{(e)} = \sqrt{(x_2 - x_3)^2 + (y_2 - y_3)^2} = \sqrt{(4-3)^2 + (0-3)^2} = \sqrt{10}\ m$$

The load vector $\mathbf{F}^{(e)}$ is evaluated from Eq. (4.74) except that q is replaced by $-hT_\infty$.

Thus,

$$F^{(e)} = \left\{ f_i^{(e)} \right\} = \left\{ \int_{\ell_{2-3}^{(e)}} hT_\infty N_i^{(e)} d\Gamma \right\} = hT_\infty \int_{\ell_{2-3}^{(e)}} \begin{bmatrix} 0 \\ L_2^1 L_3^0 \\ L_2^0 L_3^1 \end{bmatrix} d\Gamma = \frac{hT_\infty \ell_{2-3}^{(e)}}{2} \begin{bmatrix} 0 \\ 1 \\ 1 \end{bmatrix} \tag{4.83}$$

Replacing Eqs. (4.73), (4.78), and (4.83) and the problem data into Eq. (4.68), we get

$$\left\{ \frac{10}{4 \cdot 4} \begin{bmatrix} 9+1 & -6+2 & -3-3 \\ -6+2 & 4+4 & 2-6 \\ -3-3 & 2-6 & 1+9 \end{bmatrix} + \frac{5 \cdot \sqrt{10}}{6} \begin{bmatrix} 0 & 0 & 0 \\ 0 & 2 & 1 \\ 0 & 1 & 2 \end{bmatrix} \right\} \begin{bmatrix} T_1^{(e)} \\ T_2^{(e)} \\ T_3^{(e)} \end{bmatrix}$$

$$= \frac{2 \cdot 4}{3} \begin{bmatrix} 1 \\ 1 \\ 1 \end{bmatrix} + \frac{5 \cdot 100 \cdot \sqrt{10}}{2} \begin{bmatrix} 0 \\ 1 \\ 1 \end{bmatrix}$$

That reduces to

$$\begin{bmatrix} 6.25 & -2.5 & -3.75 \\ -2.5 & 10.27 & 0.135 \\ -3.75 & 0.135 & 11.52 \end{bmatrix} \begin{bmatrix} T_1^{(e)} \\ T_2^{(e)} \\ T_3^{(e)} \end{bmatrix} = \begin{bmatrix} 2.6667 \\ 793.24 \\ 793.24 \end{bmatrix} \tag{4.84}$$

Because no temperatures have been prescribed at the nodes, the linear equation, Eq. (4.84), represents a triangular region with adiabatic sides $\ell_{1-2}^{(e)}$ and $\ell_{3-1}^{(e)}$; by not imposing any conditions along these sides, we have the equivalent to a flux of the form (4.59a) with $q = 0$. This is the same situation we encountered before in the one-dimensional case. Notice that by imposing any conditions along these boundaries, the finite element method automatically introduces the condition of adiabatic boundaries for free.

The nodal temperatures for the equilibrium state between the internal heat generated in the triangle and the heat flux through the side $\ell_{2-3}^{(e)}$ is readily obtained by solving Eq. (4.84). This is left to Exercise 4.8.

At this point, let us review the major steps necessary to obtain the element diffusion matrix and load vector:

1. Define the shape functions for a linear triangular element using Eq. (4.71) with the coefficients given in Table 4.2.
2. Approximate the temperature using the shape functions in the form of Eq. (4.66), with $M = 3$ for a linear triangle.

3. Determine the gradient of T using Eq. (4.71). This yields

$$\begin{bmatrix} \dfrac{\partial T}{\partial x} \\ \dfrac{\partial T}{\partial y} \end{bmatrix} = \frac{1}{2A}\begin{bmatrix} b_1 & b_2 & b_3 \\ c_1 & c_2 & c_3 \end{bmatrix}\begin{bmatrix} T_1^{(e)} \\ T_2^{(e)} \\ T_3^{(e)} \end{bmatrix} \tag{4.85}$$

4. Evaluate the element conduction matrix using Eq. (4.72).
5. Evaluate the convection integral if convection occurs along a boundary, as done in Eq. (4.78). In general, this is obtained as

$$\mathbf{H}^{(e)} = h\int_{\Gamma_B}\begin{bmatrix} N_1^{(e)}N_1^{(e)} & N_1^{(e)}N_2^{(e)} & N_1^{(e)}N_3^{(e)} \\ N_2^{(e)}N_1^{(e)} & N_2^{(e)}N_2^{(e)} & N_2^{(e)}N_3^{(e)} \\ N_3^{(e)}N_1^{(e)} & N_3^{(e)}N_2^{(e)} & N_3^{(e)}N_3^{(e)} \end{bmatrix} d\Gamma \tag{4.86}$$

Since $L_1 = N_1$, etc., and normally only one side (e.g., the side joining nodes 1 and 2 is subjected to convection), $\mathbf{H}^{(e)}$ becomes

$$H^{(e)} = h\int_{\ell_{1-2}^{(e)}}\begin{bmatrix} L_1L_1 & L_1L_2 & 0 \\ L_2L_1 & L_2L_2 & 0 \\ 0 & 0 & 0 \end{bmatrix} d\Gamma = \frac{h\ell_{1-2}^{(e)}}{6}\begin{bmatrix} 2 & 1 & 0 \\ 1 & 2 & 0 \\ 0 & 0 & 0 \end{bmatrix} \tag{4.87}$$

6. Evaluate the source/sink term. This is given by Eq. (4.73) for a constant source Q.
7. If a heat flux is prescribed along any of the sides of the triangle, evaluate it using Eq. (4.74). A convection term, hT_∞, is also given by Eq. (4.74), with q replaced by $-hT_\infty$.

In the previous example, we assumed the region consisted of only one triangular element. We could have as easily discretized the region into any number of elements, then performed the individual calculations on each element. The resulting answers would have been assembled into an overall global matrix equation, to be solved by some form of an elimination procedure. It should be obvious by now that performing the local element calculations by hand can become quite tedious, even for simple linear approximations.

4.8 THE AXISYMMETRIC CONDUCTION EQUATION

The axisymmetric heat conduction equation for steady-state conditions and constant conductivity is expressed by the relation

$$-K\left[\frac{1}{r}\frac{\partial}{\partial r}\left(r\frac{\partial T}{\partial r}\right)+\frac{\partial^2 T}{\partial z^2}\right]=Q \tag{4.88}$$

with $T = T_A$ on the portion Γ_A of the boundary, and

$$-K\left(\frac{\partial T}{\partial r}n_r+\frac{\partial T}{\partial z}n_z\right)=q \tag{4.89}$$

over the portion Γ_B of the boundary, where n_r and n_z are the direction cosines of the outward unit vector normal to Γ.

We now construct the weak form of Eqs. (4.88) and (4.89):

$$\int_V W\left\{-K\left[\frac{1}{r}\frac{\partial}{\partial r}\left(r\frac{\partial T}{\partial r}\right)+\frac{\partial^2 T}{\partial z^2}\right]-Q\right\}dV=0 \tag{4.90}$$

where the elemental volume is assumed axisymmetric, i.e.,

$$dV = rd\theta dA = 2\pi rdz \tag{4.91}$$

The resulting integral relation becomes

$$2\pi\int_\Omega W\left[K\left(-\frac{\partial}{\partial r}\left(r\frac{\partial T}{\partial r}\right)-r\frac{\partial^2 T}{\partial z^2}\right)-Qr\right]drdz=0 \tag{4.92}$$

Note that we can replace Kr with K' and Qr with Q' if we wish to solve a Cartesian problem. Hence, we can use the same type of methodology for either Cartesian or axisymmetric cases; we only have to evaluate the radical component r in the axisymmetric case.

There are basically two ways to account for r. One way is to use an average radial distance for r based on the element centroid. This procedure is sufficient if the element size is small compared to the radial distance. A more accurate method is to express r as

$$r = N_1^{(e)}r_1 + N_2^{(e)}r_2 + N_3^{(e)}r_3 \tag{4.93}$$

where r_i, $i = 1, 2, 3$, are the value of r at the corner points. The axisymmetric conduction matrix for an individual element becomes

$$\mathbf{K}^{(e)}=\left[k_{ij}^{(e)}\right]=\left[\frac{2\pi\mathbf{K}}{4\left(A^{(e)}\right)^2}\left(b_ib_j+c_ic_j\right)\int_{A^{(e)}}\left(L_1r_1+L_2r_2+L_3r_3\right)d\Omega\right] \tag{4.94}$$

Performing the integration using Eq. (4.44b) gives

$$\left[k_{ij}^{(e)}\right] = \frac{2\pi\mathbf{K}}{12A^{(e)}}\left(r_1 + r_2 + r_3\right)\left[b_i b_j + c_i c_j\right] \tag{4.95}$$

The load vector containing the internal heat generation term for constant Q is written as

$$F_Q^{(e)} = 2\pi Q \int_{A^{(e)}} r \begin{bmatrix} N_1 \\ N_2 \\ N_3 \end{bmatrix} dA = 2\pi Q \int_{A^{(e)}} \begin{bmatrix} L_1 \\ L_2 \\ L_3 \end{bmatrix} \left(L_1 r_1 + L_2 r_2 + L_3 r_3\right) d\Omega \tag{4.96}$$

which can be integrated to give

$$F_Q^{(e)} = \frac{2\pi Q A^{(e)}}{12} \begin{bmatrix} 2r_1 + r_2 + r_3 \\ r_1 + 2r_2 + r_3 \\ r_1 + r_2 + 2r_3 \end{bmatrix} \tag{4.97}$$

The loading associated with the flux boundary condition on an element face is

$$F_{q_{1-2}}^{(e)} = -2\pi q \int_{\ell_{1-2}^{(e)}} r \begin{bmatrix} N_1 \\ N_2 \\ 0 \end{bmatrix} d\Gamma = -2\pi q \int_{\ell_{1-2}^{(e)}} r \begin{bmatrix} L_1 \\ L_2 \\ 0 \end{bmatrix} d\Gamma$$

$$= -2\pi q \int_{\ell_{1-2}^{(e)}} [L_1 r_1 + L_2 r_2] \begin{bmatrix} L_1 \\ L_2 \\ 0 \end{bmatrix} \tag{4.98a}$$

$$= -\frac{2\pi q \ell_{1-2}^{(e)}}{6} \begin{bmatrix} 2r_1 + r_2 \\ r_1 + 2r_2 \\ 0 \end{bmatrix}$$

Likewise,

$$F_{q_{2-3}}^{(e)} = -\frac{2\pi q \ell_{2-3}^{(e)}}{6} \begin{bmatrix} 0 \\ 2r_2 + r_3 \\ r_2 + 2r_3 \end{bmatrix} \tag{4.98b}$$

$$F_{q_{3-1}}^{(e)} = -\frac{2\pi q \ell_{3-1}^{(e)}}{6}\begin{bmatrix} 2r_1 + r_3 \\ 0 \\ r_1 + 2r_3 \end{bmatrix} \tag{4.98c}$$

The total force vector is then the sum of Eqs. (4.97), (4.98a), (4.98b), and (4.98c), where $\ell_{1-2}^{(e)}$, $\ell_{2-3}^{(e)}$, and $\ell_{3-1}^{(e)}$ are the lengths of the element sides given by $\ell_{i-j}^{(e)} = \sqrt{(r_j - r_i)^2 + (z_j - z_i)^2}$.

4.9 THE QUADRATIC TRIANGULAR ELEMENT

As discussed in Section 4.3.2, the quadratic triangular element consists of three vertex nodes, plus three mid-size nodes, as shown in Figure 4.4. On some occasions, the increased accuracy of the quadratic element significantly enhances overall convergence of the solution over that of the linear element. Computational savings can be achieved in spite of the larger bandwidth matrices that result due to the increased number of nodes in the element. Such situations occur in many nonlinear problems; a large mesh of finely sized linear elements may be required where a much coarser mesh of quadratic elements would have been sufficient.

Example 4.4

Let us return to the steady-state diffusion problem with convection and internal heat source previously discussed in Section 4.7. This time, we use the quadratic triangular element consisting of six nodes. The element is shown in Figure 4.11. The steady-state diffusion Eq. (4.58) with convective boundary condition (4.76) over side $\ell_{2-3}^{(e)}$ is considered. The conduction matrix, after applying the Galerkin formulation, is identical to Eq. (4.77), only i and j now range between 1 and 6. The shape functions in terms of natural area coordinates are given by Eq. (4.26). The

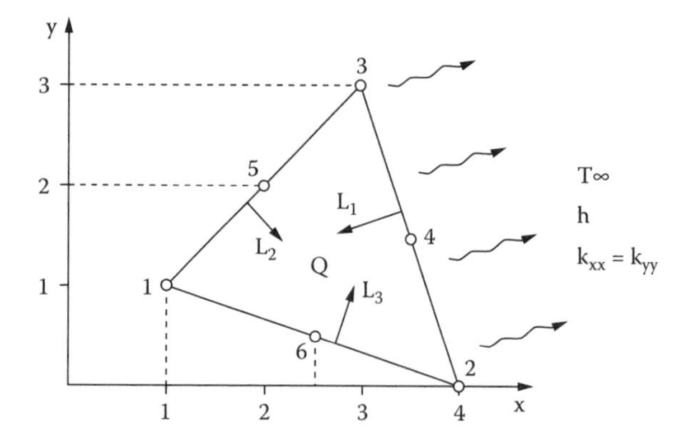

Figure 4.11 Quadratic triangular element subjected to convection cooling and internal heat generation

gradients $\partial N_i / \partial x$ and $\partial N_i / \partial y$ are obtained from Eq. (4.36). Thus, for example, we have for N_1,

$$\begin{bmatrix} \dfrac{\partial N_1}{\partial x} \\ \dfrac{\partial N_1}{\partial y} \end{bmatrix} = \mathbf{J}^{-1} \begin{bmatrix} \dfrac{\partial N_1}{\partial L_1} - \dfrac{\partial N_1}{\partial L_3} \\ \dfrac{\partial N_1}{\partial L_2} - \dfrac{\partial N_1}{\partial L_3} \end{bmatrix} = \mathbf{J}^{-1} \begin{bmatrix} 4L_1 - 1 \\ 0 \end{bmatrix} \tag{4.99}$$

The Jacobian is given by

$$\mathbf{J} = \begin{bmatrix} \dfrac{\partial x}{\partial L_1} - \dfrac{\partial x}{\partial L_3} & \dfrac{\partial y}{\partial L_1} - \dfrac{\partial y}{\partial L_3} \\ \dfrac{\partial x}{\partial L_2} - \dfrac{\partial x}{\partial L_3} & \dfrac{\partial y}{\partial L_2} - \dfrac{\partial y}{\partial L_3} \end{bmatrix} \tag{4.100}$$

For this particular element,

$$\left.\begin{aligned} x &= L_1 x_1 + L_2 x_2 + L_3 x_3 = L_1 + 4L_2 + 3L_3 \\ y &= L_1 y_1 + L_2 y_2 + L_3 y_3 = L_1 + 3L_3 \end{aligned}\right\} \tag{4.101}$$

Hence,

$$\mathbf{J} = \begin{bmatrix} -2 & -2 \\ 1 & -2 \end{bmatrix} \tag{4.102}$$

The inverse is

$$\mathbf{J}^{-1} = \frac{1}{8} \begin{bmatrix} -3 & 2 \\ -1 & -2 \end{bmatrix} \tag{4.103}$$

Consequently,

$$\begin{bmatrix} \dfrac{\partial N_1}{\partial x} \\ \dfrac{\partial N_1}{\partial y} \end{bmatrix} = \frac{1}{8} \begin{bmatrix} -3 & 2 \\ -1 & -2 \end{bmatrix} \begin{Bmatrix} 4L_1 - 1 \\ 0 \end{Bmatrix}$$

$$= \begin{bmatrix} -\dfrac{12L_1 + 3}{8} \\ -\dfrac{4L_1 + 1}{8} \end{bmatrix} = \begin{bmatrix} -\dfrac{3}{2}L_1 + \dfrac{3}{8} \\ -\dfrac{L_1}{2} + \dfrac{1}{8} \end{bmatrix}$$

Following the same procedure, we find the remaining derivative values:

$$\begin{bmatrix} \dfrac{\partial N_2}{\partial x} \\ \dfrac{\partial N_2}{\partial y} \end{bmatrix} = \begin{bmatrix} L_2 - \dfrac{1}{4} \\ -L_2 + \dfrac{1}{4} \end{bmatrix} \qquad \begin{bmatrix} \dfrac{\partial N_3}{\partial x} \\ \dfrac{\partial N_3}{\partial y} \end{bmatrix} = \begin{bmatrix} -\dfrac{1}{2} L_3 - 1 \\ \dfrac{3}{2} L_3 - \dfrac{3}{8} \end{bmatrix}$$

$$\begin{bmatrix} \dfrac{\partial N_4}{\partial x} \\ \dfrac{\partial N_4}{\partial y} \end{bmatrix} = \begin{bmatrix} -\dfrac{1}{2} L_2 + L_3 \\ \dfrac{3}{2} L_2 - L_3 \end{bmatrix} \qquad \begin{bmatrix} \dfrac{\partial N_5}{\partial x} \\ \dfrac{\partial N_5}{\partial y} \end{bmatrix} = \begin{bmatrix} -\dfrac{3}{2} L_3 + \dfrac{1}{2} L_1 \\ -\dfrac{1}{2} L_3 + \dfrac{3}{2} L_1 \end{bmatrix}$$

$$\begin{bmatrix} \dfrac{\partial N_6}{\partial x} \\ \dfrac{\partial N_6}{\partial y} \end{bmatrix} = \begin{bmatrix} \dfrac{3}{2} L_2 + L_1 \\ -\dfrac{1}{2} L_2 - L_1 \end{bmatrix}$$

The first term of the diffusion integral,

$$\int_{A^{(e)}} \mathrm{K} \frac{\partial N_i}{\partial x} \frac{\partial N_j}{\partial x} dA$$

is expressed in expanded form as

$$\begin{aligned} &\mathrm{K} \int_{A^{(e)}} \frac{\partial N_i}{\partial x} \frac{\partial N_j}{\partial x} dxdy \\ &= \mathrm{K} \int_{A^{(e)}} \begin{bmatrix} \dfrac{\partial N_1}{\partial x}\dfrac{\partial N_1}{\partial x} & \dfrac{\partial N_1}{\partial x}\dfrac{\partial N_2}{\partial y} & \cdots & \dfrac{\partial N_1}{\partial x}\dfrac{\partial N_6}{\partial x} \\ \dfrac{\partial N_2}{\partial x}\dfrac{\partial N_1}{\partial x} & & & \\ \cdot & & & \\ \cdot & \cdot & & \\ \cdot & \cdot & & \\ \cdot & & & \\ \dfrac{\partial N_6}{\partial x}\dfrac{\partial N_1}{\partial x} & \dfrac{\partial N_6}{\partial x}\dfrac{\partial N_2}{\partial x} & \cdots & \dfrac{\partial N_6}{\partial x}\dfrac{\partial N_6}{\partial x} \end{bmatrix} dxdy \end{aligned} \tag{4.104}$$

For the first term,

$$\begin{aligned} \mathrm{K} \int_{A^{(e)}} \frac{\partial N_1}{\partial x} \frac{\partial N_1}{\partial x} dxdy &= \mathrm{K} \int_{A^{(e)}} \left\{ -\frac{3}{2} L_1 + \frac{3}{8} \right\} \left[-\frac{3}{2} L_1 + \frac{3}{8} \right] dxdy \\ &= \mathrm{K} \int_{A^{(e)}} \left(\frac{9}{4} L_1^2 - \frac{18}{16} L_1 + \frac{9}{64} \right) dxdy \end{aligned} \tag{4.105}$$

Recalling Eq. (4.44b), it is a simple matter to evaluate the terms in Eq. (4.105). Thus,

$$K\int_{A^{(e)}} \frac{9}{4} L_1^2 dA = \left(\frac{9}{4}K\right) \frac{2!0!0!}{(2+0+0+2)!} 2A^{(e)}$$

$$= \frac{9}{4} K \frac{2!}{4!} 2A^{(e)}$$

$$= \frac{3}{8} K\, A^{(e)}$$

Similarly,

$$K\int_{A^{(e)}} -\frac{18}{16} L_1 dxdy = -\frac{18}{16} K \frac{1!0!0!}{(1+0+0+2)!} 2A^{(e)}$$

$$= -\frac{3}{8} KA^{(e)}$$

and

$$K\int_{A^{(e)}} -\frac{9}{64} dxdy = \frac{9}{64} K \frac{1!0!0!}{(0+0+0+2)!} 2A^{(e)}$$

$$= \frac{9}{64} KA^{(e)}$$

Thus,

$$K\int_{A^{(e)}} \frac{\partial N_1}{\partial x}\frac{\partial N_1}{\partial x} dxdy = KA^{(e)}\left(\frac{3}{8} - \frac{3}{8} + \frac{9}{64}\right) \tag{4.106}$$

$$= \frac{9}{64} KA^{(e)}$$

For the sixth term in Eq. (4.104),

$$K\int_{A^{(e)}} \frac{\partial N_1}{\partial x}\frac{\partial N_6}{\partial x} dxdy = K\int_{A^{(e)}} \left\{-\frac{3}{2}L_1 + \frac{3}{8}\right\}\left[-\frac{3}{8}L_2 + L_1\right] dxdy$$

$$= K\int_{A^{(e)}} \left(\frac{9}{4}L_1L_2 - \frac{3}{2}L_1^2 - \frac{9}{16}L_2 + \frac{3}{8}L_1\right) dxdy \tag{4.107}$$

$$= K\, A^{(e)}\left(\frac{3}{16} - \frac{4}{16} - \frac{3}{16} + \frac{2}{16}\right)$$

$$= -\frac{1}{8} KA^{(e)}$$

One proceeds in a similar fashion for 34 remaining terms. The diffusion term containing the derivatives of $\partial N / \partial y$ is also evaluated in exactly the same manner, i.e.,

$$K\int_{A^{(e)}} \frac{\partial N_i}{\partial y}\frac{\partial N_j}{\partial y}dxdy = K\int_{A^{(e)}} \begin{bmatrix} \frac{\partial N_1}{\partial y}\frac{\partial N_1}{\partial y} & \frac{\partial N_1}{\partial y}\frac{\partial N_2}{\partial y}\cdots & \frac{\partial N_1}{\partial y}\frac{\partial N_6}{\partial y} \\ \frac{\partial N_2}{\partial y}\frac{\partial N_1}{\partial y} & & \\ \cdot & & \\ \cdot & \cdot & \\ \cdot & \cdot & \\ \cdot & & \\ \frac{\partial N_6}{\partial y}\frac{\partial N_1}{\partial y} & \frac{\partial N_6}{\partial y}\frac{\partial N_2}{\partial y}\cdots & \frac{\partial N_6}{\partial y}\frac{\partial N_6}{\partial y} \end{bmatrix} \tag{4.108}$$

where

$$\begin{aligned} K\int_{A^{(e)}} \frac{\partial N_1}{\partial y}\frac{\partial N_1}{\partial y}dxdy &= K\int_{A^{(e)}} \left\{-\frac{1}{2}L_1+\frac{1}{8}\right\}\left[-\frac{1}{2}L_1+\frac{1}{8}\right]dxdy \\ &= K\int_{A^{(e)}} \left(\frac{1}{4}L_1^2-\frac{1}{8}L_1+\frac{1}{64}\right)dxdy \\ &= KA^{(e)}\left[\frac{1}{24}-\frac{1}{24}+\frac{1}{64}\right]=\frac{1}{64}KA^{(e)} \end{aligned} \tag{4.109}$$

etc. Note that the evaluation of the integrals, even using Eq. (4.44) has now become considerably more complex; in this instance, it is more convenient to use numerical integration.

The remaining convection term in the stiffness matrix, corresponding to Eq. (4.86), is evaluated as

$$\begin{aligned} H^{(e)} &= \left[\int_{\ell^{(e)}_{2-4-3}} hN_iN_j d\Gamma\right] = h\int_{\ell^{(e)}_{2-4-3}} \begin{bmatrix} N_1 \\ N_2 \\ N_3 \\ N_4 \\ N_5 \\ N_6 \end{bmatrix} \begin{bmatrix} N_1 & N_2 & N_3 & N_4 & N_5 & N_6 \end{bmatrix} d\Gamma \\ &= h\int_{\ell^{(e)}_{2-4-3}} \begin{bmatrix} N_1N_1 & N_1N_2 & \cdots & N_1N_6 \\ N_2N_1 & N_2N_2 & & \\ \cdot & & & \cdot \\ \cdot & & & \cdot \\ \cdot & & & \cdot \\ N_6N_1 & N_6N_2 & \cdots & N_6N_6 \end{bmatrix} d\Gamma \end{aligned} \tag{4.110}$$

Since one side 2–4–3 is affected, $\mathbf{H}^{(e)}$ becomes

$$H^{(e)} = h\int_{\ell^{(e)}_{2-4-3}} \begin{bmatrix} 0 & 0 & 0 & 0 & 0 & 0 \\ 0 & N_2N_2 & N_2N_3 & N_2N_4 & 0 & 0 \\ 0 & N_3N_2 & N_3N_3 & N_3N_4 & 0 & 0 \\ 0 & N_4N_2 & N_4N_3 & N_4N_4 & 0 & 0 \\ 0 & 0 & 0 & 0 & 0 & 0 \\ 0 & 0 & 0 & 0 & 0 & 0 \end{bmatrix} d\Gamma \tag{4.111}$$

which, in our example, yields

$$H^{(e)} = \frac{h\ell^{(e)}_{2-4-3}}{30} \begin{bmatrix} 0 & 0 & 0 & 0 & 0 & 0 \\ 0 & 4 & -1 & 2 & 0 & 0 \\ 0 & -1 & 4 & 2 & 0 & 0 \\ 0 & 2 & 2 & 16 & 0 & 0 \\ 0 & 0 & 0 & 0 & 0 & 0 \\ 0 & 0 & 0 & 0 & 0 & 0 \end{bmatrix} \tag{4.112}$$

where a typical term, say $h^{(e)}_{22}$, is given by

$$\begin{aligned} h^{(e)}_{22} &= h\int_{\ell^{(e)}_{2-4-3}} N_2N_2 d\Gamma = h\int_{\ell^{(e)}_{2-4-3}} \{2L_2^2 - L_2\}[2L_2^2 - L_2] d\Gamma \\ &= h\int_{\ell^{(e)}_{2-4-3}} (4L_2^4 - 4L_2^3 - L_2^2) d\Gamma \\ &= h\,\ell^{(e)}_{2-4-3}\left(4\frac{4!0!}{(1+0+4)!} - 4\frac{3!0!}{(1+0+3)!} + \frac{2!0!}{(1+0+2)!}\right) \\ &= \frac{2h\,\ell^{(e)}_{2-4-3}}{15} \end{aligned} \tag{4.113}$$

To evaluate Eq. (4.111), we had to use Eq. (4.44a) because only the side containing nodes 2, 4 and 3, i.e., L_2L_3 coordinates, is considered.

The integral of the source term, Q, is evaluated as

$$Q\int_{A^{(e)}} N_i dxdy = Q\int_{A^{(e)}} \begin{bmatrix} N_1 \\ N_2 \\ N_3 \\ N_4 \\ N_5 \\ N_6 \end{bmatrix} dxdy = \frac{Q\,A^{(e)}}{3} \begin{bmatrix} 0 \\ 0 \\ 0 \\ 1 \\ 1 \\ 1 \end{bmatrix} \tag{4.114}$$

where use is made of Eq. (4.44). Notice that the source term is distributed equally among the three midsize nodes.

The last remaining term to consider is the heat flux, q, or boundary condition flux if present. In the quadratic element, the term is evaluated as

$$q\int_{\Gamma_B} N_i d\Gamma = q\int_{\Gamma_B} \begin{bmatrix} N_1 \\ N_2 \\ N_3 \\ N_4 \\ N_5 \\ N_6 \end{bmatrix} d\Gamma$$

$$= \begin{cases} \dfrac{q\ell_{1-6-2}^{(e)}}{6} \begin{bmatrix} 1 \\ 1 \\ 0 \\ 0 \\ 0 \\ 4 \end{bmatrix} \\ \dfrac{q\ell_{2-4-3}^{(e)}}{6} \begin{bmatrix} 0 \\ 1 \\ 1 \\ 4 \\ 0 \\ 0 \end{bmatrix} \\ \dfrac{q\ell_{3-5-1}^{(e)}}{6} \begin{bmatrix} 1 \\ 0 \\ 1 \\ 0 \\ 4 \\ 0 \end{bmatrix} \end{cases} \tag{4.115}$$

where Eq. (4.44a) is utilized for each element side.

4.10 TIME-DEPENDENT DIFFUSION EQUATION

The time-dependent equation for transport of a scalar variable $\phi(x, y, t)$ due to diffusion is written in its two-dimensional form as

$$\frac{\partial \phi}{\partial t} = \frac{\partial}{\partial x}\left(\alpha \frac{\partial \phi}{\partial y}\right) + \frac{\partial}{\partial y}\left(\alpha \frac{\partial \phi}{\partial y}\right) + S \tag{4.116}$$

for a region Ω such as depicted in Figure 4.9, with boundary conditions

$$\phi = \phi \qquad \text{in } \Gamma_A \tag{4.117a}$$

$$-\alpha \frac{\partial \phi}{\partial n} = a\phi + b \qquad \text{in } \Gamma_B \tag{4.117b}$$

and initial condition:

$$\phi(x, y, 0) = \phi_0(x, y) \tag{4.118}$$

Equations (4.116) through (4.118) are two-dimensional extensions of Eqs. (3.87) through (3.90), which describe one-dimensional transport of heat. The variable ϕ now represents any unknown scalar quantity; α is the diffusion coefficient, e.g., thermal diffusivity, which can be spatially dependent; and S is source/sink term. The values of a and b in (4.117b) are prescribed for each problem.

The weighted residuals form of Eq. (4.116) is

$$\int_\Omega W\left[\frac{\partial \phi}{\partial t} - \frac{\partial}{\partial x}\left(\alpha \frac{\partial \phi}{\partial x}\right) - \frac{\partial}{\partial y}\left(\alpha \frac{\partial \phi}{\partial y}\right) - S\right] d\Omega = 0 \tag{4.119}$$

and, after applying Green's theorem, as was done before to obtain Eq. (4.65),

$$\int_\Omega \left\{W\frac{\partial \phi}{\partial t} + \alpha\left(\frac{\partial W}{\partial x}\frac{\partial \phi}{\partial x} + \frac{\partial W}{\partial y}\frac{\partial \phi}{\partial y}\right) - W\, S\right\} d\Omega + \int_{\Gamma_B} W\left(-\alpha \frac{\partial \phi}{\partial n}\right) d\Gamma = 0 \tag{4.120}$$

where $d\Gamma$ denotes the surface element of Γ_B over which the normal gradients (fluxes) are applied. We approximate the variables in terms of shape functions as before:

$$\left.\begin{aligned}\phi(x,y,t) &\cong \sum_{i=1}^{N} N_i(x,y)\phi_i(t)\\ \alpha(x,y) &\cong \sum_{i=1}^{N} N_i(x,y)\alpha_i\\ S(x,y,t) &\cong \sum_{i=1}^{N} N_i(x,y)S_i(t)\end{aligned}\right\} \tag{4.121}$$

Substituting Eq. (4.121) into Eq. (4.120) and setting $W_i = N_i$, we obtain the semi-discrete Galerkin approximation as

$$\begin{aligned}&\left[\int_\Omega N_i N_j d\Omega\right]\dot{\phi}_j + \left[\int_\Omega \alpha\left(\frac{\partial N_i}{\partial x}\frac{\partial N_j}{\partial x} + \frac{\partial N_i}{\partial y}\frac{\partial N_i}{\partial y}\right)dA\right]\phi_j\\ &+\left[\int_B aN_iN_jd\Gamma\right]\phi_j = \int_\Omega N_i S d\Omega - \int_{\Gamma_B} N_i b d\Gamma\end{aligned} \tag{4.122}$$

where the normal derivative term along Γ_B has been replaced using Eq. (4.117b), and summation in the index j is implied on the left-hand side of the equation. Notice the similarity between Eq. (4.122) and Eq. (3.97). As mentioned in Chapter 3, development of the finite element algorithms for the one-dimensional element is also applicable for multidimensional problem geometries. For a linear triangular element, the element mass matrix is evaluated as

$$\begin{aligned}\mathbf{M}^{(e)} &= \int_{A^{(e)}} N_i N_j dA = \int_{A^{(e)}} \begin{bmatrix} L_1 \\ L_2 \\ L_3 \end{bmatrix} [L_1\ L_2\ L_3] dA\\ &= \int_{A^{(e)}} \begin{bmatrix} L_1L_1 & L_1L_2 & L_1L_3 \\ L_2L_1 & L_2L_2 & L_2L_3 \\ L_3L_1 & L_3L_2 & L_3L_3 \end{bmatrix} dA = \frac{A^{(e)}}{12}\begin{bmatrix} 2 & 1 & 1 \\ 1 & 2 & 1 \\ 1 & 1 & 2 \end{bmatrix}\end{aligned} \tag{4.123}$$

The diffusion (stiffness) matrix is given by Eq. (4.72) with K replaced by α when the diffusion coefficient is constant. If $\alpha = \alpha(x, y)$, we have an expression similar to Eq. (4.94), that is

$$\mathbf{k}^{(e)} = \left[k_{ij}^{(e)}\right] = \left[\frac{b_i b_j + c_i c_j}{4\left(A^{(e)}\right)^2} \int_{A^{(e)}} \left(L_1\alpha_1 + L_2\alpha_2 + L_3\alpha_3\right) dxdy\right] \tag{4.124}$$

The source term yields an expression similar to the mass matrix, i.e.,

$$\mathbf{F}_S^{(e)} = [f_{Si}] = \frac{A^{(e)}}{12}\begin{bmatrix} 2 & 1 & 1 \\ 1 & 2 & 1 \\ 1 & 1 & 2 \end{bmatrix}\begin{bmatrix} S_1 \\ S_2 \\ S_3 \end{bmatrix} = \mathbf{MS} \tag{4.125}$$

where $\mathbf{S}$ is the vector of nodal values of $S(x, y, t)$. If $\mathbf{S}$ is a constant function, we get the vector

$$\mathbf{F}_S^{(e)} = \frac{SA^{(e)}}{3}\begin{bmatrix} 1 \\ 1 \\ 1 \end{bmatrix} \tag{4.126}$$

which is what we obtain for steady-state heat conduction, Eq. (4.73).

The convective term $a\phi$ generates a matrix identical to Eq. (4.86), with h replaced by a, so if convection occurs on side 1–2, we get

$$\mathbf{H}_{1-2}^{(e)} = a\ell_{1-2}^{(e)}\begin{bmatrix} 2 & 1 & 0 \\ 1 & 2 & 0 \\ 0 & 0 & 0 \end{bmatrix} \tag{4.127}$$

and similarly for the other two sides. Finally, the flux b acts as a boundary condition applied to an element side (surface). Assuming that the flux is constant over the side, the flux boundary relations are given by Eq. (4.74) with q replaced by b. We use the notation:

$$\mathbf{F}_{b_{1-2}}^{(e)} = -\frac{b\ell_{1-2}^{(e)}}{2}\begin{bmatrix} 1 \\ 1 \\ 0 \end{bmatrix} \tag{4.128}$$

and similarly for the other two sides.

Evaluating all matrices over an element, we obtain the general element matrix equation, as we did before for the one-dimensional element, Eq. (3.108).

$$\mathbf{M}^{(e)}\dot{\phi} + \mathbf{K}^{(e)}\phi = \mathbf{F}^{(e)} \tag{4.129}$$

where

$$\mathbf{K}^{(e)} = \mathbf{k}^{(e)} + \mathbf{H}^{(e)} \tag{4.130}$$

and

$$\mathbf{F}^{(e)} = \mathbf{F}_Q^{(e)} + \mathbf{F}_b^{(e)} \tag{4.131}$$

The matrix $\mathbf{H}^{(e)}$ and the vector $\mathbf{F}_b^{(e)}$ contain the contributions of the mixed boundary conditions (4.117b) along one or more sides of the element.

In this instance, the matrix coefficient terms were obtained from a linear triangular two-dimensional element. Remember that the finite element procedure and resulting matrix equation are applicable to any type of element and any number of spatial dimensions.

Example 4.5*

Derive the matrix-equivalent relation for unsteady diffusion of heat with internal heat source $\bar{Q} = 1\,C/s$, where $\bar{Q} = Q/\rho c_p$, over the linear two-dimensional element shown in Figure 4.12. For illustrative purposes, assume α = 10 m²/s with convective heat loss out of face 2–3, $\bar{h} = 1\,m/s$ ($\bar{h} = h/\rho c_p$), and T_∞ = 100°C. Initially, $T(x, y, 0)$ = 500°C. The time-dependent diffusion equation is written as

$$\frac{\partial T}{\partial t} = \alpha\left(\frac{\partial^2 T}{\partial x^2} + \frac{\partial^2 T}{\partial y^2}\right) + \bar{Q} \tag{4.132}$$

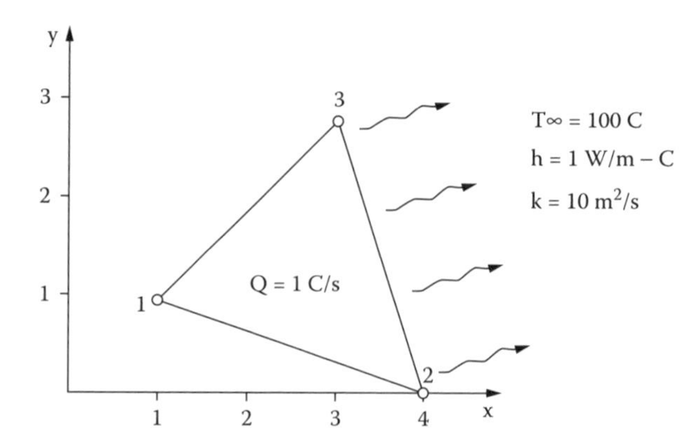

Figure 4.12 Transport of heat in a triangular element

* FEM-2D and COMSOL files are available.

with

$$-\alpha\frac{\partial T}{\partial n}=\bar{h}(T-T_{\infty}) \qquad on\,\Gamma_{2-3} \tag{4.133}$$

$$-\alpha\frac{\partial T}{\partial n}=0 \qquad on\,\Gamma_{1-2}\ and\ \Gamma_{3-1} \tag{4.134}$$

and

$$T(x,y,0)=500 \tag{4.135}$$

Referring to Eq. (4.129), we immediately have

$$\mathbf{M}\dot{\mathbf{T}}+(\mathbf{k}+\mathbf{H}_{2-3})\mathbf{T}=\mathbf{F}_{q2-3}+\mathbf{F}_{Q} \tag{4.136}$$

The area of the element is calculated from Eq. (4.6) and yields A = 4.

The mass matrix from Eq. (4.123) is

$$\mathbf{M}=\frac{1}{3}\begin{bmatrix}2 & 1 & 1\\ 1 & 2 & 1\\ 1 & 1 & 2\end{bmatrix} \tag{4.137}$$

The matrices $\mathbf{k}$ and $\mathbf{H}_{2-3}$ are given by Eqs. (4.72) and (4.127), respectively, and yield

$$\mathbf{k}=\frac{5}{8}\begin{bmatrix}10 & -4 & -6\\ -4 & 8 & -4\\ -6 & -4 & 10\end{bmatrix} \tag{4.138}$$

$$\mathbf{H}_{2-3}=\frac{\sqrt{10}}{6}=\begin{bmatrix}0 & 0 & 0\\ 0 & 2 & 1\\ 0 & 1 & 2\end{bmatrix} \tag{4.139}$$

Finally, the vectors F_{q2-3} and F_{Q} are given by Eqs. (4.128) and (4.126), respectively, and are

$$F_{q2-3}=50\sqrt{10}\begin{bmatrix}0\\ 1\\ 1\end{bmatrix} \tag{4.140}$$

$$F_Q = \frac{8}{3}\begin{bmatrix} 1 \\ 1 \\ 1 \end{bmatrix} \tag{4.141}$$

Substituting Eqs. (4.137)–(4.141) into Eq. (4.136), we obtain

$$\frac{1}{3}\begin{bmatrix} 2 & 1 & 1 \\ 1 & 2 & 1 \\ 1 & 1 & 2 \end{bmatrix}\begin{bmatrix} \dot{T}_1 \\ \dot{T}_2 \\ \dot{T}_3 \end{bmatrix} + \begin{bmatrix} \frac{25}{4} & -\frac{5}{2} & -\frac{15}{4} \\ -\frac{5}{2} & 5+\frac{\sqrt{10}}{3} & -\frac{5}{2}+\frac{\sqrt{10}}{6} \\ -\frac{15}{4} & -\frac{5}{2}+\frac{\sqrt{10}}{6} & \frac{25}{4}+\frac{\sqrt{10}}{3} \end{bmatrix}\begin{bmatrix} T_1 \\ T_2 \\ T_3 \end{bmatrix} = \begin{bmatrix} \frac{4}{3} \\ \frac{4}{3}+50\sqrt{10} \\ \frac{4}{3}+50\sqrt{10} \end{bmatrix} \tag{4.142}$$

We can now approximate the time derivative using the θ method described in Section 3.6 and defined by Eqs. (3.98) and (3.99). If we use $\theta = 1$, a backward implicit method, we have

$$\dot{T} = \frac{T^{n+1} - T^n}{\Delta t} \tag{4.143}$$

At $t = 0$, $T_1 = T_2 = T_3 = 500°\text{C}$. The last step is to choose Δt, substitute Eq. (4.143) into Eq. (4.142), and solve Eq. (4.142) for the unknown temperature using a suitable matrix solution algorithm.

4.11 BANDWIDTH

The concept of bandwidth becomes important when dealing with two- and three-dimensional elements. As briefly discussed in Chapter 3, the bandwidth of a matrix is determined by the sequence of node numbering. In a one-dimensional problem, the elements and hence, node numbering, are sequential, as shown in Figure 4.13.

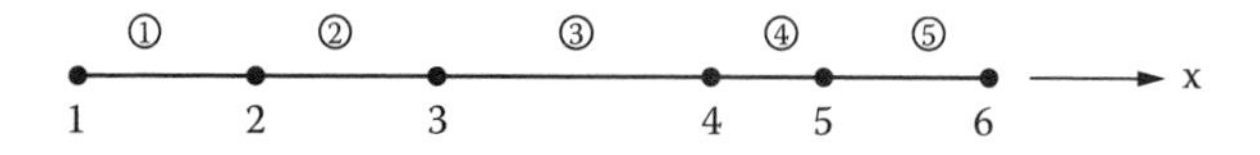

Figure 4.13 Node numbering and element connectivity in a one-dimensional domain; O denotes element number

The "global" node numbering associated with an element is termed "element connectivity." In a one-dimensional linear element, two "local" nodes are associated with the element, i.e., every element has a node located at each end of the element. For example, referring to Figure 4.13, the two local nodes associated with element 3 are globally numbered 3 and 4. Table 4.3 lists the element number and connectivity for the one-dimensional mesh shown in Figure 4.13.

When the element matrix equations derived from the weak statement formulation are "assembled" over the entire problem domain, i.e., all elements are incorporated, the resulting global system of linear equations, $\mathbf{A}\phi = \mathbf{F}$, involves an $n \times n$ square matrix, where n is the total number of nodes. In Figure 4.13, the global matrix is 6 × 6 square matrix. As shown in Figure 3.7, assemblage over two adjacent one-dimensional linear elements produces a tri-diagonal matrix that is generic over the entire mesh. Thus, the 6 × 6 is actually a sparse matrix with only a main diagonal and an upper and lower adjacent diagonal, as shown in Figure 4.14.

The a_{ij} terms represent the individual coefficients that make up the 6 × 6 array. Notice the sparseness of the matrix in Figure 4.14. The largest number of nonzero coefficients in a row is 3. This is also the bandwidth of the matrix. The distance (number of columns) from the main diagonal (and including the main diagonal) to

Table 4.3 One-Dimensional Element Connectivity

Element	Local Node	Local Node
1	1[a]	2[a]
2	2	3
3	3	4
4	4	5
5	5	6

[a] Global node number.

$$A = \begin{bmatrix} a_{11} & a_{12} & 0 & 0 & 0 & 0 \\ a_{21} & a_{22} & a_{23} & 0 & 0 & 0 \\ 0 & a_{32} & a_{33} & a_{34} & 0 & 0 \\ 0 & 0 & a_{43} & a_{44} & a_{45} & 0 \\ 0 & 0 & 0 & a_{54} & a_{55} & a_{56} \\ 0 & 0 & 0 & 0 & a_{65} & a_{66} \end{bmatrix}$$

Figure 4.14 A 6 × 6 sparse matrix with bandwidth 3 and half-bandwidth 2

the farthest most nonzero coefficient is the half-bandwidth (or semi-bandwidth) is 2. We can calculate the half-bandwidth of a matrix from the simple relation:

$$\text{HBW} = \text{NOF} \times (N_{\text{max}} - N_{\text{diag}} + 1) \tag{4.144}$$

where NOF is the number of degrees of freedom per node, i.e., the number of nodal unknowns such as temperature, velocity, concentration, etc., and N_{max} and N_{diag} are the largest node number assigned to any nonzero element in a row and the number of the row, respectively. For the one-dimensional domain consisting of six elements, and solving for only one unknown (NOF = 1), the half-bandwidth is

$$\text{HBW} = 1(1+1) = 2 \tag{4.145}$$

where $N_{\text{max}} - N_{\text{diag}} = 1$ for any row. The coefficients outside the bandwidth are zero; these coefficients need not be stored. Efficient computer programming (matrix solution algorithm) deals only with the coefficients within the bandwidth; this is particularly important when using Gaussian elimination methods. Every effort should be made to reduce the bandwidth to a minimum because a reduction in the bandwidth directly results in a reduction in computer storage and computing time. For the one-dimensional element, the bandwidth is already at its minimum.

In a two-dimensional mesh, the labeling of node numbers becomes important with regard to the bandwidth. Figure 4.15 shows a two-dimensional domain with two different node numberings using linear triangular elements. Each element consists of three local nodes (1,2,3) with assigned global numbers. Element connectivities for both mesh configurations are listed in Table 4.4 (locally, we number in a counterclockwise direction beginning with the lower left node in each element). Notice that poor nodal labeling in Figure 4.15a creates a half-bandwidth which is 17 wide; in Figure 4.15b, an improved node-numbering scheme creates a half-bandwidth of only 6, a considerable improvement.

Example 4.6

Determine the half-bandwidth of the mesh shown in Figure 4.16, which uses quadratic triangular elements. We create a table listing the element connectivities and local node numbering (there are 6 local nodes per element). The mesh in Figure 4.16 produces a half-bandwidth 11; in this case, every element produces a half-bandwidth of 11; i.e., the mesh numbering is optimized. If we number across instead of down, beginning with the upper left node, what is the maximum bandwidth?

The user has direct control over the semi-bandwidth; the node numbering can be minimized by visual inspection for simple problems using linear elements. However, for higher-degree elements or large three-dimensional problems, bandwidth minimization by inspection is very difficult. Commercially available software packages that perform mesh generation and reorder node numbering to minimize the

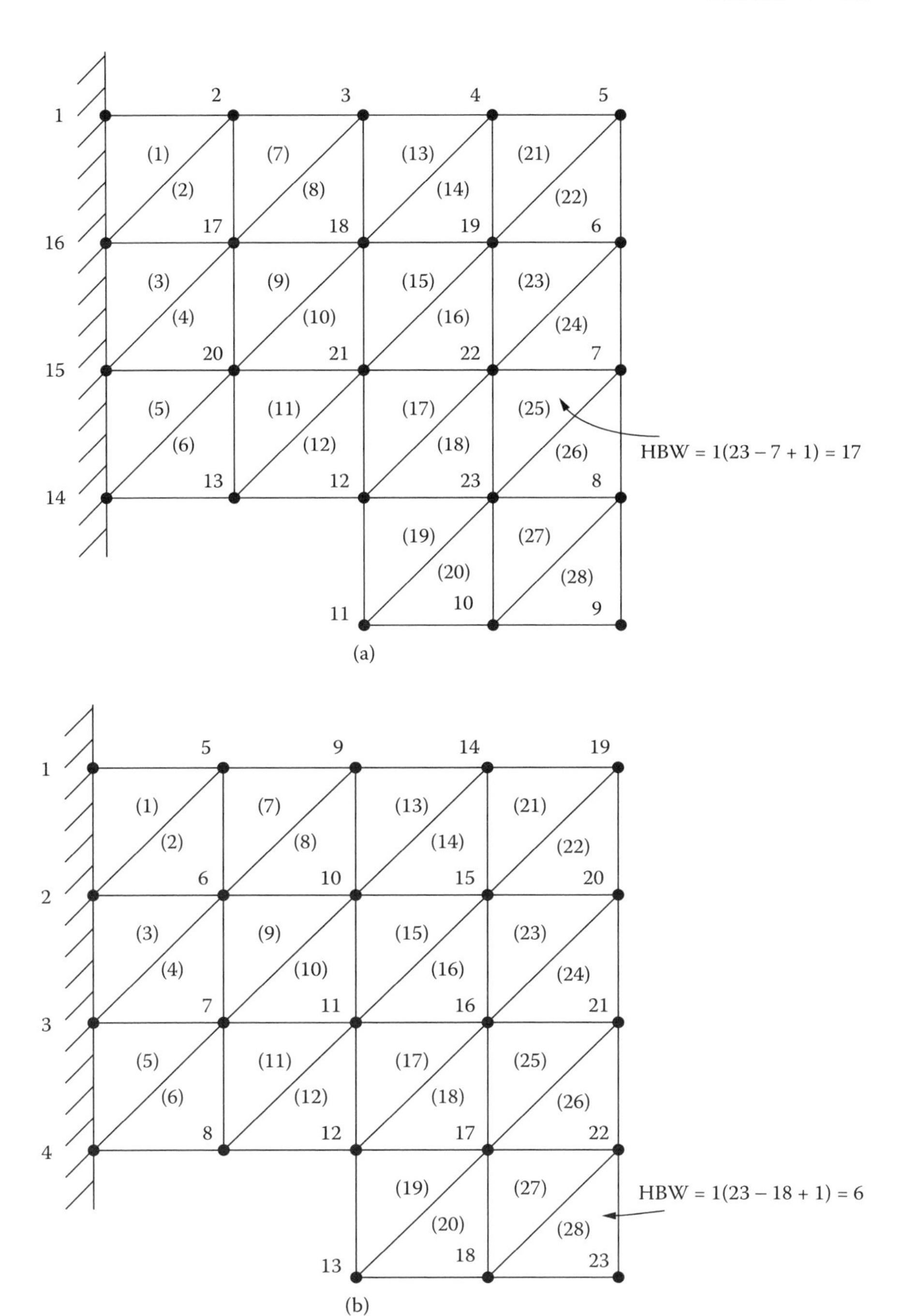

Figure 4.15 Node numbering in a two-dimensional domain: (a) numbering across and (b) numbering down

Table 4.4 Two-Dimensional Element Connectivity for Mesh Configurations in Figure 4.15

	(a)				(b)			
Element	1	2	3	HBW	1	2	3	HBW
1	16	2	1	16	2	5	1	5
2	16	17	2	16	2	6	5	5
3	15	17	16	3	3	6	2	5
4	15	20	17	6	3	7	6	5
5	14	20	15	7	4	7	3	5
6	14	13	20	8	4	8	7	5
7	17	3	2	16	6	9	5	5
8	17	18	3	16	6	10	9	5
9	20	18	17	4	7	10	6	5
10	20	21	18	4	7	11	10	5
11	13	21	20	9	8	11	7	5
12	13	12	21	10	8	12	11	5
13	18	4	3	16	10	14	9	6
14	18	19	4	16	10	15	14	6
15	21	19	18	4	11	15	10	6
16	21	22	19	4	11	16	15	6
17	12	22	21	11	12	16	11	6
18	12	23	22	12	12	17	16	6
19	11	23	12	13	13	17	12	6
20	11	10	23	14	13	18	17	6
21	19	5	4	16	15	19	14	6
22	19	6	5	15	15	20	19	6
23	22	6	19	17	16	20	15	6
24	22	7	6	17	16	21	20	6
25	23	7	22	17	17	21	16	6
26	23	8	7	17	17	22	21	6
27	10	8	23	16	18	22	17	6
28	10	9	8	3	18	23	22	6

semi-bandwidth are available for the PC. The programs typically command a premium price. MESH-2D uses a simple nodal renumbering scheme to reorder the node numbers to produce the smallest bandwidth. COMSOL and other commercial finite element codes do a much better job at reordering node numbers; many of these optimization schemes are based on the Cuthill–McGee (1969) algorithm introduced many years ago. The reader is referred to the mesh generation text by Carey (1997) and to Heinrich and Pepper (1999) for more details on mesh optimization.

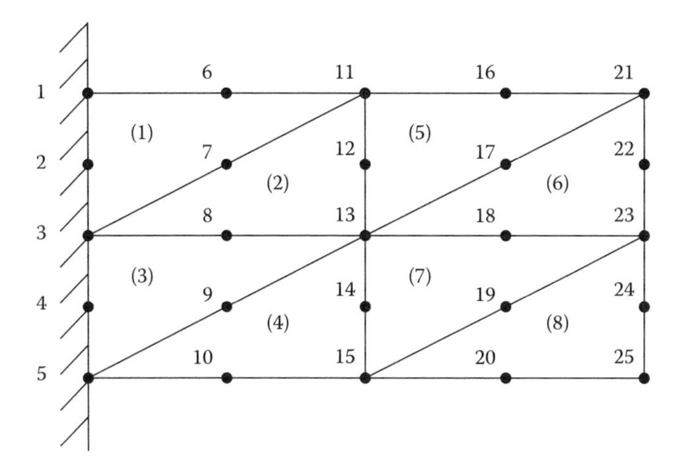

Figure 4.16 Two-dimensional mesh consisting of quadratic triangular elements

4.12 MASS LUMPING

In Chapter 3, we briefly touched on the subject of mass lumping when we discussed the possibility of using explicit time-marching algorithms, i.e., the θ method with $\theta = 0$. Application of the θ method with $\theta = 0$ to Eq. (4.136) yields

$$\mathbf{MT}^{n+1} = \Delta t\left[\mathbf{F}_{q_{2-3}} + \mathbf{F}_Q - (\mathbf{k} + \mathbf{H}_{2-3})\mathbf{T}^n\right] + \mathbf{MT}^n \tag{4.146}$$

To calculate $\mathbf{T}^{n+1}$ requires $\mathbf{M}^{-1}$ to be computed or some elimination method to be used to solve for $\mathbf{T}^{n+1}$ due to the coupling created by the mass matrix.

To obtain a fully explicit scheme, we diagonalize the mass matrix **M**, or "lump" the masses. For the type of elements under consideration, this is simply achieved by adding all elements in each row of **M** and putting the sum in the diagonal. The mass matrix **M** is then replaced by the lumped mass matrix $\mathbf{M}^{\ell}$, defined by

$$\mathbf{M}^{\ell} = \left[m_{ij}^{\ell}\right] \tag{4.147}$$

where

$$m_{ij}^{\ell} = \begin{cases} 0 & \text{if } i \neq j \\ \displaystyle\sum_{j=1}^{n} m_{ij} & \text{if } i = j \end{cases} \tag{4.148}$$

which is a diagonal matrix. Equation (4.146) can then be modified to

$$T^{n+1} = \Delta t\left(\mathbf{M}^{\ell}\right)^{-1}\left[F_{q_{2-3}} + F_Q - \left(k + H_{2-3}\right)T^n\right] + T^n \tag{4.149}$$

which if fully explicit. Furthermore, $\left(\mathbf{M}^{\ell}\right)^{-1}$ is trivial to compute since it is given by

$$(\mathbf{M}_{ij}^{\ell})^{-1} = \begin{cases} 0 & if\ i \neq j \\ \dfrac{1}{m_{ij}^{\ell}} & if\ i = j \end{cases} \tag{4.150}$$

For example, the lumped mass matrix corresponding to the mass matrix in Eq. (4.137) is

$$\mathbf{M}^{\ell} = \frac{1}{3}\begin{bmatrix} 4 & 0 & 0 \\ 0 & 4 & 0 \\ 0 & 0 & 4 \end{bmatrix}$$

The question that now arises is whether a consistent mass matrix or mass lumping should be used when solving time-dependent problems. There is no definitive answer to this question, and the situation must be examined on a problem-by-problem basis. As a general rule, if we are interested in a transient solution where the time evolution must be accurately tracked, a consistent mass formulation is best. However, if we wish to rapidly achieve steady-state conditions by means of a time integration scheme (in lieu of iteration), then mass lumping is preferred. Unfortunately there are other considerations that come into play when deciding which formulation to use.

A consistent mass formulation requires the assembly and storage of the mass matrix; in the case where $\theta = 0$ (see Eq. 3.112a), a system of equations in which the global matrix is the mass matrix must be solved at each time step. As a result, an increase in storage and computing time takes place, which can be significant.

Additional considerations arise when we address the question of accuracy of the solution and the choice of the time step size. In this case, stability limitations and the correct representation of the physics involved must be evaluated. The analyses necessary to answer these questions are not trivial and require the computation of the eigenvalues of the matrices involved. Results from such studies show that the time step size is more limiting when using the consistent formulation than when using mass lumping.

For a more detailed discussion of the issues related to mass lumping, we refer the reader to Segerlind (1984). For an advanced treatment of time-dependent problems in dynamics and field problems, the reader should consult the books by Zienkiewicz and Taylor (1989) and Hughes (1987).

4.13 CLOSURE

We have extended the finite element procedure to two dimensions based on the fundamental concepts originally derived for one-dimensional elements. While more detail is required in establishing a two-dimensional mesh and formulating the matrices, the underlying principles are the same for any number of spatial dimensions. We have concentrated on the two-dimensional triangular element in this chapter, i.e., the linear three-noded element and the six-noded quadratic element. Most of the early developmental work on the finite element method revolved around the triangular element; extensive application of triangular elements in structural problems occurred during the 1960s and early 1970s. Triangular elements are still used today; they are simple to construct, easy to use, and can approximate irregular boundaries because they can be oriented as desired.

The reader is encouraged to become acquainted with either a commercial finite element code or the accompanying codes on the Web (see femcodes.nscee.edu) to learn about the process of mesh generation and boundary condition input. Several problems in the exercises will require the use of a computer program—working out the problems by hand is only for the stout hearted!

EXERCISES

A set of exercises follow, some of which require the user to generate two-dimensional meshes using linear and quadratic triangular elements. Once a mesh is generated, the user loads the two-dimensional finite element solver and inputs specific data as requested by the computer program. As in the one-dimensional programs MESH-1D and FEM-1D used in Chapter 3, MESH-2D and FEM-2D are modular driven and can be used to solve many of the exercises. COMSOL is also an easy program to use; once familiar with how the code operates, the user will find that it can easily and quickly solve a wide range of problems. The simple 2-D mesh generator written by Persson and Strang (2004) in MATLAB will also produce high-quality meshes.

4.1 Derive the expressions in Eq. (4.5) and use these to obtain the shape functions (4.8) to (4.10).

4.2 As in the one-dimensional case show that for the linear triangle

$$\sum_{i=1}^{3} N_i\left(x, y\right) \equiv 1 \text{ and } \sum_{i=1}^{3} \frac{\partial N_i}{\partial x} = \sum_{i=1}^{3} \frac{\partial N_i}{\partial y} = 0$$

4.3 Show that the shape function for the linear triangle satisfies $N_i(x_j, y_j) = \delta_{ij}$.

4.4 The temperature along the sides of the triangle in the figure below is prescribed as shown.

(a) Interpolate using one linear element and find T and its partial derivatives at (0.25, 0.25).

(b) Repeat part (a) using one quadratic element.
(c) Compare the results and discuss.

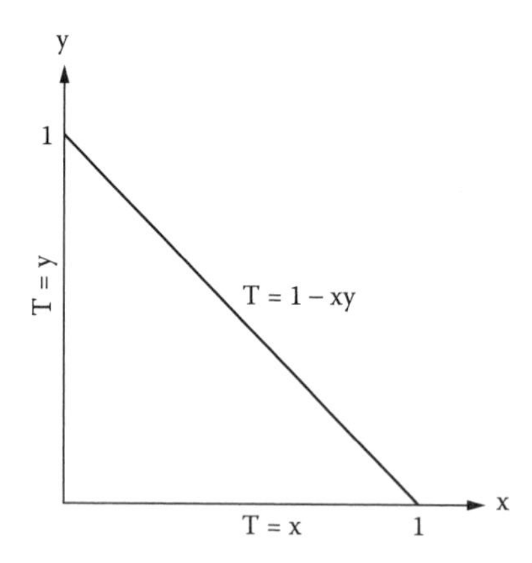

4.5 Find the temperature and the temperature gradients at point (0.25, 0.25) of a linear triangular element defined by nodes at (0,0), (1,0), and (0,1) with temperatures –10°C, 50°C, and 15°C, respectively.

4.6 Find the derivatives $\delta N_i/\partial x$ and $\partial N_i/\partial y$, where N_i is given in Eqs. (4.26), for the element of Figure 4.8 at the point (1,5,2).

4.7 Evaluate the integral of Example 4.3 using formulae (4.44).

4.8 Solve the systems of equations in Eq. (4.84) for the nodal temperatures in the triangle of Figure 4.10.

4.9 Find Eq. (4.97) from (4.96).

4.10 Use Eq. (4.44a) and Eq. (4.111) to derive Eq. (4.112).

4.11 Use Eq. (4.44b) to verify (4.114) and (4.115).

4.12 Subdivide the following geometrical figure into ten linear triangular elements. Label the nodes and elements and determine the bandwidth.

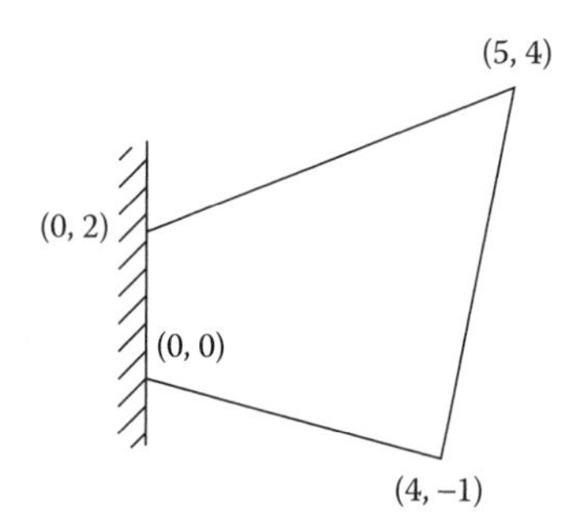

4.13 Calculate the area of each element in Exercise 4.12. Reconfigure the mesh such that all areas are approximately the same.

4.14 Find $\partial N_3/\partial x$ and $\partial N_3/\partial y$ in Example 4.2.

4.15 Generate a mesh consisting of 20 triangular elements using MESH-2D or COMSOL for Exercise 4.1.

4.16 Calculate the shape functions for the following elements:

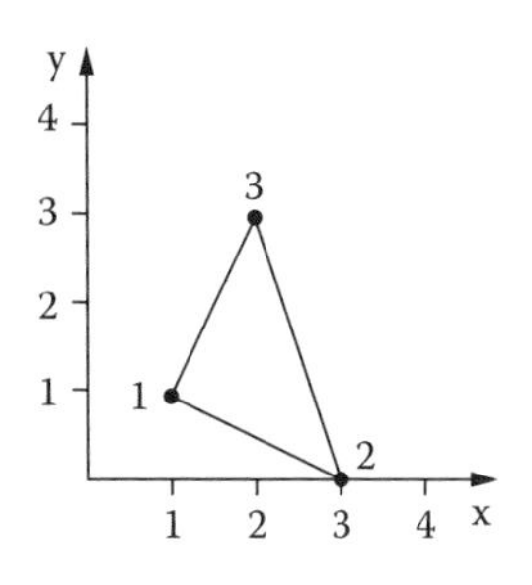

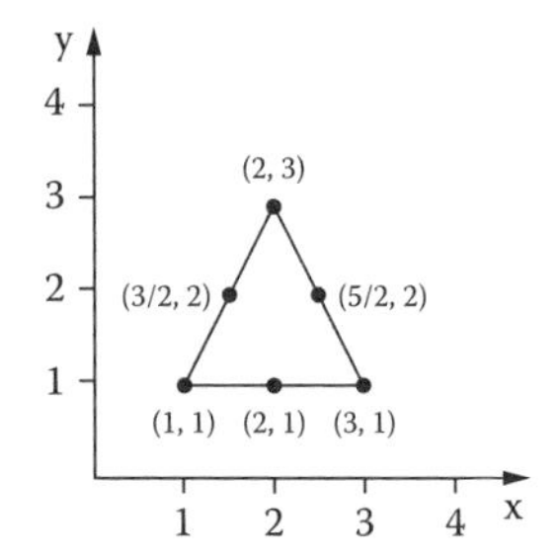

4.17 Determine the amount of heat generated by the source, $Q(6,3) = 50\ \text{W/m}^3$, allocated to each node. (The point source can be written as $Q = 50\delta(x-6)\ \delta(y-3)$, where $\delta(x-a)$ is the Dirac δ function).

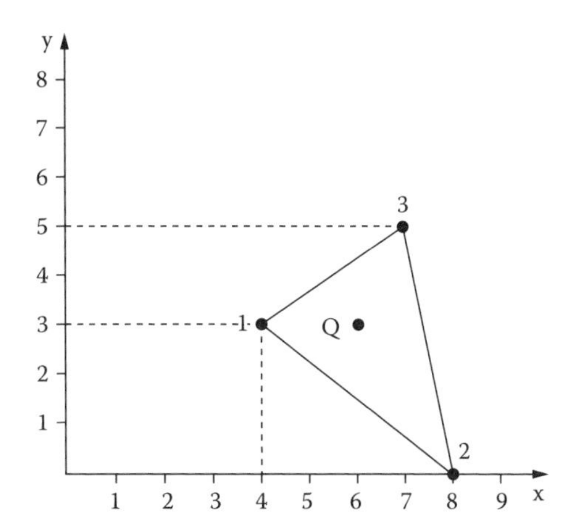

4.18 Find the stiffness matrix for a linear triangle with nodes at (0,0), (3,0), and (0.2). The heat conduction coefficient is constant and distances are in centimeters.

4.19 Repeat Exercise 4.18 for a linear triangle with nodes at (1,0), (3,0), and (2.2). Compare the matrices and discuss.

4.20 Calculate the element matrices for the element shown below if $K = 5$ W/mC and $h = 1$ W/m²C, with $Q = 100\delta(x - 2)\delta(y - 3/2)$ and $T_\infty - 20°C$.

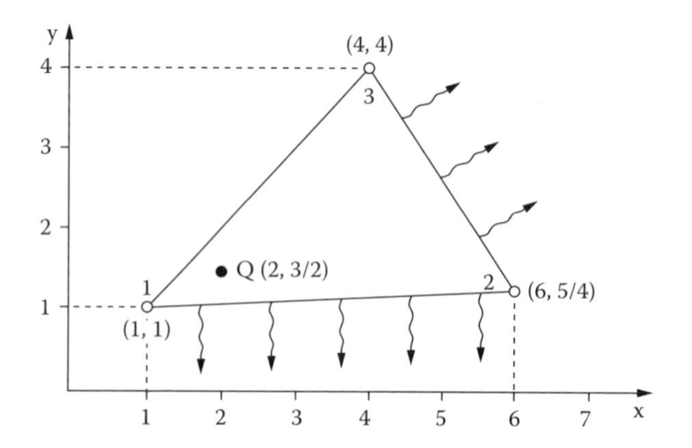

4.21 Calculate the element equations for the axisymmetric triangular element shown below if $K_{rr} = 1$ W/mC,* $K_{zz} = 0.5$ W/mC, h = 1 W/m²C, and $T_\infty = 100°C$.

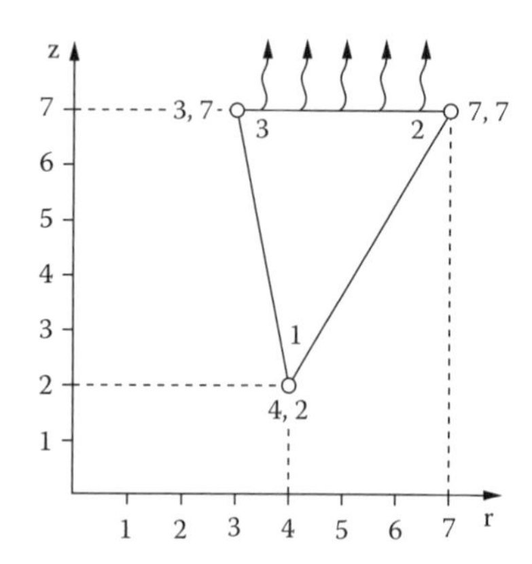

* In this problem, the material is not isotropic and different diffusion coefficients are needed in the x- and z- (or x- and y-) directions. It should be clear how to introduce this in Eq. (4.116).

4.22 Find the temperature distribution in the square plate of length 1 m shown in the figure below, with $K_{xx} = 10$ W/mK, $K_{yy} = 20$ W/mK, $T_1 = 100°C$ and $T_4 = 50°C$. The upper and lower surfaces are insulated and at the right surface $h = 30$ W/m²K and $T_\infty = 0°C$.

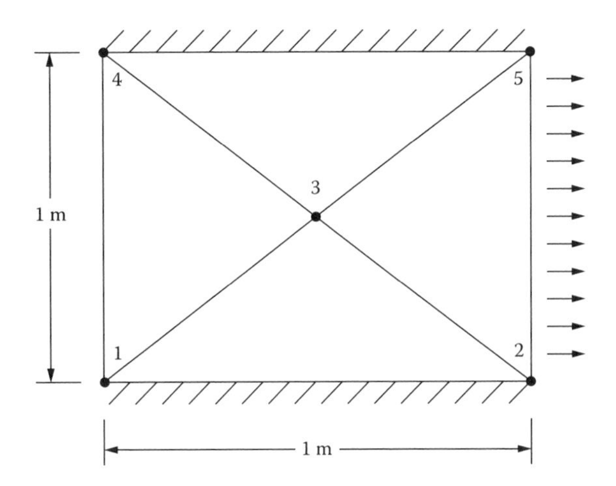

4.23 Find the relationship between the stiffness matrices of two similar triangles with proportional sides; $\ell_{12} = \alpha\ell_{45}$, $\ell_{22} = \alpha\ell_{56}$, and $\ell_{31} = \alpha\ell_{64}$.

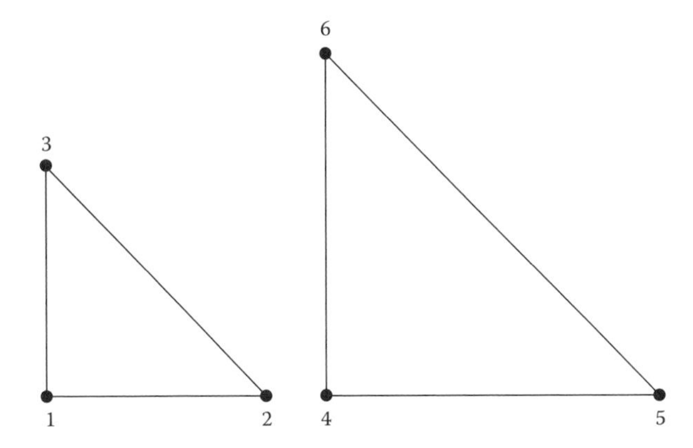

4.24 Calculate the time-dependent diffusion of concentration emitted into the atmosphere from a point source in the domain shown below if $K_{xx} = 10$ cm²/s, $K_{yy} = 1$ cm²/s, and $S = 100$ g/s. Use 50 linear triangular elements and let the time step (Δt) = 1 s. End the calculation with time = 30 s.

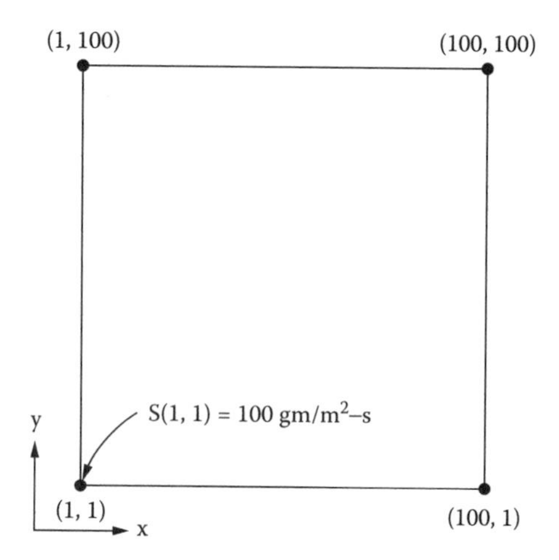

4.25 Determine the steady-state temperature distribution in the square plate shown below if one edge of the plate is subjected to a temperature defined by the relation $T = 100\cos(\pi x/2)$ and the remaining edges are maintained at $T = 0$. Let $K_{xx} = K_{yy} = 1.0$ and use 25 node points. Compare the results with the analytical solution at $x = 0.5$, $y = 0.5$ (notice that the units are dimensionless in this problem):

$$T(x,y) = \frac{100}{\sinh(\pi)}\sinh\left(\frac{\pi y}{2}\right)\cos\left(\frac{\pi x}{2}\right)$$

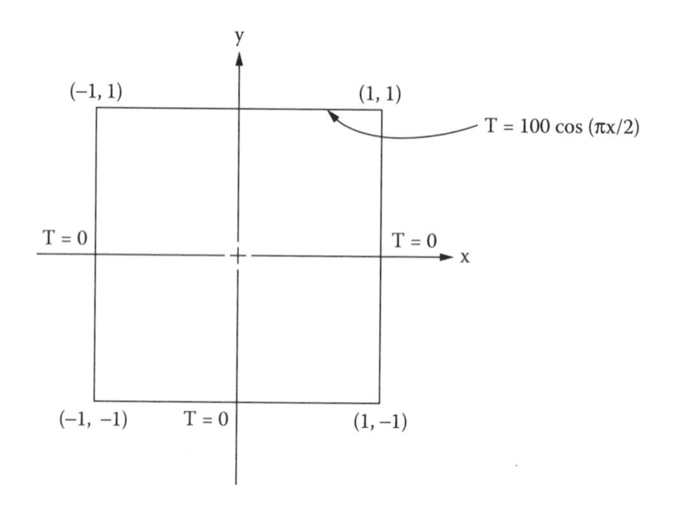

4.26 Using four quadratic triangular elements, obtain the element matrices for the diffusion of a scalar through the problem domain shown below if $K = 10$ cm^2/s and $S = 100$ g/s. Assume steady-state conditions.

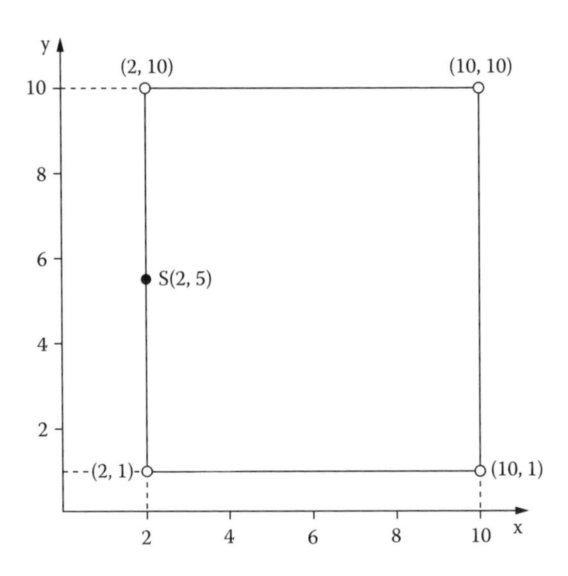

4.27 The axisymmetric cross section of a cylindrical duct is shown in the figure, together with a finite element mesh. The top and bottom ends are insulated, and the left-hand side carries a fluid at 100°C and has a convective coefficient $\bar{h}_L = 2 \times 10^{-4}\, m/s$. The walls at the right-hand side are exposed to air at 25°C and the convective heat transfer coefficient is $\bar{h}_R = 5 \times 10^{-5}\, m/s$. Find the steady-state temperature field if $\alpha = 3 \times 10^{-5}$ m^{2}s.

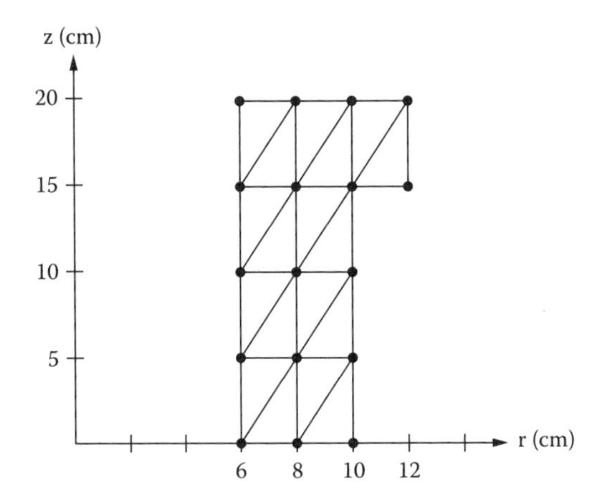

4.28 Find the matrices (4.104) and (4.108) for the element in Figure 4.11.

4.29 Assemble the finite element equation for the problem shown in the figure with a constant heat source Q acting everywhere in the domain.

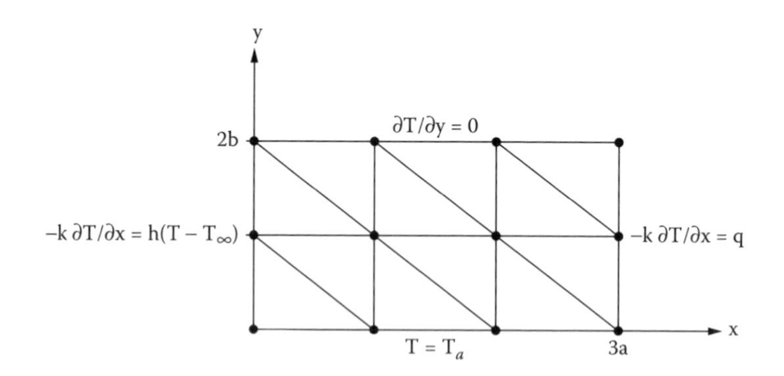

4.30 Solve Exercise 4.29 when Q = 20,000 W/m³, a = 20 cm, b = 25 cm, h = 5 W/m²C, T_∞ = 200°C, q = 500 W/m², and K = 50 W/mC.

4.31 The irrotational flow of an ideal fluid can be written in terms of the velocity potential function, ϕ. The governing equation is written as

$$\frac{\partial^2\phi}{\partial x^2}+\frac{\partial^2\phi}{\partial y^2}=0$$

with $u = \partial\phi/\partial x$ and $v = \partial\phi/\partial y$. The velocity normal to a fixed boundary is zero; likewise, $\partial\phi/\partial n = 0$, or

$$\frac{\partial\phi}{\partial x}n_x+\frac{\partial\phi}{\partial y}n_y=0$$

Calculate the potential flow over the triangular-shaped object shown in the figure using linear triangular elements (at least ten elements recommended).

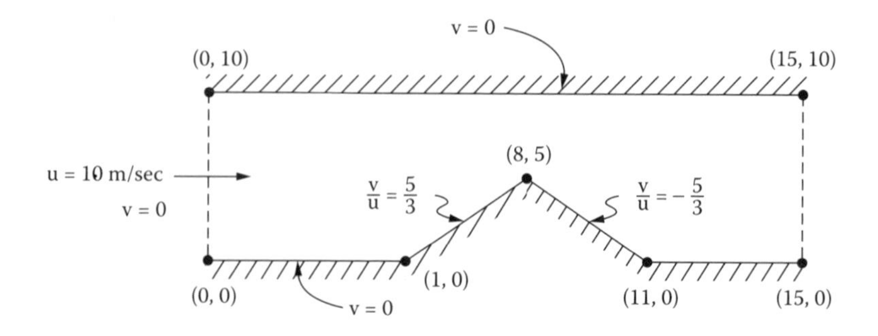

REFERENCES

Carey, G. F. (1997). *Computational Grids: Generation, Adaptation, and Solution Strategies.* Washington, D.C.: Taylor & Francis.

Cowper, G. R. (1973). "Gaussian Quadrature Formulas for Triangles," *Int. J. Num. Mech. Eng.* 7:405–408.

Cuthill, E. and McGee, J. (1969). "Reducing the Bandwidth of Sparse Symmetric Matrices." ACM Publ. P-69, pp. 157–172.

FEMLAB 3 (2004). *User's Manual.* Burlington, Mass.: COMSOL, Inc.

GAMBIT 2.2 (2004). *User's Guide*, Fluent, Inc., Lebanon, N.H.

Hammer, P. C., Marlowe, O. P., and Stroud, A. H. (1956). "Numerical Integration over Simplexes and Cones." *Mathematical Tables Aids Comp.* 10:130–137.

Heinrich, J. C. and Pepper, D. W. (1999). *Intermediate Finite Element Method, Application to Fluid Flow and Heat Transfer.* Philadelphia: Taylor & Francis.

Huebner, K. H. and Thornton, E. A. (1982). *The Finite Element Method for Engineers*, 2nd ed. New York: Wiley-Interscience.

Hughes, R. J. R. (1987). *The Finite Element Method.* Englewood Cliffs, N.J.: Prentice-Hall.

Oden, J. T. (1972). *Finite Elements of Nonlinear Continua.* New York: McGraw-Hill.

pcGRIDGEN (1996). *Numerical Technologies*, ver. 1.1.

Persson, P.-O. and Strang, G. (2004). "A Simple Mesh Generator in MATLAB." *SIAM Rev.* 46(2):329–345.

Segerlind, L. J. (1976). *Applied Finite Element Analysis.* New York: John Wiley & Sons.

Segerlind, L. J. (1984). *Applied Finite Element Analysis.* 2nd ed. New York: John Wiley & Sons.

Sorenson, R. L. (1989). "The 3DGRAPE Book: Theory, User's Manual Examples." NASA Technical Memorandum 102223. Moffett Field, Calif.: Ames Research Center.

Thompson, J. F. (1987). "Program EAGLE Numerical Grid Generation System User's Manual." Volume II: "Surface Generation System." Report AFATL-TR-87-15, Vol. II. Eglin Air Force Base, Fla.: Air Force Armament Laboratory.

TrueGrid (2004). Livermore, Calif.: XYZ Applications, Inc.

Zienkiewicz, O. C. (1977). *The Finite Element Method,* 3rd ed. London: McGraw-Hill.

Zienkiewicz, O. C. and Taylor, R. L. (1989). *The Finite Element Method,* 4th ed. Maidenhead, U.K.: McGraw-Hill.

CHAPTER

5

THE TWO-DIMENSIONAL QUADRILATERAL ELEMENT

5.1 BACKGROUND

The quadrilateral element is a four-sided polygon. The most commonly used shapes contain four nodes located at the vertices and are bilinear in a rectangular configuration. Other commonly used rectangular elements involve 8-noded quadratics, 9-noded biquadratics, and 12-noded cubic configurations. Originally, the finite element method used triangular elements exclusively. However, many researchers now prefer quadrilateral elements. In the bilinear approximation, quadrilateral elements add one more node, compared to only three nodes for the triangle. In addition, gradients are linear functions of the coordinate directions, compared to the gradients being constant in triangular element. For higher-order elements, more complex representations are achieved that are increasingly more accurate from an approximation point of view. However, care must be taken to evaluate the benefits of increased accuracy against the increased computational cost associated with more sophisticated elements.

5.2 ELEMENT MESH

Like the two-dimensional triangular mesh, the quadrilateral mesh subdivides a region into a set of small domains, i.e., elements, only now these are four-sided. Figure 5.1

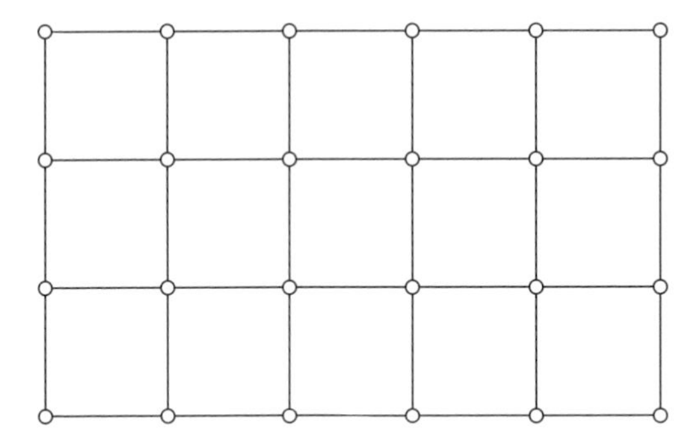

Figure 5.1 Rectangular region divided into bilinear quadrilateral elements

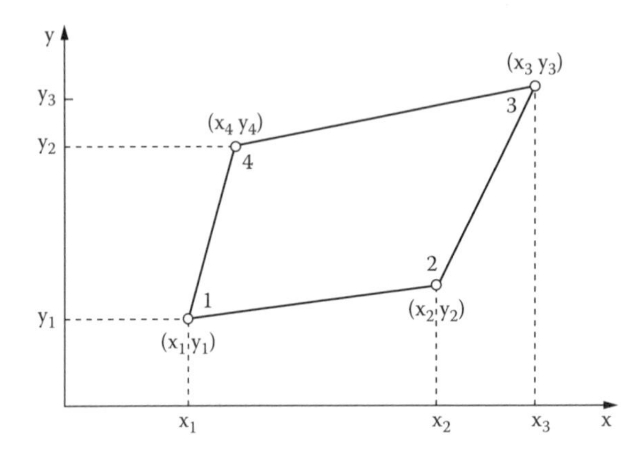

Figure 5.2 Two-dimensional four-noded quadrilateral element

shows a rectangular computational domain discretized into a set of quadrilateral elements.

Note that if we connect a pair of diagonally opposite nodes, we obtain two triangular elements. The bilinear quadrilateral element contains four local nodes, which are designated as 1, 2, 3, and 4 in a counterclockwise order. A general quadrilateral element containing four local nodes is shown in Figure 5.2.

The quadrilateral mesh more closely resembles the standard two-dimensional finite difference mesh, as opposed to the triangular mesh. However, the quadrilateral is just as versatile as the triangle in discretizing a region. The difference between a finite difference mesh and a quadrilateral mesh is that the finite difference mesh must be orthogonal; i.e., all line intersections (node point locations) must be at right angles to one another. In the finite element quadrilateral, each element is unique—each face of the element can have a difference slope. In other words, a finite difference mesh is formed globally and must be perpendicular at all surfaces and intersection; the finite element mesh is formed locally and need not be orthogonal (in physical space)—however, it must be geometrically compatible in the computational domain to ensure continuity between elements.

In this instance, continuity means that the interpolation functions must be continuous between elements. If a governing equation contains a second derivative (e.g., heat diffusion), the approximating function must be continuous between elements. If we wish to integrate $d^n\phi/dx^n$, then ϕ must have a continuous $(n-1)$-order derivative so that only a finite jump discontinuity exists in the nth derivative. A more detailed discussion of this requirement for finite elements and the effect on triangular and quadrilateral elements is given by Carey and Oden (1983) and Heinrich and Pepper (1999).

Although an additional node is required to define the quadrilateral, the number of elements is reduced to half that of a mesh consisting of triangular elements for the same number of nodal points. In discretizing a geometrical domain, the region is generally divided into quadrilaterals first. If triangular elements are desired, the quadrilaterals are easily subdivided. In nearly all instances, a mesh consisting of quadrilateral elements is sufficient and usually more accurate than a mesh consisting of triangular elements. Extension of the quadrilateral mesh to three dimensions is easily obtained, i.e., a box, and conceptually easier to visualize than a group of three-dimensional tetrahedrons.

5.3 SHAPE FUNCTIONS

As in the two-dimensional triangular element, the two-dimensional quadrilateral element is based on linear, quadratic, cubic, or higher approximations. The linear triangular element is classified as a simplex element (approximation containing a constant and linear terms) and the quadratic triangle as a complex element (constant-, linear-, and quadratic-order terms). The element boundaries of the simplex and complex elements do not require the element sides to be parallel to a coordinate axis. However, because it is a multiplex element, for a simple polynomial representation to be possible, the quadrilateral element requires that its sides be parallel to the coordinate system. As we shall see later, this requirement is easily relaxed through a local coordinate transformation when the sides are not parallel to the global axial system.

5.3.1 Bilinear Rectangular Element

In its simplest form, the quadrilateral becomes a rectangular element with the boundaries of the element parallel to a coordinate system (in physical space). Further extension of the element, using a local natural coordinate system, results in a generalized quadrilateral of which the rectangle is a subset. We begin with the rectangular element.

The interpolation function for a four-noded bilinear rectangle is given as

$$\phi = \alpha_1 + \alpha_2 x + \alpha_3 y + \alpha_4 xy \tag{5.1}$$

The xy term was chosen in lieu of x^2 or y^2 since xy does not break the symmetry and assumes a linear variation in ϕ along x or y. For lines of constant x, the element is linear in y, and vice versa; for lines of constant y, the element is linear in x, therefore the name bilinear element. The gradient of ϕ is seen to be a linear function in one coordinate direction, i.e.,

$$\left.\begin{aligned}\frac{\partial\phi}{\partial x} &= \alpha_2 + \alpha_4 y \\ \frac{\partial\phi}{\partial y} &= \alpha_3 + \alpha_4 x\end{aligned}\right\} \tag{5.2}$$

Figure 5.3 shows the rectangular element and nodal configuration. If the distances to the coordinate origin, located at the midpoint of the element, are denoted by a and b, the node values are given as

$$\left.\begin{aligned}\phi &= \phi_1 && at\ x = -b, y = -a \\ \phi &= \phi_2 && at\ x = \ \ b, y = -a \\ \phi &= \phi_3 && at\ x = \ \ b, y = \ \ a \\ \phi &= \phi_4 && at\ x = -b, y = \ \ a\end{aligned}\right\} \tag{5.3}$$

Substituting these values into Eq. (5.1) gives, for the α_i terms, $i = 1, 4$,

$$\left.\begin{aligned}\alpha_1 &= \frac{\phi_1 + \phi_2 + \phi_3 + \phi_4}{4} \\ \alpha_2 &= \frac{\phi_1 - \phi_2 - \phi_3 + \phi_4}{4a} \\ \alpha_3 &= \frac{\phi_1 + \phi_2 - \phi_3 - \phi_4}{4b} \\ \alpha_4 &= \frac{\phi_1 - \phi_2 + \phi_3 - \phi_4}{4ab}\end{aligned}\right\} \tag{5.4}$$

If we now approximate ϕ in terms of the nodal values and shape functions as we did in Eq. (4.7), we obtain

$$\phi = N_1\phi_1 + N_2\phi_2 + N_3\phi_3 + N_4\phi_4 \tag{5.5}$$

where the shape functions are given as

$$\left.\begin{aligned}N_1 &= \frac{1}{4ab}(b - x)(a - y) \\ N_2 &= \frac{1}{4ab}(b + x)(a - y) \\ N_3 &= \frac{1}{4ab}(b + x)(a + y) \\ N_4 &= \frac{1}{4ab}(b - x)(a + y)\end{aligned}\right\} \tag{5.6}$$

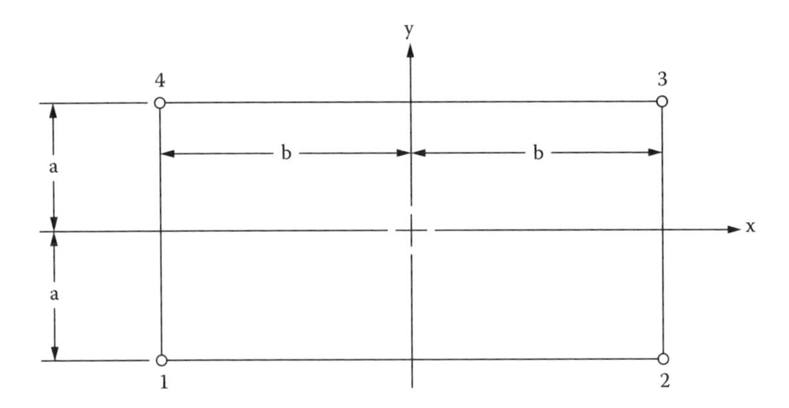

Figure 5.3 Four-noded rectangular element

We see in Eq. (5.6) that the shape functions can be expressed in terms of length ratios, x/b and y/a. For example,

$$N_1 = \frac{1}{4ab}(b-x)(a-y) = \frac{1}{4}\left(1-\frac{x}{b}\right)\left(1-\frac{y}{a}\right) \tag{5.7}$$

where $-1 \le (x/b) \le 1$ and $1 \le (y/a) \le 1$.

5.3.2 Quadratic Rectangular Element

We can also develop a two-dimensional, higher-order, quadrilateral element, just as we did for the triangular element. There are two ways that this can be done, which lead to two different quadratic elements: one containing eight nodes one on each corner and one on the midpoint of each side; the other containing nine nodes, with the ninth in the center of the element. We discuss the eight-noded element here; the nine-noded element is left to Exercise 5.1. The four-noded bilinear and eight-noded quadratic quadrilaterals, along with a nine-noded biquadratic element, are used predominantly in industry and have been found to be sufficient for most computational purposes.

The interpolating polynomial for the eight-noded quadratic element, as shown in Figure 5.4, consists of eight terms. The unknown variable, ϕ, is expressed as

$$\phi = \alpha_1 + \alpha_2 x + \alpha_3 y + \alpha_4 xy + \alpha_5 x^2 + \alpha_6 y^2 + \alpha_6 y^2 + \alpha_7 x^2 y + \alpha_8 xy^2 \tag{5.8}$$

If nodal values are defined as in Figure 5.4,

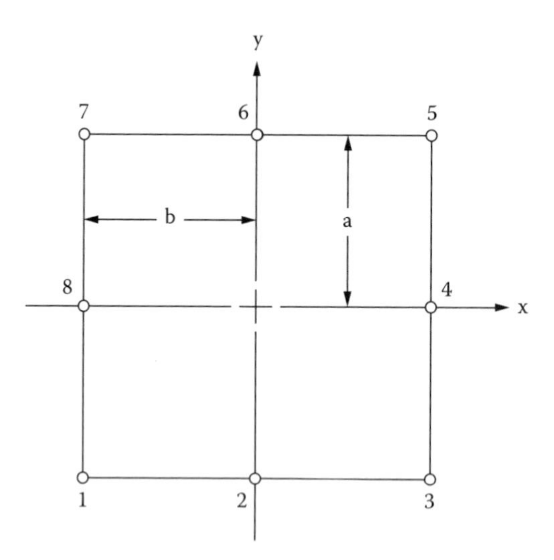

Figure 5.4 Eight-noded element

$$\left.\begin{array}{ll}
\phi = \phi_1 & at\ x = -b, y = -a \\
\phi = \phi_2 & at\ x = \ \ 0, y = -a \\
\phi = \phi_3 & at\ x = \ \ b, y = -a \\
\phi = \phi_4 & at\ x = \ \ b, y = \ \ 0 \\
\phi = \phi_5 & at\ x = \ \ b, y = \ \ a \\
\phi = \phi_6 & at\ x = \ \ 0, y = \ \ a \\
\phi = \phi_7 & at\ x = -b, y = \ \ a \\
\phi = \phi_8 & at\ x = -b, y = \ \ 0
\end{array}\right\} \tag{5.9}$$

Substituting the values from Eq. (5.9) into Eq. (5.8) yields the following eight equations (expressed in matrix form):

$$\begin{bmatrix}
1 & -b & -a & ab & b^2 & a^2 & -ab^2 & -a^2b \\
1 & 0 & -a & 0 & 0 & a^2 & 0 & 0 \\
1 & b & -a & -ab & b^2 & a^2 & -ab^2 & a^2b \\
1 & b & 0 & 0 & b^2 & 0 & 0 & 0 \\
1 & b & a & ab & b^2 & a^2 & ab^2 & a^2b \\
1 & 0 & a & 0 & 0 & a^2 & 0 & 0 \\
1 & -b & a & -ab & b^2 & a^2 & ab^2 & -a^2b \\
1 & -b & 0 & 0 & b^2 & 0 & 0 & 0
\end{bmatrix}
\begin{bmatrix} \alpha_1 \\ \alpha_2 \\ \alpha_3 \\ \alpha_4 \\ \alpha_5 \\ \alpha_6 \\ \alpha_7 \\ \alpha_8 \end{bmatrix}
=
\begin{bmatrix} \phi_1 \\ \phi_2 \\ \phi_3 \\ \phi_4 \\ \phi_5 \\ \phi_6 \\ \phi_7 \\ \phi_8 \end{bmatrix} \tag{5.10}$$

from which the α_i terms, $i = 1, \ldots, 8$, can be obtained.

Approximating ϕ as a function of nodal values and shape functions gives

$$\phi = N_1\phi_1 + N_2\phi_2 + N_3\phi_3 + N_4\phi_4 + N_5\phi_5 + N_6\phi_6 + N_7\phi_7 + N_8\phi_8 \qquad (5.11)$$

The shape functions can be found to be (Zienkiewicz, 1977; Heinrich and Pepper, 1999)

$$\left.\begin{aligned}
N_1 &= -\frac{1}{4(ab)^2}(b-x)(a-y)(ab+xa+yb) \\
N_2 &= \frac{1}{2ab^2}(b^2-x^2)(a-y) \\
N_3 &= \frac{1}{4(ab)^2}(b-x)(a-y)(ax-by-ab) \\
N_4 &= \frac{1}{2a^2b}(a^2-y^2)(b+x) \\
N_5 &= \frac{1}{4(ab)^2}(b+x)(a+y)(ax+by-ab) \\
N_6 &= \frac{1}{2ab^2}(b^2-x^2)(a+y) \\
N_7 &= \frac{1}{4(ab)^2}(b-x)(a+y)(ab+ax-by) \\
N_8 &= \frac{1}{2a^2b}(a^2-y^2)(b-x)
\end{aligned}\right\} \qquad (5.12)$$

The gradients of ϕ now become

$$\left.\begin{aligned}
\frac{\partial\phi}{\partial x} &= \alpha_2 + \alpha_4 y + 2\alpha_5 x + 2\partial_7 xy + \partial_8 y^2 \\
\frac{\partial\phi}{\partial y} &= \alpha_3 + \alpha_4 x + 2\alpha_6 y + \partial_7 x^2 + \partial_8 xy
\end{aligned}\right\} \qquad (5.13)$$

Notice that $\partial\phi/\partial x$ is linear in the x-direction but quadratic in the y-direction; $\partial\phi/\partial y$ is the opposite. In terms of length ratios, the shape function for N_1, for example, can be rewritten as

$$N_1 = -\frac{1}{4}\left(1-\frac{x}{b}\right)\left(1-\frac{y}{a}\right)\left(1+\frac{x}{a}+\frac{y}{a}\right) \qquad (5.14)$$

5.4 NATURAL COORDINATE SYSTEM

In most situations, the physical domain to be modeled does not consist of straight-sided, orthogonal boundaries. A region with curved boundaries can be discretized using quadrilateral elements with curved sides to obtain a more accurate solution. The transformation from straight to curved sides is achieved by expressing the x,y coordinates in terms of curvilinear coordinates:

$$\left.\begin{aligned} x &= x(\xi,\eta) \\ y &= y(\xi,\eta) \end{aligned}\right\} \tag{5.15}$$

The choice of ξ,η depends on the element geometry. The ξ,η coordinate system is normally referred to as the "natural" coordinate system when the coordinate variables lie within the range $-1 \leq \xi \leq \eta \leq 1$. For the rectangular case, we can find the set of dimensionless coordinates ξ,η such that

$$\left.\begin{aligned} \xi &= \frac{x}{b} \\ \eta &= \frac{y}{a} \end{aligned}\right\} \tag{5.16}$$

where ξ,η effectively replace the x and y global coordinates with local coordinates for an individual element; we previously referred to these specific ratios as length ratios (Figure 5.3 and Figure 5.4). Thus, utilizing our definition in Eq. (5.16), we rewrite our bilinear shape functions (5.6) in the natural coordinate system as

$$\left.\begin{aligned} N_1 &= \frac{1}{4}(1-\xi)(1-\eta) \\ N_2 &= \frac{1}{4}(1+\xi)(1-\eta) \\ N_3 &= \frac{1}{4}(1+\xi)(1+\eta) \\ N_4 &= \frac{1}{4}(1-\xi)(1+\eta) \end{aligned}\right\} \tag{5.17}$$

We see that our local coordinate system always varies between –1 and 1. We take advantage of this fact when we formulate the derivative terms. Figure 5.5 shows the ξ,η coordinate system centered within the element. Notice that it is no longer important for this new coordinate system to be orthogonal—it can also be curvilinear. However, derivatives with respect to x and y are not obtainable directly and require an additional transformation. We illustrate this point later when dealing with isoparametric elements; for now, we assume a simple Cartesian system of coordinates.

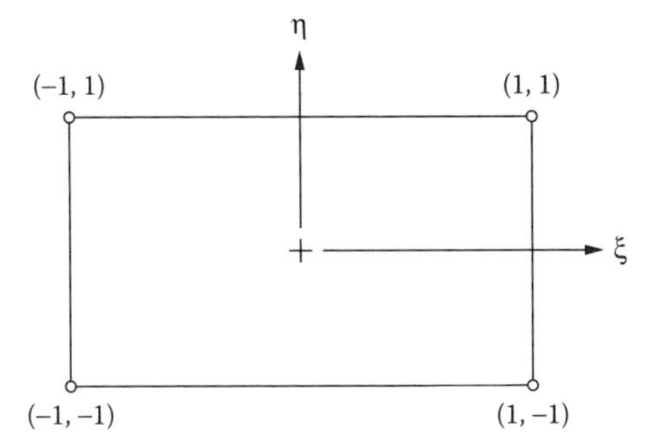

Figure 5.5 The ξ,η coordinate system for a rectangular element

The evaluation of the derivative terms $\partial N_i/\partial x$ and $\partial N_i/\partial y$ is similar to that for the triangular elements in Chapter 4. The Jacobian matrix relates the derivatives through

$$\begin{bmatrix} \dfrac{\partial N_i}{\partial \xi} \\ \dfrac{\partial N_i}{\partial \eta} \end{bmatrix} = \begin{bmatrix} \dfrac{\partial x}{\partial \xi} & \dfrac{\partial y}{\partial \xi} \\ \dfrac{\partial x}{\partial \eta} & \dfrac{\partial y}{\partial \eta} \end{bmatrix} \begin{bmatrix} \dfrac{\partial N_i}{\partial x} \\ \dfrac{\partial N_i}{\partial y} \end{bmatrix} = \mathrm{J} \begin{bmatrix} \dfrac{\partial N_i}{\partial x} \\ \dfrac{\partial N_i}{\partial y} \end{bmatrix} \tag{5.18}$$

from which the inverse Jacobian, $\mathbf{J}^{-1}$, can be used of find the values of the derivatives with respect to x and y and the determinant of the transformation.

Example 5.1

As an illustrative example find the values for $\partial N_1/\partial x$ and $\partial N_1/\partial y$ at $x = 2$ and $y = 2$ for the element shown in Figure 5.6. The x and y transformations are expressed in the form:

$$\left.\begin{aligned} x &= \frac{1}{2}(3\xi + 5) \\ y &= \eta + 2 \end{aligned}\right\} \tag{5.19}$$

We first evaluate the Jacobian matrix. The Jacobian is obtained from Eq. (5.18) as

$$\mathrm{J} = \begin{bmatrix} \dfrac{\partial}{\partial \xi} & \dfrac{\partial y}{\partial \xi} \\ \dfrac{\partial}{\partial \eta} & \dfrac{\partial y}{\partial \eta} \end{bmatrix} \tag{5.20}$$

This gives

$$\mathrm{J} = \frac{1}{2}\begin{bmatrix} 3 & 0 \\ 0 & 2 \end{bmatrix} \tag{5.21}$$

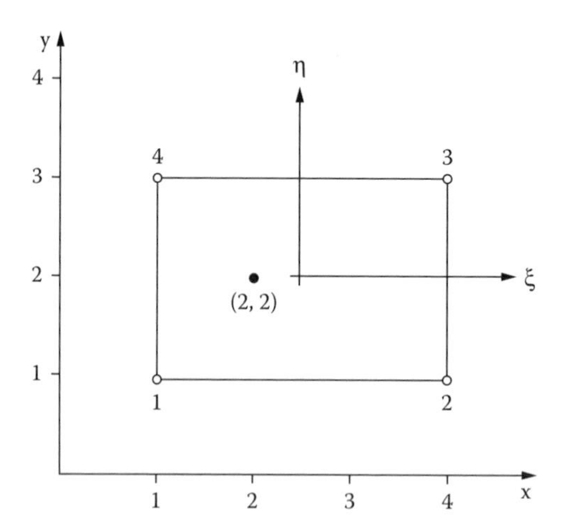

Figure 5.6 Example problem to find $\partial N_1/\partial x$ and $\partial N_1/\partial y$

The inverse is

$$J^{-1} = \frac{1}{3}\begin{Bmatrix} 2 & 0 \\ 0 & 3 \end{Bmatrix} \tag{5.22}$$

The derivatives $\partial N_1/\partial x$ and $\partial N_1/\partial y$ can be calculated once the derivatives $\partial N_1/\partial \xi$ and $\partial N_1/\partial \eta$ are known. We know that

$$N_1 = \frac{(1-\xi)(1-\eta)}{4} \tag{5.23}$$

The derivatives are given as

$$\left.\begin{aligned} \frac{\partial N_1}{\partial \xi} &= -\left(\frac{1-\eta}{4}\right) \\ \frac{\partial N_1}{\partial \eta} &= -\frac{(1-\xi)}{4} \end{aligned}\right\} \tag{5.24}$$

We evaluate these derivatives at $x = 2$ and $y = 2$ using Eq. (5.10):

$$\left.\begin{aligned} \xi &= \frac{1}{3}(2x-5) = -\frac{1}{3} \\ \eta &= y - 2 = 0 \end{aligned}\right\} \tag{5.25}$$

Thus,

$$\left.\begin{aligned} \frac{\partial N_1}{\partial \xi} &= -\frac{(1-0)}{4} = -\frac{1}{4} \\ \frac{\partial N_1}{\partial \eta} &= -\frac{\left(1 = \frac{1}{3}\right)}{4} = -\frac{1}{3} \end{aligned}\right\} \tag{5.26}$$

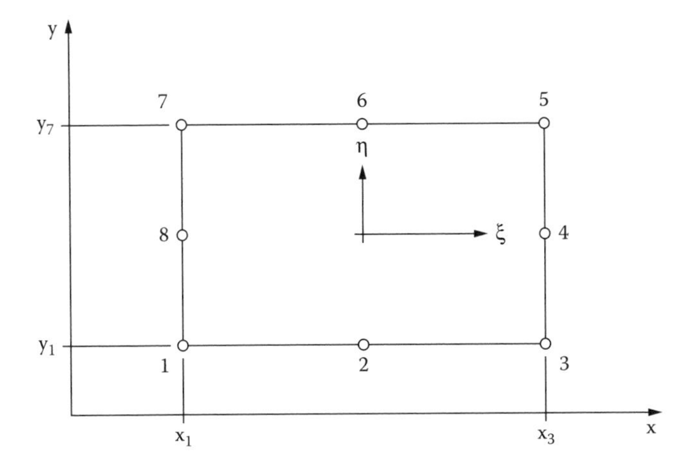

Figure 5.7 Quadratic quadrilateral element in ξ,η coordinates

Matrix multiplication yields the derivatives

$$\begin{bmatrix} \dfrac{\partial N_1}{\partial x} \\ \dfrac{\partial N_1}{\partial y} \end{bmatrix} = \frac{1}{3}\begin{bmatrix} 2 & 0 \\ 0 & 3 \end{bmatrix}\begin{bmatrix} -\dfrac{1}{4} \\ -\dfrac{1}{3} \end{bmatrix} = \begin{bmatrix} -\dfrac{1}{6} \\ -\dfrac{1}{3} \end{bmatrix} \tag{5.27}$$

Let us now consider the interpolation polynomial for the eight-noded quadratic quadrilateral element, which can be expressed in terms of the local coordinates, ξ,η as

$$\phi = \alpha_1 + \alpha_2\xi + \alpha_3\eta + \alpha_4\xi\eta + \alpha_5\xi^2 + \alpha_6\eta^2 + \alpha_7\xi^2\eta + \alpha_8\xi\eta^2 \tag{5.28}$$

The shape functions are obtained in the same manner as before, only this time there are eight functions, denoted by numbers 1-8. Figure 5.7 shows the eight-noded quadrilateral in ξ,η coordinates. The shape functions are given as

$$\left.\begin{aligned} N_1 &= -\frac{(1-\xi)(1-\eta)(1+\xi+\eta)}{4} \\ N_2 &= -\frac{(1-\xi)(1-\eta)}{2} \\ N_3 &= -\frac{(1+\xi)(1-\eta)(\xi-\eta-1)}{4} \\ N_4 &= -\frac{(1-\eta^2)(1+\xi)}{2} \\ N_5 &= -\frac{(1+\xi)(1-\eta)(\xi+\eta-1)}{4} \\ N_6 &= -\frac{(1-\xi^2)(1+\eta)}{2} \\ N_7 &= -\frac{(1-\xi)(1+\eta)(1+\xi-\eta)}{4} \\ N_8 &= -\frac{(1-\eta^2)(1-\xi)}{2} \end{aligned}\right\} \tag{5.29}$$

Evaluation of the quadratic shape function derivatives follow in exactly the same manner as in the bilinear case. From Eq. (5.18),

$$\begin{bmatrix} \dfrac{\partial N_i}{\partial \xi} \\ \dfrac{\partial N_i}{\partial \eta} \end{bmatrix} = \mathrm{J} \begin{bmatrix} \dfrac{\partial N_i}{\partial \xi} \\ \dfrac{\partial N_i}{\partial y} \end{bmatrix} \tag{5.30}$$

The Jacobian is

$$\mathrm{J} = \begin{bmatrix} \dfrac{\partial x}{\partial \xi} & \dfrac{\partial y}{\partial \xi} \\ \dfrac{\partial x}{\partial \eta} & \dfrac{\partial y}{\partial \eta} \end{bmatrix} \tag{5.31}$$

The derivatives $\partial x/\partial \xi$, $\partial y/\partial \eta$ are obtained from

$$\left.\begin{aligned} x &= \frac{1}{2}(x_3 - x_1)\xi + \frac{1}{2}(x_1 + x_3) \\ y &= \frac{1}{2}(y_7 - y_1)\eta + \frac{1}{2}(y_1 + y_7) \end{aligned}\right\} \tag{5.32}$$

for an element such as the one given in Figure 5.7, and

$$\left.\begin{aligned} \frac{\partial N_1}{\partial \xi} &= \frac{(1-\eta)(2\xi+\eta)}{4} \\ \frac{\partial N_2}{\partial \xi} &= \xi(1-\eta) \\ \frac{\partial N_3}{\partial \xi} &= \frac{(1-\eta)(2\xi-\eta)}{4} \\ \frac{\partial N_4}{\partial \xi} &= \frac{(1-\eta^2)}{2} \\ \frac{\partial N_5}{\partial \xi} &= \frac{(1+\eta)(2\xi+\eta)}{4} \\ \frac{\partial N_6}{\partial \xi} &= -\xi(1+\eta) \\ \frac{\partial N_7}{\partial \xi} &= \frac{(1-\eta)(\eta-2\xi)}{4} \\ \frac{\partial N_8}{\partial \xi} &= -\frac{(1-\eta^2)}{2} \end{aligned}\right\} \tag{5.33}$$

with similar expressions for the derivatives with respect to η. These are found using Eqs. (5.29). The derivatives are then evaluated depending on the shape function and location in question. For example, if we wanted to find $\partial N_1/\partial\xi$ and $\partial N_1/\partial\eta$ at $\xi = (1/2)$ and $\eta = (1/2)$, then

$$\left.\begin{aligned}\frac{\partial N_1}{\partial\xi} &= \frac{(1-\eta)(2\xi+\eta)}{4} = \frac{\left(1-\frac{1}{2}\right)\left(2\left(\frac{1}{2}\right)+\frac{1}{2}\right)}{4} = \frac{3}{16}\\ \frac{\partial N_1}{\partial\eta} &= \frac{(1-\xi)(2\eta+\xi)}{4} = \frac{\left(1-\frac{1}{2}\right)\left(2\left(\frac{1}{2}\right)+\frac{1}{2}\right)}{4} = \frac{3}{16}\end{aligned}\right\} \tag{5.34}$$

5.5 NUMERICAL INTEGRATION USING GAUSSIAN QUADRATURES

The formulation of integrals in terms of ξ and η yields simple integration limits. This is especially advantageous when we are dealing with curved boundaries or boundaries that are not parallel to the coordinate axis. The integrals are defined as

$$\int_{-a}^{a}\int_{-b}^{b} F(x,y)dxdy = \int_{-1}^{1}\int_{-1}^{1} f(\xi,\eta)|\mathrm{J}|d\xi d\eta \tag{5.35}$$

The conduction matrix thus becomes

$$\begin{aligned}\mathrm{K}^{(e)} &= \iint_{\Omega^{(e)}} \mathrm{K}\left[\frac{\partial N_i}{\partial x}\frac{\partial N_j}{\partial x} + \frac{\partial N_i}{\partial y}\frac{\partial N_j}{\partial y}\right]dxdy\\ &= \int_{-1}^{1}\int_{-1}^{1} \mathrm{K}\left[\frac{\partial N_i}{\partial x}\frac{\partial N_j}{\partial x} + \frac{\partial N_i}{\partial y}\frac{\partial N_j}{\partial y}\right]|\mathrm{J}|d\xi d\eta\end{aligned} \tag{5.36}$$

where the derivatives $\partial N_i/\partial x$ and $\partial N_i/\partial y$ can be obtained from Eq. (5.18) as a function of ξ, η, $\partial N_i/\partial\xi$, and $\partial N_i/\partial\eta$. All area integral terms containing the product N_iN_j become

$$\iint_{\Omega^{(e)}} N_iN_j dxdy = \int_{-1}^{1}\int_{-1}^{1} N_iN_j|\mathrm{J}|d\xi d\eta \tag{5.37}$$

Integration of the term on the right-hand integral in Eq. (5.36) no longer involves simple polynomials, due to the term $1/|\mathbf{J}|$ appearing in the expressions for the derivatives with respect to ξ and η can become troublesome. The most common practice is to use the numerical integration procedure known as Gaussian quadrature. The procedure is easy to program* and gives nearly exact values for the integrations.

* See FEM-2D and the subroutine Gauss.

A Gaussian quadrature approximates a definite integral with a weighted sum over a finite set of points. For example, in one dimension,

$$\int_{-1}^{1} f(\xi)d\xi \cong \sum_{i=1}^{N} W_i f(\xi_i) \tag{5.38}$$

where N is the number of integration points, W_i are the weight factors, and ξ_i the Gauss points. With N Gauss points, we can integrate exactly a polynomial of degree $2N - 1$. Thus, if $f(\xi)$ is a cubic polynomial $f(\xi) = a + b\xi + c\xi^2 + d\xi^3$, then choosing $N = 2$, we can integrate it exactly. The ξ_i values are defined as (Heinrich and Pepper, 1999):

$$\left.\begin{aligned} \xi_1 &= -1/\sqrt{3} \\ \xi_2 &= -1/\sqrt{3} \end{aligned}\right\} \tag{5.39}$$

the weights are $W_1 = W_2 = 1$, and the function will be integrated exactly. For the double integrals in Eq. (5.37),

$$\int_{-1}^{1}\int_{-1}^{1} f(\xi, \eta)d\xi d\eta \cong \sum_{i=1}^{N}\sum_{j=1}^{N} W_i W_j f(\xi_i, \eta_j) \tag{5.40}$$

The Gauss points are defined by the values:

$$\left.\begin{aligned} \xi_1 &= -1/\sqrt{3} \\ \xi_2 &= 1/\sqrt{3} \\ \eta_1 &= -1/\sqrt{3} \\ \eta_2 &= 1/\sqrt{3} \end{aligned}\right\} \tag{5.41}$$

Four Gauss points are used with the four-noded quadrilateral element shown in Figure 5.8 and Table 5.1.

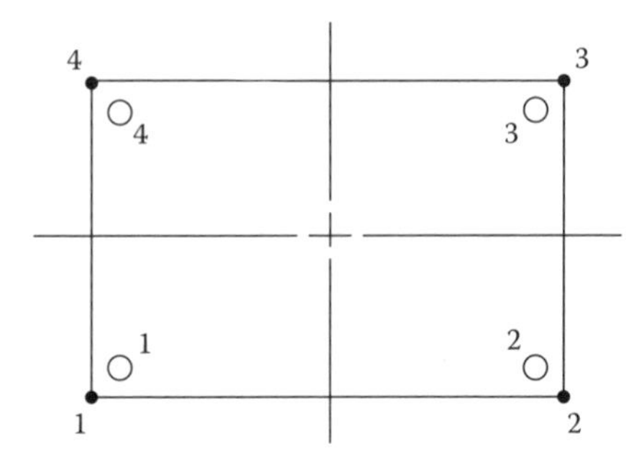

Figure 5.8 A 2 × 2 Gauss quadrature for bilinear elements

Table 5.1 Gauss Point Values in Two Dimensions for $N = 2$

Gauss Point	ξ_i	η_j
1	$-1/\sqrt{3}$	$-1/\sqrt{3}$
2	$-1/\sqrt{3}$	$-1/\sqrt{3}$
3	$1/\sqrt{3}$	$1/\sqrt{3}$
4	$1/\sqrt{3}$	$1/\sqrt{3}$

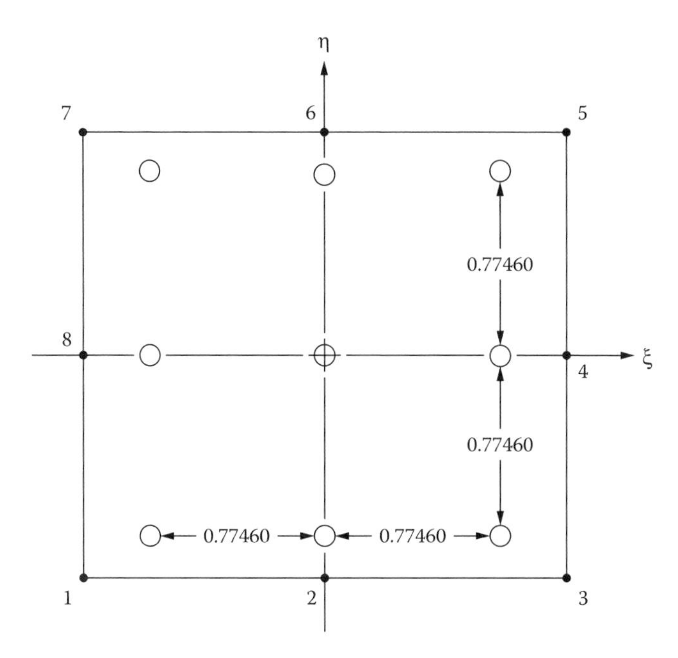

Figure 5.9 Sampling locations for numerical integration involving nine Gauss points

When eight-noded quadratic or nine-noded biquadratic elements are used, the integral appearing in Eq. (5.36) is obtained numerically as

$$\int_{-1}^{1}\int_{-1}^{1} f(\xi, \eta)d\xi d\eta = \sum_{i=1}^{3}\sum_{j=1}^{3} W_i W_j f(\xi_i, \eta_j) \tag{5.42}$$

The nine integration points are depicted in Figure 5.9. The points and corresponding weights are given in Table 5.2

A solution obtained with bilinear quadrilaterals involves 4 × 4 matrix multiplications for each element; a quadratic quadrilateral deals with an 8 × 8 matrix, or four times more product calculations per element than the linear element. Furthermore, bilinear elements require evaluation at four Gauss points, while quadratics involve nine Gauss points. Therefore, it is nine times less expensive in terms of

Table 5.2 Gauss Points and Weights for $N = 3$

Gauss Point	ξ_i	η_i	W_i	W_j
1	–0.77460	0.77460	0.55556	0.55556
2	0.0	–0.77460	0.88889	0.55556
3	0.77460	–0.77460	0.55556	0.55556
4	0.77460	0.0	0.55556	0.88889
5	0.77460	0.77460	0.55556	0.55556
6	0.0	0.77460	0.88889	0.55556
7	–0.77460	0.77460	0.55556	0.55556
8	–0.77460	0.0	0.55556	0.88889
9	0.0	0.0	0.88889	0.88889

operations to construct the element stiffness matrix for a four-noded bilinear element than to obtain the element stiffness matrix for a quadratic eight-noded element. This can be an important consideration when choosing the type of element to be used, because it means a difference of almost an order of magnitude in CPU time.

5.6 STEADY-STATE CONDUCTION EQUATION

We direct our attention once more to determining the steady-state conduction of heat within a two-dimensional domain. At this point, we also generalize our heat conduction matrix to allow for anisotropy materials, assuming that the coordinate axis coincides with the principal axis of the conductivity tensor, i.e., that the heat conduction is given in the x–y coordinate system by

$$\mathrm{K} = \begin{bmatrix} K_{xx} & 0 \\ 0 & K_{yy} \end{bmatrix} \tag{5.43}$$

The basic equation to be solved can then be written, as was done in Exercises 4.21 and 4.24, as

$$-\frac{\partial}{\partial x}\left(K_{xx}\frac{\partial T}{\partial x}\right) - \frac{\partial}{\partial y}\left(K_{yy}\frac{\partial T}{\partial y}\right) = Q \tag{5.44}$$

A heat flux condition now takes the form:

$$-\left(K_{xx}\frac{\partial T}{\partial x}n_x + K_{yy}\frac{\partial T}{\partial y}n_y\right) = q \tag{5.45}$$

where n_x and n_y are defined as in Eq. (4.60). Following the same procedure as was previously with Eq. (4.58), the weighted residuals form of Eq. (5.44) is

$$\int_{\Omega} W\left[-\frac{\partial}{\partial x}\left(K_{xx}\frac{\partial T}{\partial x}\right)-\left(K_{yy}\frac{\partial T}{\partial y}\right)-Q\right]d\Omega=0 \tag{5.46}$$

and after application of Green's theorem,

$$\begin{aligned}&\int_{\Omega}\left[K_{xx}\frac{\partial W}{\partial x}\frac{\partial T}{\partial x}+K_{yy}\frac{\partial W}{\partial y}\frac{\partial T}{\partial y}-WQ\right]dxdy\\&+\int_{\Gamma_B} W\left(-K_{xx}\frac{\partial T}{\partial x}n_x-K_{yy}\frac{\partial T}{\partial y}n_y\right)d\Gamma=0\end{aligned} \tag{5.47}$$

We now approximate T using the bilinear shape functions, i.e.,

$$T=\sum_{i=1}^{4}N_iT_i$$

and set $W = N_i$. The Galerkin formulation of Eq. (5.44) becomes

$$\begin{aligned}&\left[\int_{\Omega}\left(K_{xx}\frac{\partial N_i}{\partial x}\frac{\partial N_j}{\partial x}+K_{yy}\frac{\partial N_i}{\partial y}\frac{\partial N_j}{\partial y}\right)dxdy\right]T_j=\\&\left[\int_{\Omega}N_iQdxdy-\int_{\Gamma_B}N_iqd\Gamma\right]\end{aligned} \tag{5.48}$$

where Eq. (5.45) has been used to replace the applied heat flux in the boundary integral.

We now transform Eq. (5.48) to the natural ξ,η coordinate system. We define the temperature in that system as

$$T(\xi,\eta)=\sum_{i=1}^{4}N_i(\xi,\eta)T_i \tag{5.49}$$

where the shape functions $N_i(\xi,\eta)$ are given in Eq.(5.17), defined on the generic element shown in Figure 5.5. The first derivative values are obtained from

$$\begin{bmatrix}\dfrac{\partial N_i}{\partial x}\\[2ex]\dfrac{\partial N_i}{\partial y}\end{bmatrix}=\mathrm{J}^{-1}\begin{bmatrix}\dfrac{\partial N_i}{\partial \xi}\\[2ex]\dfrac{\partial N_i}{\partial \eta}\end{bmatrix} \tag{5.50}$$

where

$$J = \begin{bmatrix} \dfrac{\partial x}{\partial \xi} & \dfrac{\partial y}{\partial \xi} \\ \dfrac{\partial x}{\partial \eta} & \dfrac{\partial y}{\partial \eta} \end{bmatrix} = \begin{bmatrix} J_{11} & J_{21} \\ J_{12} & J_{22} \end{bmatrix} = \frac{1}{2}\begin{bmatrix} x_2 - x_1 & 0 \\ 0 & y_4 - y_1 \end{bmatrix} \tag{5.51}$$

is obtained using Eq. (5.32). The inverse Jacobian is given by

$$J^{-1} = \frac{1}{|J|}\begin{bmatrix} J_{22} & -J_{22} \\ -J_{21} & J_{11} \end{bmatrix} \tag{5.52}$$

with

$$|J| = J_{11}J_{22} - J_2 J_{21} \tag{5.53}$$

Thus, Eq. (5.48) becomes

$$\begin{gathered} \int_{-1}^{1}\int_{-1}^{1}\left[K_{xx}\frac{\partial N_i}{\partial x}\frac{\partial N_j}{\partial x} + K_{yy}\frac{\partial N_i}{\partial y}\frac{\partial N_j}{\partial y}\right]|J|\,d\xi d\eta T_j \\ = -\int_{\Gamma_B} qN_i d\Gamma + \int_{-1}^{1}\int_{-1}^{1} QN_i |J|\,d\xi d\eta \end{gathered} \tag{5.54}$$

where $\partial N_i/\partial x$ and $\partial N_i/\partial y$ are given in terms of $\partial N_i/\partial \xi$ and $\partial N_i/\partial \eta$ by Eq. (5.50).

Equation (5.54) is the general finite element algorithm for steady-state conduction of heat with internal source/sink over a generic element. One only has to input the appropriate x,y values and boundary conditions associated with the nodes of the particular element under consideration.

Example 5.2

Determine the matrix-equivalent equation for steady-state conduction of heat in the rectangular domain shown in Figure 5.10. Assume $K_{xx} = K_{yy} = K$ = constant and heat flux occurs only at the right-hand face. Create a computational mesh consisting of only one four-noded bilinear element. We begin by discretizing the problem domain into a computational domain (mesh) consisting of one element, as shown in Figure 5.11. In this instance, we set the ξ,η coordinate at local node 1: the limits of integration will vary from 0 to 1; for a coordinate system located at the centroid of the element, the limits of integration vary from −1 to 1. We are free to establish our local coordinate system wherever we please. Generally, the coordinate system is located either at the centroid of the element or at the bottom left corner node.

The shape functions are defined as

$$\left.\begin{aligned} N_1 &= (1-\xi)(1-\eta) \\ N_2 &= \xi(1-\eta) \\ N_3 &= \xi\eta \\ N_4 &= (1-\xi)\eta \end{aligned}\right\} \tag{5.55}$$

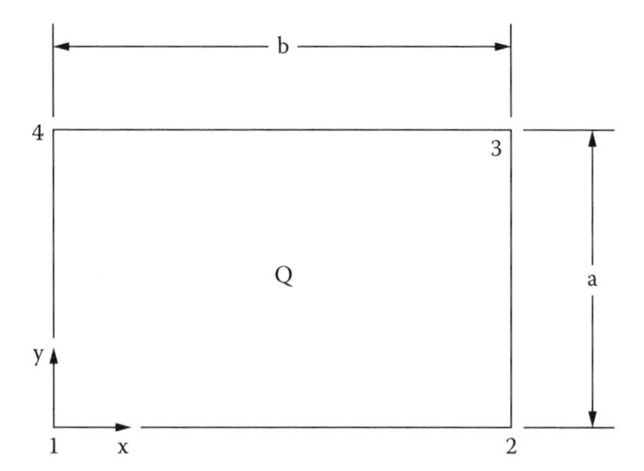

Figure 5.10 Rectangular domain for steady-state heat conduction

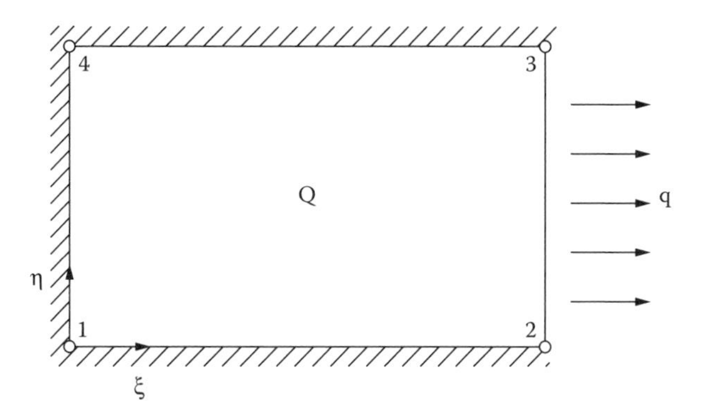

Figure 5.11 One-element mesh

where $\xi = x/b$ and $\eta = y/a$. We wish to solve an equation of the form similar to Eq. (5.54), whereby

$$\int_0^1\int_0^1\left[K_{xx}\frac{\partial N_i}{\partial x}\frac{\partial N_j}{\partial x}+K_{yy}\frac{\partial N_i}{\partial y}\frac{\partial N_j}{\partial y}\right]|\mathrm{J}|\,d\xi\, d\eta\, T_j = -\int_0^1 qN_i d\eta+\int_0^1\int_0^1 QN_i|\mathrm{J}|\,d\xi\, d\eta \tag{5.56}$$

The first derivative values of the shape functions are obtained from Eq. (5.50) where

$$\mathrm{J}=\begin{bmatrix} b & 0 \\ 0 & a \end{bmatrix} \tag{5.57}$$

The inverse Jacobian is

$$\mathrm{J}^{-1}=\frac{1}{ab}\begin{bmatrix} a & 0 \\ 0 & b \end{bmatrix} \tag{5.58}$$

where $|\mathbf{J}| = ab$. Thus,

$$\begin{bmatrix} \dfrac{\partial N_i}{\partial x} \\ \dfrac{\partial N_i}{\partial y} \end{bmatrix} = \frac{1}{ab}\begin{bmatrix} a & 0 \\ 0 & b \end{bmatrix}\begin{bmatrix} \dfrac{\partial N_i}{\partial \xi} \\ \dfrac{\partial N_i}{\partial \eta} \end{bmatrix} = \begin{bmatrix} \dfrac{1}{b}\dfrac{\partial N_i}{\partial \xi} \\ \dfrac{1}{a}\dfrac{\partial N_i}{\partial \eta} \end{bmatrix} \tag{5.59}$$

Substituting Eq. (5.59) into Eq. (5.56), we obtain, for the first term

$$\begin{aligned} \int_0^1\int_0^1 K_{xx}\frac{\partial N_i}{\partial x}\frac{\partial N_j}{\partial x}|\mathrm{J}|\,d\xi d\eta &= K_{xx}ab\int_0^1\int_0^1 \frac{\partial N_i}{\partial x}\frac{\partial N_j}{\partial x}\,d\xi d\eta \\ &= K_{xx}\left(\frac{1}{b^2}\right)\int_0^1\int_0^1 \begin{bmatrix} (1-\eta)^2 & -(1-\eta)^2 & -\eta(1-\eta) & \eta(1-\eta) \\ -(1-\eta)^2 & (1-\eta)^2 & \eta(1-\eta) & -\eta(1-\eta) \\ -\eta(1-\eta) & \eta(1-\eta) & \eta^2 & -\eta^2 \\ \eta(1-\eta) & -\eta(1-\eta) & -\eta^2 & \eta^2 \end{bmatrix} d\xi d\eta \\ &= \frac{1}{6}K_{xx}\left(\frac{a}{b}\right)\begin{bmatrix} 2 & -2 & -1 & 1 \\ -2 & 2 & 1 & -1 \\ -1 & 1 & 1 & -2 \\ 1 & -1 & -2 & 2 \end{bmatrix} \end{aligned} \tag{5.60}$$

Similarly, the second term is

$$\int_0^1\int_0^1 K_{yy}\frac{\partial N_i}{\partial y}\frac{\partial N_j}{\partial y}|\mathrm{J}|\,d\xi d\eta = \frac{1}{6}K_{yy}\left(\frac{b}{a}\right)\begin{bmatrix} 2 & 1 & -1 & -2 \\ 1 & 2 & -2 & -1 \\ -1 & -2 & 2 & 1 \\ -2 & -1 & 1 & 2 \end{bmatrix} \tag{5.61}$$

For a constant heat source term, we get

$$\int_0^1\int_0^1 QN_i|\mathrm{J}|\,d\xi d\eta = abQ\int_0^1\int_0^1 \begin{bmatrix} (1-\xi)(1-\eta) \\ \xi(1-\eta) \\ \xi\eta \\ (1-\xi)\eta \end{bmatrix} d\xi d\eta = \frac{abQ}{4}\begin{bmatrix} 1 \\ 1 \\ 1 \\ 1 \end{bmatrix} \tag{5.62}$$

As in the linear triangular element, the heat source is distributed equally among the four nodes.

The surface heat flux acts only over element face 2–3. On the straight side containing nodes 2 and 3, we evaluate the shape functions at $\xi = 1$ (Figure 5.11). We obtain

$$\left.\begin{aligned} N_1 &= (1-\xi)(1-\eta) = 0 \\ N_2 &= \xi(1-\eta) = (1-\eta) \\ N_3 &= \xi\eta = \eta \\ N_4 &= (1-\xi)\eta = 0 \end{aligned}\right\} \tag{5.63}$$

The surface differential is evaluated by relating $d\Gamma$ to $d\eta$ in order to take advantage of the limits of integration along the $\xi = 1$ side. The term $d\Gamma$ is related to $d\eta$ by

$$d\Gamma = |J|\,d\eta \tag{5.64}$$

where, for the side with nodes 2 and 3, Γ can be defined as

$$\Gamma_{2-3} = \eta \ell_{2-3} \tag{5.65}$$

with ℓ_{2-3} the length of the side between nodes 2 and 3. From Eq. (5.65),

$$|J| = \frac{d\Gamma}{d\eta} = \ell_{2-3} = a$$

or

$$d\Gamma = a\,d\eta \tag{5.66}$$

Thus, at $\xi = 1$,

$$\int_{\Gamma_{B=q}} qN_i d\Gamma \int_0^1 \begin{bmatrix} 0 \\ 1-\eta \\ \eta \\ 0 \end{bmatrix} a\,d\eta = \frac{aq}{2}\begin{bmatrix} 0 \\ 1 \\ 1 \\ 0 \end{bmatrix} \tag{5.67}$$

Combining Eqs. (5.60) through (5.63) and replacing terms in Eq. (5.56) yields

$$\left\{ K_{xx}\left(\frac{a}{b}\right)\frac{1}{6}\begin{bmatrix} 2 & -2 & -1 & 1 \\ -2 & 2 & 1 & -1 \\ -1 & 1 & 2 & -2 \\ 1 & -1 & -2 & 2 \end{bmatrix} \right.$$

$$\left. + K_{yy}\left(\frac{b}{a}\right)\frac{1}{6}\begin{bmatrix} 2 & 1 & -1 & -2 \\ 1 & 2 & -2 & -1 \\ -1 & -2 & 2 & 1 \\ -2 & -1 & 1 & 2 \end{bmatrix} \right\} \begin{bmatrix} T_1 \\ T_2 \\ T_3 \\ T_4 \end{bmatrix} \tag{5.68}$$

$$= \frac{abQ}{4}\begin{bmatrix} 1 \\ 1 \\ 1 \\ 1 \end{bmatrix} - \frac{aq}{2}\begin{bmatrix} 0 \\ 1 \\ 1 \\ 0 \end{bmatrix}$$

It is a simple matter (for the computer) to solve for the unknown temperatures at the four nodes once the a, b, K_{xx}, K_{yy}, Q, and q values are substituted in Eq. (5.68).

5.7 STEADY-STATE CONDUCTION WITH BOUNDARY CONVECTION

As discussed previously in Chapter 4, the addition of convection occurring over an element face alters the boundary conditions, i.e., load vector, and the stiffness matrix.

The steady-state conduction equation is identical to Eq. (5.44), but with the boundary condition along Γ_B given by

$$-\left(K_{xx}\frac{\partial T}{\partial x}n_x + K_{yy}\frac{\partial T}{\partial y}n_y\right) = h\left(T - T_\infty\right) \tag{5.69}$$

where the term $h\ (T - T_\infty)$ accounts for surface flux due to convection; otherwise, the equation set is analogous to the problem of steady-state conduction just discussed in Section 5.6 with the heat flux given by Eq. (5.45).

Applying the Galerkin procedure and integrating by parts, Eq. (5.44) becomes

$$\begin{aligned}&\iint_\Omega\left[K_{xx}\frac{\partial N_i}{\partial x}\frac{\partial N_j}{\partial x} + K_{yy}\frac{\partial N_i}{\partial y}\frac{\partial N_j}{\partial y}\right]dxdyT_j\\&= -\int_{\Gamma_B} h(T - T_\infty)N_i d\Gamma + \iint_\Omega QN_i dxdy\end{aligned} \tag{5.70}$$

where Eq. (5.69) has been used to define the first-order gradient boundary condition terms. The procedure used to evaluate Eq. (5.70) is exactly the same as before; in this instance, we must account for the surface integral containing the convective heat flux. The evaluation of these integrals when the element sides are straight can be performed analytically. When the elements are no longer rectangular (or have curved faces, as we shall see later in the quadratic element), numerical integration must be performed.

Assume that we wish to use an element-centered coordinate system for ξ, η and that side $\ell_{2\text{-}3}$ experiences convection. Using Eq. (5.17), the surface integrals are

$$\begin{aligned}\int_{\ell_{2-3}} hN_iTd\Gamma &= \left[h\int_{-1}^{1}\frac{1}{2}\begin{bmatrix}0\\1-\eta\\1+\eta\\0\end{bmatrix}\frac{1}{2}[0 \quad 1-\eta \quad 1+\eta]\frac{\ell_{2\text{-}3}}{2}d\eta\right]\begin{bmatrix}T_1\\T_2\\T_3\\T_4\end{bmatrix}\\
&= \left[\frac{h}{4}\int_{-1}^{1}\begin{bmatrix}0 & 0 & 0 & 0\\0 & (1-\eta)^2 & 1-\eta^2 & 0\\0 & 1-\eta^2 & (1+\eta)^2 & 0\\0 & 0 & 0 & 0\end{bmatrix}\frac{\ell_{2-3}}{2}d\eta\right]\begin{bmatrix}T_1\\T_2\\T_3\\T_4\end{bmatrix}\\
&= \frac{h\ell_{2-3}}{6}\begin{bmatrix}0 & 0 & 0 & 0\\0 & 2 & 1 & 0\\0 & 1 & 2 & 0\\0 & 0 & 0 & 0\end{bmatrix}\begin{bmatrix}T_1\\T_2\\T_3\\T_4\end{bmatrix}\end{aligned} \tag{5.71}$$

$$hT_\infty \int_{\Gamma_B} N_i d\Gamma = \frac{hT_\infty}{2}\int_{-1}^{1}\begin{bmatrix}0\\1-\eta\\1+\eta\\0\end{bmatrix}\frac{\ell_{2-3}}{2} = \frac{hT_\infty \ell_{2-3}}{2}\begin{bmatrix}0\\1\\1\\0\end{bmatrix} \tag{5.72}$$

If a different face contains the boundary condition for the convective flux, the 2 and 1 terms in the matrices in Eqs. (5.71) and (5.72) shift to the appropriate nodal locations. These expressions can be used as general relations and easily programmed to account for the boundary convection effects.*

Example 5.3†

A simple example is used to illustrate matrix manipulations in the global assembly process, and ultimate matrix solution for steady-state heat conduction with convective boundary conditions. Consider a rectangular domain discretized into elements, as shown in Figure 5.12, with $K_{xx} = K_{yy} = 1$ W/mC, convective heat transfer occurring along the right boundary, the top and bottom walls are insulated, and the left end held at a constant temperature $T = 10$°C. For simplicity, let $a = b = 1$ m and $Q = 0$; we carry a and b in the expressions until the final global assembly over both elements.

The Galerkin form, Eq. (5.48), in this particular case, reduces to

$$\left[\int_\Omega\left(K_{xx}\frac{\partial N_i}{\partial x}\frac{\partial N_j}{\partial x} + K_{yy}\frac{\partial N_i}{\partial y}\frac{\partial N_j}{\partial y}\right)dxdy\right]T_j = -\int_{\ell_{5-6}} N_i h(T - T_\infty)d\Gamma \tag{5.73}$$

We now evaluate Eq. (5.73) over the first element of the two-element domain (shown in Figure 5.13) using local coordinates at the lower left corner of the elements as follows (Figure 5.13): The temperature is defined in terms of the shape functions by Eq. (5.49) in the local coordinate

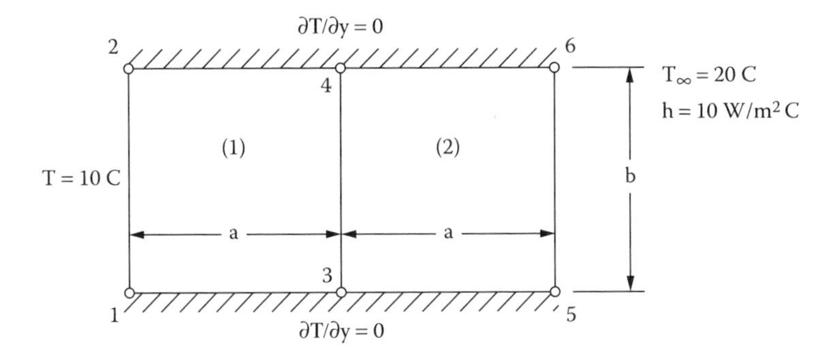

Figure 5.12 Two-element heat conduction with convective boundary conditions

* See FEM-2D, subroutine BNDCON.

† FEM-2D and COMSOL files are available.

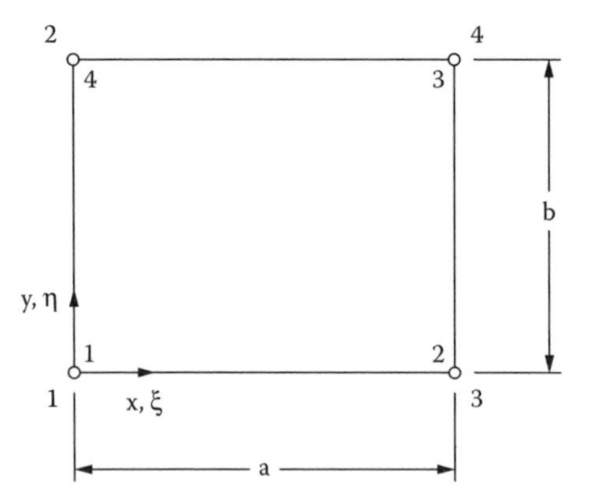

Figure 5.13 Local and global coordinate systems and nodal numbering for element 1

system with the shape functions given by Eq. (5.63). The coordinate transformation, in this case, is

$$\xi = \frac{x}{a}, \qquad \eta = \frac{y}{b} \tag{5.74}$$

So the first derivatives of temperature are

$$\begin{bmatrix} \dfrac{\partial T}{\partial x} \\ \dfrac{\partial T}{\partial y} \end{bmatrix} = \begin{bmatrix} \dfrac{1}{a}\dfrac{\partial T}{\partial \xi} \\ \dfrac{1}{b}\dfrac{\partial T}{\partial \eta} \end{bmatrix} \tag{5.75}$$

Element 1

In discretized form, Eq. (5.75) becomes

$$\begin{bmatrix} \dfrac{\partial T}{\partial x} \\ \dfrac{\partial T}{\partial y} \end{bmatrix} = \begin{bmatrix} -\dfrac{1}{a}(1-\eta) & \dfrac{1}{a}(1-\eta) & \dfrac{1}{a}\eta & -\dfrac{1}{a}\eta \\ -\dfrac{1}{b}(1-\xi) & -\dfrac{1}{b}\xi & \dfrac{1}{b}\xi & \dfrac{1}{b}(1-\xi) \end{bmatrix} \begin{bmatrix} T_1 \\ T_2 \\ T_3 \\ T_4 \end{bmatrix} \tag{5.76}$$

Thus,

$$\begin{aligned} \int_{e_1} K_{xx} \frac{\partial N_i}{\partial x}\frac{\partial N_j}{\partial x} dxdy &= ab\, K_{xx} \int_0^1\int_0^1 \left(\frac{1}{a^2}\frac{\partial N_i}{\partial \xi}\frac{\partial N_j}{\partial \xi} \right) d\xi d\eta \\ &= \frac{b}{6a} K_{xx} \begin{bmatrix} 2 & -2 & -1 & 1 \\ -2 & 2 & 1 & -1 \\ -1 & 1 & 2 & -2 \\ 1 & -1 & -2 & 2 \end{bmatrix} \end{aligned} \tag{5.77}$$

Similarly,

$$\int_{e_1} K_{yy} \frac{\partial N_i}{\partial y}\frac{\partial N_j}{\partial y} dxdy = \frac{a}{6b} K_{yy} \begin{bmatrix} 2 & 1 & -1 & -2 \\ 1 & 2 & -2 & -1 \\ -1 & -2 & 2 & 1 \\ -2 & -1 & 1 & 2 \end{bmatrix} \tag{5.78}$$

Notice that the line integral along the side 2–1 is ignored because a Dirichlet boundary condition is imposed along that side; the equations for T_1 and T_2 can be used to calculate the heat flux across that boundary once the rest of the temperatures are known, as we did in Section 3.3 for the one-dimensional problem. The line integrals along side 1–3 and 4–2 vanish because $q = 0$ (adiabatic walls). The line integral along the interior line 3–4 will cancel out once the contribution from the second element is added, which is a statement of continuity of fluxes across inter-element boundaries, exactly the same as in the one-dimensional problem. Hence, no line integrals need be calculated except on side 5–6 in the second element, as suggested by Eq. (5.73).

Replacing the problem data in Eq. (5.77) and (5.78), the element stiffness matrix for element 1 is given by

$$K^{(e^1)} = \frac{1}{6} \begin{bmatrix} 4 & -1 & -2 & -1 \\ -1 & 4 & -1 & -2 \\ -2 & -1 & 4 & -1 \\ -1 & -2 & -1 & 4 \end{bmatrix} \tag{5.79}$$

The procedure and relations are similar for element 2 (Figure 5.14). The only other contributions to evaluate are those due to convection. The boundary integrals for convection are given by

$$\left[-h\int_{\ell_{5-6}} N_i N_j d\Gamma\right] T_j + \left[hT_\infty \int_{\ell_{5-6}} N_i d\Gamma\right] \tag{5.80}$$

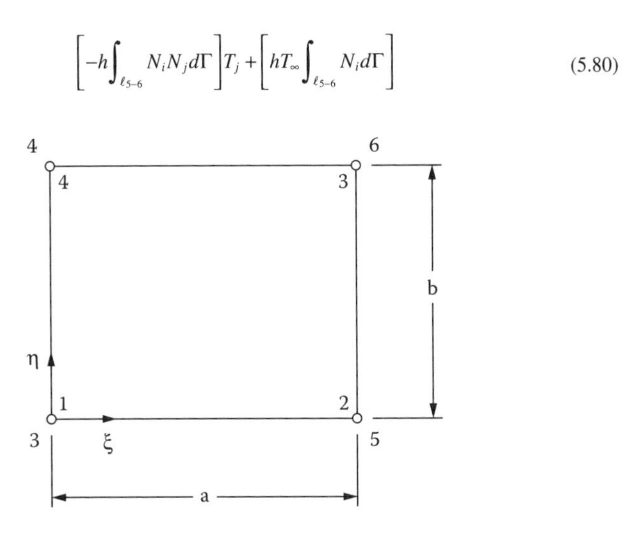

Figure 5.14 Local and global nodal numbering and local coordinate system

Element 2

When evaluated for $\xi = 1$, the first integral yields

$$-hb\int_0^1 \begin{bmatrix} 0 & 0 & 0 & 0 \\ 0 & (1-\eta)^2 & \eta(1-\eta) & 0 \\ 0 & \eta(1-\eta) & (1-\eta)^2 & 0 \\ 0 & 0 & 0 & 0 \end{bmatrix} d\eta = -\frac{10}{6}\begin{bmatrix} 0 & 0 & 0 & 0 \\ 0 & 2 & 1 & 0 \\ 0 & 1 & 2 & 0 \\ 0 & 0 & 0 & 0 \end{bmatrix} \qquad (5.81)$$

where $h = 10$ W/m^{2}C. Using $T_\infty = 20$°C. the second integral becomes

$$hT_\infty \int_{\ell_{5-6}} N_i d\Gamma = hT_\infty b\int_0^1 \begin{bmatrix} 0 \\ 1-\eta \\ \eta \\ 0 \end{bmatrix} d\eta = 100\begin{bmatrix} 0 \\ 1 \\ 1 \\ 0 \end{bmatrix} \qquad (5.82)$$

If convection were to occur on another face of the element, contributions would appear in different locations in matrix (5.81) and vector (5.82), but the values would be the same. These relations also hold for any element (triangular or rectangular) that uses two nodes to define an element face.

Evaluating the matrices from each element, we can now assemble them into global form for the final solution. The global stiffness matrix in this example is a 6 × 6 square matrix, as shown in Figure 5.15.

We begin by assembling the matrix with the contributions obtained from the element 1 calculations. The values are given in Figure 5.16. Notice that the entries corresponding to nodes 2, 3 and 4 are changed from the local to the global system.

The contributions of element 2 are shown in Figure 5.17. The additional values due to convection appear at locations i = 5, 6 and j = 5, 6 in the coefficients matrix and right-hand-side vector.

Adding together the element contributions shown in Figures 5.16 and 5.17, we obtain the full system of equations

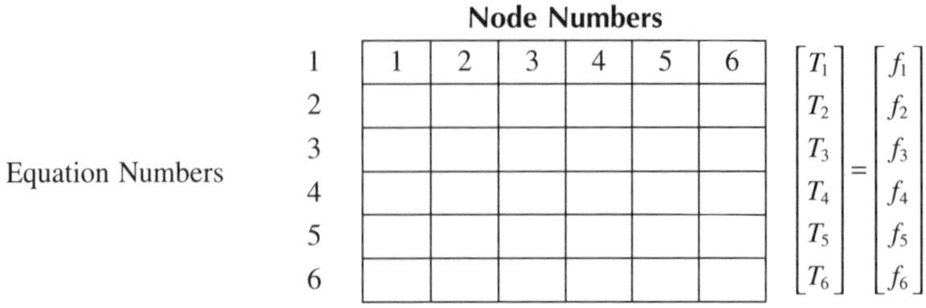

Node Numbers

Equation Numbers	1	2	3	4	5	6	T		f
1							T_1		f_1
2							T_2		f_2
3							T_3	=	f_3
4							T_4		f_4
5							T_5		f_5
6							T_6		f_6

Figure 5.15 Global matrix assembly

$$
\begin{array}{c} \\ |1| \\ |2| \\ |3| \\ |4| \\ |5| \\ |6| \end{array}
\frac{1}{6}
\begin{array}{c}
\begin{array}{cccccc} 1 & 2 & 3 & 4 & 5 & 6 \end{array} \\
\begin{bmatrix}
4 & -1 & -1 & -2 & 0 & 0 \\
-1 & 4 & -2 & -1 & 0 & 0 \\
-1 & -2 & 4 & -1 & 0 & 0 \\
-2 & -1 & -1 & 4 & 0 & 0 \\
0 & 0 & 0 & 0 & 0 & 0 \\
0 & 0 & 0 & 0 & 0 & 0
\end{bmatrix}
\end{array}
\begin{bmatrix} T_1 = 10 \\ T_2 = 10 \\ T_3 \\ T_4 \\ T_5 \\ T_6 \end{bmatrix}
=
\begin{bmatrix} 0 \\ 0 \\ 0 \\ 0 \\ 0 \\ 0 \end{bmatrix}
$$

Figure 5.16 Global matrix equation for element 1 values

$$
\begin{array}{c} \\ |1| \\ |2| \\ |3| \\ |4| \\ |5| \\ |6| \end{array}
\frac{1}{6}
\begin{array}{c}
\begin{array}{cccccc} 1 & 2 & 3 & 4 & 5 & 6 \end{array} \\
\begin{bmatrix}
0 & 0 & 0 & 0 & 0 & 0 \\
0 & 0 & 0 & 0 & 0 & 0 \\
0 & 0 & 4 & -1 & -1 & -2 \\
0 & 0 & -1 & 4 & -2 & -1 \\
0 & 0 & -1 & -2 & 4+20 & -1+10 \\
0 & 0 & 0 & 0 & -1+10 & 4+20
\end{bmatrix}
\end{array}
\begin{bmatrix} T_1 = 10 \\ T_2 = 10 \\ T_3 \\ T_4 \\ T_5 \\ T_6 \end{bmatrix}
=
\begin{bmatrix} 0 \\ 0 \\ 0 \\ 0 \\ 100 \\ 100 \end{bmatrix}
$$

Figure 5.17 Contributions of element 2 to the global matrix equation

$$
\frac{1}{6}
\begin{bmatrix}
4 & -1 & -1 & -2 & 0 & 0 \\
-1 & 4 & -2 & -1 & 0 & 0 \\
-1 & -2 & 8 & -2 & -1 & -2 \\
-2 & -1 & -2 & 8 & -2 & -1 \\
0 & 0 & -1 & -2 & 24 & 9 \\
0 & 0 & -2 & -1 & 9 & 24
\end{bmatrix}
\begin{bmatrix} T_1 = 10 \\ T_2 = 10 \\ T_3 \\ T_4 \\ T_5 \\ T_6 \end{bmatrix}
=
\begin{bmatrix} 0 \\ 0 \\ 0 \\ 0 \\ 100 \\ 100 \end{bmatrix}
\tag{5.83}
$$

$$
\frac{1}{6}
\begin{bmatrix}
8 & -2 & -1 & -2 \\
-2 & 8 & -2 & -1 \\
-1 & -2 & 24 & 9 \\
-2 & -1 & 9 & 24
\end{bmatrix}
\begin{bmatrix} T_1 \\ T_2 \\ T_3 \\ T_4 \end{bmatrix}
=
\begin{bmatrix} 5 \\ 5 \\ 100 \\ 100 \end{bmatrix}
\tag{5.84}
$$

Solution of the 4 × 4 system of linear equations, which can be achieved by any of the methods already discussed in Chapter 3, yields the unknown values of the temperature at nodes 3 through 6. These are

$$
\left.
\begin{aligned}
T_3 = T_4 = 14.76\ C \\
T_5 = T_6 = 19.52\ C
\end{aligned}
\right\}
\tag{5.85}
$$

■

5.8 THE QUADRATIC QUADRILATERAL ELEMENT

As discussed in Section 5.3.2, the quadratic quadrilateral most commonly used consists of eight nodes—four corner nodes plus four mid-side nodes—as shown in Figure 5.7. The shape functions are defined in Eq. (5.29). Although use of quadratic elements requires considerably more computation work, the increase in accuracy, or approach to a converted solution for steady-state problems, is significant. The rate of error decrease for a bilinear element is $O(h^2)$, where h is the element size; for a quadratic element, the error decreases as $O(h^3)$.

The procedure for obtaining a solution with quadratic elements follows in exactly the same manner as for bilinear elements. To demonstrate the use of the quadratic quadrilateral, the following example problem dealing with heat conduction in a rectangular region with boundary convection is analyzed.

EXAMPLE 5.4*

Obtain a solution for simple heat conduction with boundary convection on one face and internal heat source for the region shown in Figure 5.18. Let $K_{xx} = K_{yy} = 1.0$ W/mC, $h = 10$ W/m²C, $T_\infty = 150$°C, and $Q = 1.0$ W/m³; the left vertical face of the region is fixed at a constant temperature, $T_o = 100$°C. The top and bottom walls are insulated. For simplicity, discretize the square region into a single eight-noded quadratic element, as shown in Figure 5.19. The Galerkin formulation is given by Eq. (5.70), which is written in matrix from as ($K_{xx} = K_{yy} = 1.0$),

$$\mathrm{KT} = \mathrm{F} \tag{5.86}$$

where

$$\mathrm{K} = \iint_\Omega \left(\frac{\partial N_i}{\partial x}\frac{\partial N_j}{\partial x} + \frac{\partial N_i}{\partial y}\frac{\partial N_j}{\partial y} \right) dxdy + \int_{\Gamma_{3-4-5}} hN_iN_j d\Gamma \tag{5.87}$$

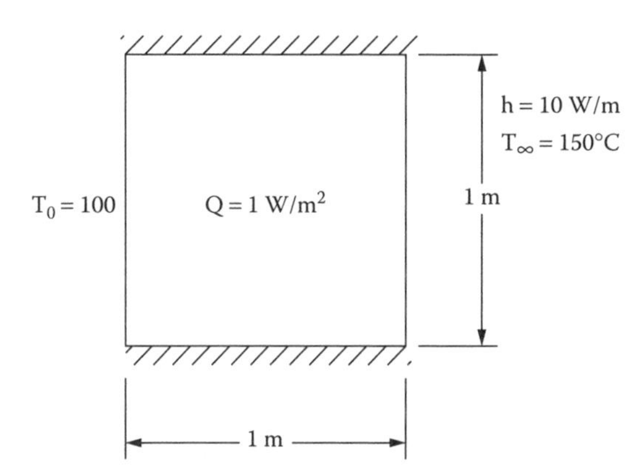

Figure 5.18 Square region with problem data for quadratic element

* FEM-2D and COMSOL files are available.

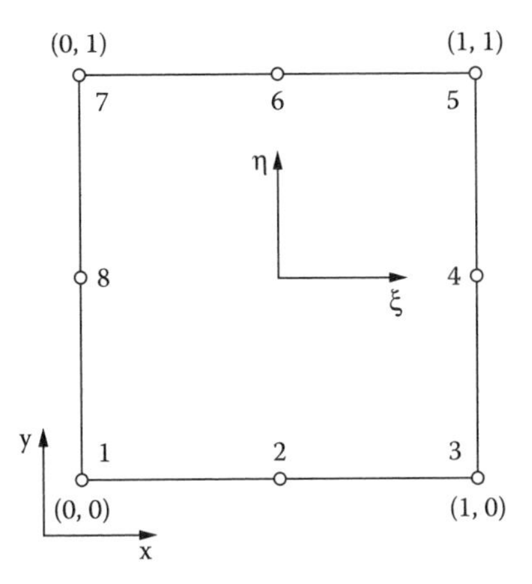

Figure 5.19 Square region discretized into a quadratic quadrilateral

and

$$F = \int_{\Gamma_{3-4-5}} hT_\infty N_i d\Gamma + \int_\Omega QN_i dxdy \tag{5.88}$$

Converting to the natural coordinate system indicated in Figure 5.19, Eqs. (5.87) and (5.88) become

$$K = \int_{-1}^{1}\int_{-1}^{1}\left(\frac{\partial N_i}{\partial x}\frac{\partial N_j}{\partial x} + \frac{\partial N_i}{\partial y}\frac{\partial N_j}{\partial y}\right)|J|\,d\xi d\eta + \int_{-1}^{1} hN_iN_j\frac{\ell_{3-5}}{2}d\eta \tag{5.89}$$

$$F = \int_{-1}^{1} hT_\infty N_i\frac{\ell_{3-5}}{2}d\eta + \int_{-1}^{1}\int_{-1}^{1} QN_i|J|\,d\xi d\eta \tag{5.90}$$

where the derivatives are given by Eq. (5.50). For quadratic quadrilaterals, the Jacobian is evaluated with reference to Figure 5.7 and Eqs. (5.32) as

$$J = \begin{bmatrix} \frac{\partial x}{\partial \xi} & \frac{\partial x}{\partial \eta} \\ \frac{\partial y}{\partial \xi} & \frac{\partial y}{\partial \eta} \end{bmatrix} = \frac{1}{2}\begin{bmatrix} x_3 - x_1 & 0 \\ 0 & y_7 - y_1 \end{bmatrix} \tag{5.91}$$

Evaluating the terms creates the 2 × 2 matrix for the Jacobian, from which the inverse can be found.

As one can see, the algebra becomes quite lengthy for quadratic functions, and calculations are best left to the computer. Table 5.3 lists the local node number, shape function values, and derivatives of the shape function in both Cartesian and natural coordinates for the one element shown in Figure 5.19, evaluated at the element midpoint.

Table 5.3 Shape Function Values

Local Node, i	N_i	$\partial N_i/\partial x$	$\partial N_i/\partial y$	$\partial N_i/\partial \xi$	$\partial N_i/\partial \eta$
1	−1/4	0	0	0	0
2	1/2	0	−1	0	−1/2
3	−1/4	0	0	0	0
4	1/2	1	0	1/2	0
5	−1/4	0	0	0	0
6	1/2	0	1	0	1/2
7	−1/4	0	0	0	0
8	1/2	−1	0	−1/2	0

Table 5.4 Gauss Integration Order for Two-Dimensional Elements (ξ, η)

Element	N_i	N_iN_j	$(\partial N_i/\partial \xi)(\partial N_j/\partial \eta)$
Linear	1, 1	2, 2	2, 2
Quadratic	2, 2	3, 3	3, 3

Once we know the values of the shape functions and their derivatives, it is a simple matter to numerically integrate Eqs. (5.89) and (5.90) using Gaussian quadrature. The order of the Gauss integration (number of Gauss points) for the two-dimensional elements under consideration is given in Table 5.4.

Evaluating Eq. (5.89) for the stiffness matrix, **K** becomes

$$K = \frac{1}{4}\int_{-1}^{1}\int_{-1}^{1}\left(\frac{\partial N_i}{\partial x}\frac{\partial N_j}{\partial x} + \frac{\partial N_i}{\partial y}\frac{\partial N_j}{\partial y}\right)d\xi\, d\eta + \int_{-1}^{1} hN_iN_j \frac{\ell_{3-5}}{2}\, d\eta$$

$$= \begin{bmatrix} 1.16 & -0.82 & 0.50 & -0.51 & 0.51 & -0.51 & 0.50 & -0.82 \\ -0.82 & 2.31 & -0.82 & 0.00 & -0.51 & 0.36 & -0.51 & 0.00 \\ 0.50 & -0.82 & 1.16 & -0.82 & 0.50 & =0.51 & 0.51 & -0.51 \\ -0.51 & 0.00 & -0.82 & 2.31 & -0.82 & 0.00 & -0.51 & 0.36 \\ 0.51 & -0.51 & 0.50 & -0.82 & 1.16 & -0.82 & 0.50 & -0.51 \\ -0.51 & 0.36 & -0.51 & 0.00 & -0.82 & 2.31 & -0.82 & 0.00 \\ 0.50 & -0.51 & 0.51 & -0.51 & 0.50 & -0.82 & 1.16 & -0.82 \\ -0.82 & 0.00 & -0.51 & 0.36 & -0.51 & 0.00 & -0.82 & 2.31 \end{bmatrix}$$

$$+ \begin{bmatrix} 0 & 0 & 0 & 0 & 0 & 0 & 0 & 0 \\ 0 & 0 & 0 & 0 & 0 & 0 & 0 & 0 \\ 0 & 0 & 1.33 & 0.66 & -0.33 & 0 & 0 & 0 \\ 0 & 0 & .066 & 5.33 & 0.66 & 0 & 0 & 0 \\ 0 & 0 & -0.33 & 0.66 & 1.33 & 0 & 0 & 0 \\ 0 & 0 & 0 & 0 & 0 & 0 & 0 & 0 \\ 0 & 0 & 0 & 0 & 0 & 0 & 0 & 0 \\ 0 & 0 & 0 & 0 & 0 & 0 & 0 & 0 \end{bmatrix} \quad (5.92)$$

where $|\mathbf{J}| = 1/4$ and $\ell_{3-5} = 1$. In similar fashion, the terms, which comprise the load vector, become

$$\mathrm{F} = \int_{-1}^{1} hT_\infty N_i \frac{\ell_{3-5}}{2} d\eta + \int_{-1}^{1}\int_{-1}^{1} QN_j |\mathrm{J}| d\xi d\eta = \begin{bmatrix} 0 \\ 0 \\ 250 \\ 1000 \\ 250 \\ 0 \\ 0 \\ 0 \end{bmatrix} + \frac{1}{12}\begin{bmatrix} -1 \\ 4 \\ -1 \\ 4 \\ -1 \\ 4 \\ -1 \\ 4 \end{bmatrix} \tag{5.93}$$

Notice that for the corner nodes in the source term integral, the integral yields a negative value, i.e., for N_1,

$$\begin{aligned} \int_{-1}^{1}\int_{-1}^{1} QN_1 |\mathrm{J}| d\xi d\eta &= \frac{Q}{4}\int_{-1}^{1}\int_{-1}^{1} -\frac{1}{4}(1-\xi\eta-\xi^2+\xi^2\eta-\eta^2+\xi\eta^2) d\xi d\eta \\ &= -\frac{Q}{16}\left(\xi\eta - \frac{\xi^2\eta^2}{4} - \frac{\xi^3\eta}{3} + \frac{\xi^3\eta^2}{6} - \frac{\xi\eta^3}{3} + \frac{\xi^2\eta^3}{6}\right)\Bigg|_{-1}^{1}\Bigg|_{-1}^{1} \\ &= -\frac{Q}{16}\left(\frac{4}{3}\right) = -\frac{q}{12} \end{aligned} \tag{5.94}$$

For N_2 and the rest of the mid-side nodes, we obtain

$$\begin{aligned} \int_{-1}^{1}\int_{-1}^{1} QN_2 |\mathrm{J}| d\xi d\eta &= \frac{Q}{4}\int_{-1}^{1}\int_{-1}^{1} -\frac{1}{2}(1-\xi\eta-\xi^2-\eta+\xi^2\eta) d\xi d\eta \\ &= \frac{Q}{8}\left(\xi\eta - \frac{\xi^3\eta}{3} + \frac{\xi\eta^2}{2} + \frac{\xi^3\eta^2}{6}\right)\Bigg|_{-1}^{1}\Bigg|_{-1}^{1} \\ &= \frac{Q}{8}\left(\frac{8}{3}\right) = -\frac{Q}{3} \end{aligned} \tag{5.95}$$

Combining all terms, Eq. (5.86) becomes

$$\begin{bmatrix} 1.16 & -0.82 & 0.50 & -0.51 & 0.51 & -0.51 & 0.50 & -0.82 \\ -0.82 & 2.31 & -0.82 & 0.00 & -0.51 & 0.36 & -0.51 & 0.00 \\ 0.50 & -0.82 & 2.49 & -0.16 & 0.17 & -0.51 & 0.51 & -0.51 \\ -0.51 & 0.00 & -0.16 & 7.64 & -0.16 & 0.00 & -0.51 & 0.36 \\ 0.51 & -0.51 & 0.17 & -0.16 & 2.49 & -0.82 & 0.50 & -0.51 \\ -0.51 & 0.36 & -0.51 & 0.00 & -0.82 & 2.31 & -0.82 & 0.00 \\ 0.50 & -0.51 & 0.51 & -0.51 & 0.50 & -0.82 & 1.16 & -0.82 \\ -0.82 & 0.00 & -0.51 & 0.36 & -0.51 & 0.00 & -0.82 & 2.31 \end{bmatrix}\begin{bmatrix} T_1 \\ T_2 \\ T_3 \\ T_4 \\ T_5 \\ T_6 \\ T_7 \\ T_8 \end{bmatrix} = \begin{bmatrix} -0.083 \\ 0.333 \\ 249.92 \\ 1000.33 \\ 249.92 \\ 0.33 \\ -0.083 \\ 0.333 \end{bmatrix} \tag{5.96}$$

Solving for the unknown temperatures, T_2 to T_6, since $T_1 = T_7 = T_8 = 100°C$, we obtain

$$\begin{bmatrix} T_1 \\ T_2 \\ T_3 \\ T_4 \\ T_5 \\ T_6 \\ T_7 \\ T_8 \end{bmatrix} = \begin{bmatrix} 100.00 \\ 122.88 \\ 145.50 \\ 145.50 \\ 145.50 \\ 122.88 \\ 100.00 \\ 100.00 \end{bmatrix} \tag{5.97}$$

Notice the symmetry about the main diagonal associated with **K** in Eq. (5.92). Advanced matrix solution algorithms typically take advantage of this symmetry, reducing in half both the storage requirements for the various terms in **K** and the computational operations.

5.9 TIME-DEPENDENT DIFFUSION

If we wish to solve the problem in a transient, time-dependent mode, the mass matrix **M** associated with the time derivative term would become

$$\int_\Omega \frac{\partial T}{\partial t} N_i dxdy = \int_{-1}^{1}\int_{-1}^{1} N_i N_j |\mathrm{J}| d\xi d\eta \left[\frac{\partial T_i}{\partial t}\right] = \mathrm{M}\left[\frac{\partial T_i}{\partial t}\right] \tag{5.98}$$

where, for the eight-noded quadratic element of Figure 5.19, we have

$$\mathrm{M} = \int_{-1}^{1}\int_{-1}^{1} N_i N_j |\mathrm{J}|$$

$$= \begin{bmatrix} 0.03 & -0.03 & 0.01 & -0.04 & 0.02 & -0.04 & 0.01 & -0.03 \\ -0.03 & 0.18 & -0.03 & 0.11 & -0.04 & 0.09 & -0.04 & 0.11 \\ 0.01 & -0.03 & 0.03 & -0.03 & 0.01 & -0.04 & 0.02 & -0.04 \\ -0.04 & 0.11 & -0.03 & 0.18 & -0.03 & 0.11 & -0.04 & 0.09 \\ 0.02 & -0.04 & 0.01 & -0.03 & 0.03 & -0.03 & 0.01 & -0.04 \\ -0.04 & 0.09 & -0.04 & 0.11 & -0.03 & 0.18 & -0.03 & 0.11 \\ 0.01 & -0.04 & 0.02 & 0.04 & 0.01 & -0.03 & 0.03 & -0.03 \\ -0.03 & 0.11 & -0.04 & 0.09 & -0.04 & 0.11 & -0.03 & 0.18 \end{bmatrix} \tag{5.99}$$

In this instance, we note from Table 5.4 that nine Gauss points must be used for the product $N_i N_j$ (Figure 5.9); hence,

$$\mathrm{M} = \left[m_{ij}\right] = \left[\sum_{k=1}^{3}\sum_{\ell=1}^{3} w_k w_\ell N_i\left(\xi_k,\ \eta_\ell\right) N_j\left(\xi_k,\ \eta_\ell\right)\left|\mathrm{J}\left(\xi_k,\ \eta_\ell\right)\right|\right] \quad (5.100)$$

The time-dependent diffusion equation, Eq. (4.116) in Chapter 4, yields a system of ordinary differential equations of the form:

$$\mathbf{M\dot{T}} + \mathbf{KT} = \mathbf{F} \quad (5.101)$$

where the matrices **K**, **F**, and **M** are given by Eqs. (5.92), (5.93), and (5.99), respectively, and can be integrated in time using the θ method.

Those readers feeling industrious can derive the general matrix equivalent relations for the eight- and/or nine-noded rectangular quadratic and biquadratic elements. Be prepared for a great number of algebraic operations!

5.10 COMPUTER PROGRAM EXERCISES*

The first example deals with steady-state heat conduction in a square, as shown in Figure 5.20. The conduction coefficients are constant, $K_{xx} = K_{yy} = 15.0$ W/mC. A total of 12 nodes with bilinear elements are arranged as shown in Figure 5.20. The left wall, nodes 1, 2, and 3, is set to a constant value of 1000°C; the right wall, nodes 10, 11, and 12, is set to 0 C. The top and bottom walls are insulated, i.e., $\partial T/\partial y = 0$. The exact solution can be obtained by inspection and serves as a test of program accuracy. Both exact and calculated values are given in Table 5.5. Model output shows values essentially identical to exact values.

The second example deals with uniform potential flow in a rectangular domain, which is shown in Figure 5.21. The right wall sets $\phi = 10$ m^2/s; the left wall specifies a Neumann condition for the gradient, in this case $\partial\phi/\partial x = -1$. At the top and bottom walls, $\partial\phi/\partial y = 0$. Nine nodes are used with four bilinear elements. The diffusion coefficients in this example are set to unity, $K_{xx} = K_{yy} = 1.0$ m^2/s. This problem depicts a uniform flow through a rectangular duct, with a constant source on the right and a constant outflow at the left boundary. Table 5.6 lists both the exact values and the calculated values.

The third example problem pertains to time-dependent heat conduction in a rectangular fin, which is attached to a wall at constant temperature. The fin is exposed to ambient convection heat transfer, as shown in Figure 5.22. The same geometry is used as in Figure 5.20. The three sides are exposed to the ambient temperature, $T_\infty = 1500$°C. The left wall is constant at 1000°C; the initial temperatures within the problem domain (at the remaining nodes) are set to zero. The conduction coefficients are $K_{xx} = K_{yy} = 15.0$ W/mC, and the ambient convection heat transfer coefficient is $h = 120$ W/m$_2$C. The time step is $\Delta t = 0.01$ h. The problem is to calculate how long it takes for the fin to reach steady-state conditions and to

* FEM-2D and COMSOL files are available.

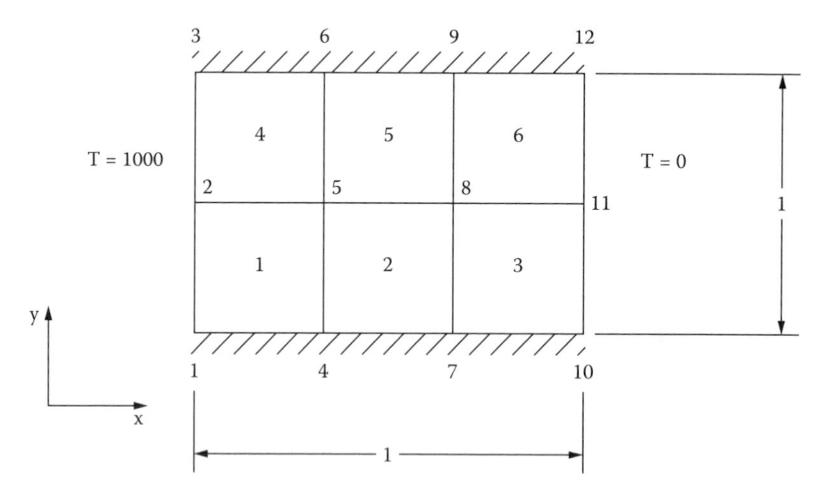

Figure 5.20 Steady-state temperature in a square

Table 5.5 Heat Conduction in a Square

Exact Values (nodes)				Calculated Values (COMSOL)			
10000	666 2/3	333 1/3	0	10000	666.666	333.333	0
10000	666 2/3	333 1/3	0	10000	666.666	333.333	0
10000	666 2/3	333 1/3	0	10000	666.666	333.333	0

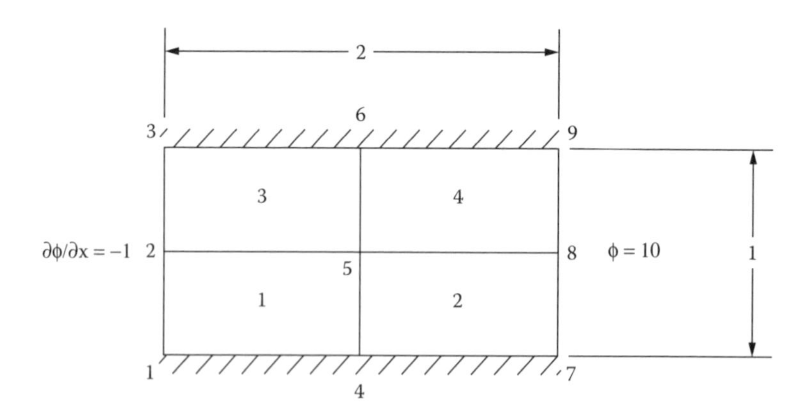

Figure 5.21 Potential flow in a rectangular domain

determine the temperature distribution at steady state. In this case, the analytical solution is not calculated; instead, numerical values are compared with values obtained from Huebner (1975), who used three-noded triangular elements. Results

Table 5.6 Potential Flow within a Rectangle

Exact Values (nodes)			Calculated Values (COMSOL)		
8	9	10	7.9999+	8.999+	10
8	9	10	7.9999+	8.999+	10
8	9	10	7.9999+	8.999+	10

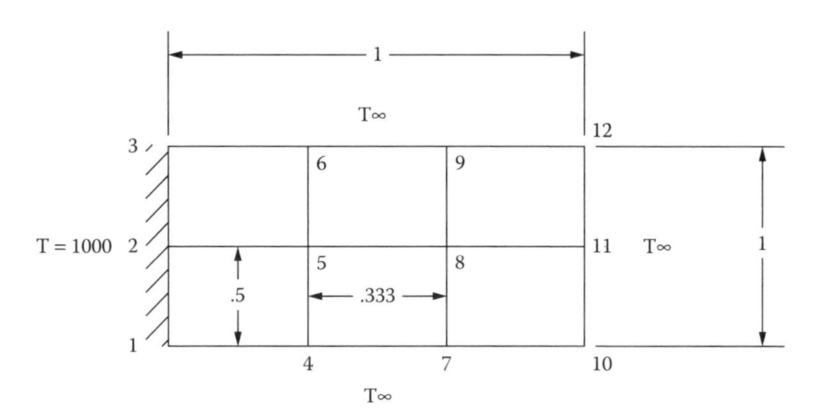

Figure 5.22 Heat conduction in a fin

Table 5.7 Heat Conduction in a Fin

Huebner's Values (nodes)				Calculated Values (COMSOL)			
1000	1420.2	1477.5	1501.0	1000	1447.18	1467.48	1493.81
1000	1268.5	1402.5	1469.6	1000	1207.86	1401.87	1477.81
1000	1420.2	1477.5	1501.0	1000	1447.18	1467.48	1493.81

are shown in Table 5.7 for steady-state conditions. Steady state should be reached in 21 time steps.

In this fourth example, uniform heat generation is analyzed in a unit square. Temperature is set to zero around the perimeter of the square, $K_{xx} = K_{yy}$ = W/mC, and Q = 8.0 W/m^3, which is applied over each individual element. The problem is shown in Figure 5.23a. Because the problem is symmetric, only one-fourth of the square needs to be analyzed, as shown in Figure 5.23b. Nine nodes are used with four bilinear elements. Both Dirichlet (fixed) and Neumann (flux) conditions are applied, as shown in Figure 5.23b. A closed-form analytical solution yields a value

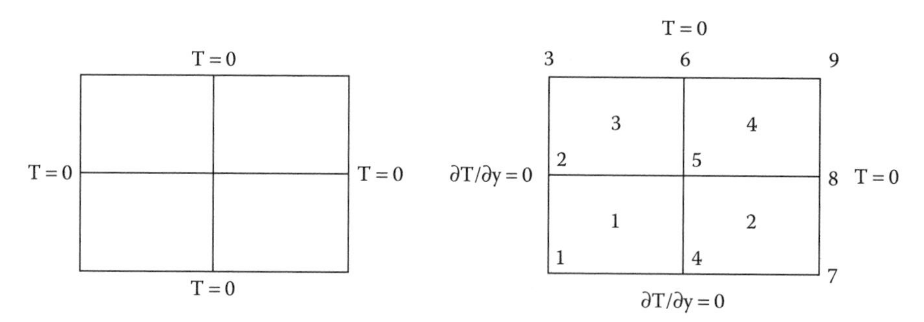

Figure 5.23 Internal heat generation in a unit square

Table 5.8 Heat Generation in a Square

Akin's Values (nodes)			Calculated Values (COMSOL)		
0.2×10^{-12}	0.2×10^{-12}	0.1×10^{-12}	0	0	0
0.46439	0.36763	0.2×10^{-12}	0.48214	0.38571	0
0.59678[a]	0.46439	0.2×10^{-12}	0.62142	0.48214	0

[a] Analytical value = 0.5894.

for node 1 of 0.5894°C at steady-state conditions. Table 5.8 lists the steady-state nodal values obtained by Akin (1981) using quadratic elements and the calculated values using bilinear quadrilaterals. The temperature at node 1 using linear elements yields 0.6214°C, which is within 6% of the analytical value. The accuracy is rather surprising considering that only four elements were used. Tests with a finer grid of 12 elements give results within 2%.

5.11 CLOSURE

We have added a second set of elements to the two-dimensional element family, the quadrilateral elements. The two-dimensional quadrilateral elements, like the triangular elements discussed in Chapter 4, are simple elements that can also fit any nonregular geometry; although this is not apparent yet, it will be clearly shown in the next chapter. The two principal differences between quadrilateal and triangular elements are (1) an additional fourth node is required in the bilinear quadrilateral element, thereby adding one more degree-of-freedom to define the element, and (2) the gradients are linear functions of one of the coordinate directions, whereas gradients in the three-noded triangular (simplex) elements are constant. A constant gradient requires that many small elements be used in regions where rapid changes occur in a variable. Since the (rectangular) quadrilateral is a multiplex element, the boundaries of the element must be parallel to a coordinate system to maintain continuity between elements. However, there is no reason a natural coordinate system has to be aligned with the Cartesian (or cylindrical) orthogonal coordinate system—it can be a curvilinear system as well.

In most instances, the quadrilateral element may yield more accurate results than triangular elements (for the same number of elements). A more extensive discussion on accuracy and convergence properties of various elements can be found in Carey and Oden (1983), Oden and Reddy (1976), Zienkiewicz and Taylor (1989), and Carey (1997).

In the next chapter, we utilize a coordinate transformation that accurately represents nonrectangular regions and regions with curved boundaries. The transformation requires only one set of relations to account for the interpolation functions and the element geometry in terms of x and y.

EXERCISES

5.1 Derive Eq. (5.4) and find the bilinear shape functions (5.6).

5.2 Find the shape functions for a bilinear element defined by (0, 0), (0.1, 0), (0.1, 0.2), and (0, 0.2).

5.3 Show that the bilinear quadratic and biquadratic shape functions satisfy

$$\sum_{i=1}^{N} N_i \equiv 1$$

and

$$\sum_{i=1}^{N} \frac{\partial N_i}{\partial x} = \sum_{i=1}^{N} \frac{\partial N_i}{\partial y} = 0$$

for $N = 4$, 8, and 9, respectively.

5.4 For the shape functions in Exercise 5.1, perform the coordinate transformation to the (ξ, η) system, $-1 \le \xi, \eta \le 1$, and obtain the shape functions. Notice that now these products on one-dimensional functions, as given in Eq. (3.73).

5.5 Repeat Example 5.1 using the nine-noded biquadratic element.

5.6 Subdivide the following geometry into 6 bilinear quadrilateral elements; label all nodes and elements. What is the bandwidth?

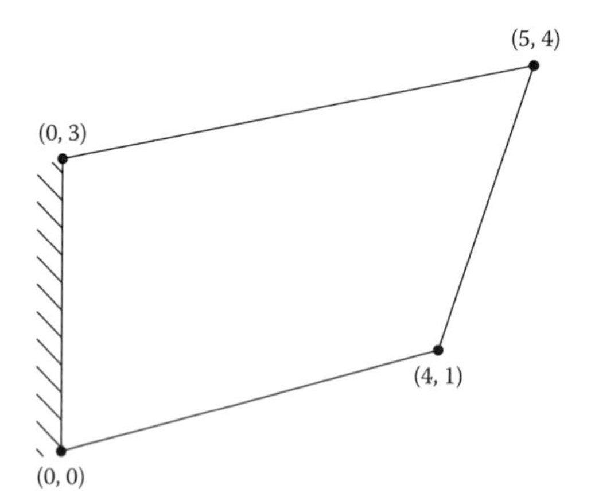

5.7 Calculate the area of the element in Exercise 5.4 that contains the point (5, 4).

5.8 Calculate $\int_\Gamma N_i d\Gamma$ along $\eta = 1$ for the quadratic eight-noded element.

5.9 Define the interpolating polynomial for a 12-noded cubic quadrilateral in ξ, η coordinates. Determine the shape function N_2; i.e., show that

$$N_2 = (1-\xi^2)(1-\eta)(\alpha_1 + \alpha_2\xi + \alpha_3\eta + \alpha_4\xi^2 + \alpha_5\eta^2)$$

and find the coefficients α_i, 1 = 1,2,...,5.

5.10 Compare results using two-dimensional linear triangular elements with results using bilinear quadrilaterals for the two meshes shown below if

$$\frac{\partial^2\phi}{\partial x^2} + \frac{\partial^2\phi}{\partial y^2} = 0$$

and at nodes 1 and 3, $\phi = 10$, and at node 2, $\phi = 20$. If the other boundaries are insulated, what are the values of ϕ at nodes 7, 8, and 9?

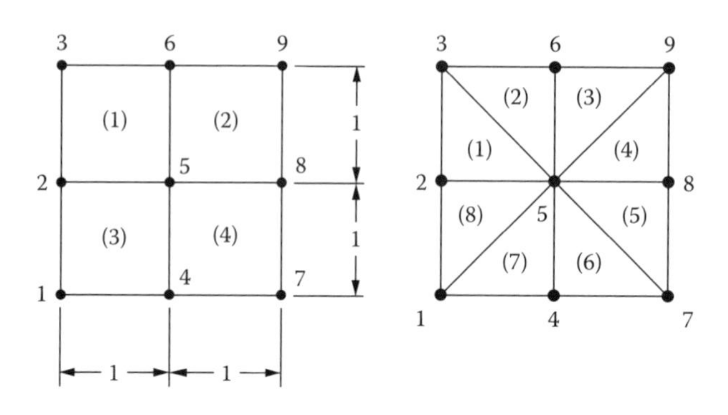

5.11 Calculate the steady-state heat distribution within the two-dimensional body shown below. Use $K_{xx} = K_{yy} = 10$ W/mC. At least 12 bilinear quadrilateral elements should be used to model the region.

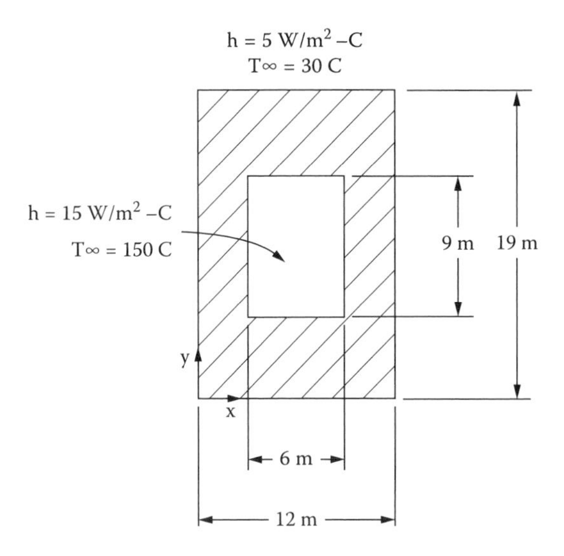

5.12 For the unit square region shown calculate the stiffness matrix using one bilinear element and two triangular elements. Compare them and discuss the differences.

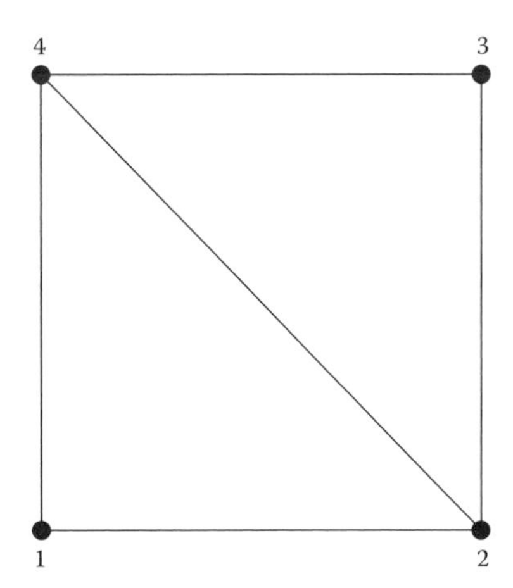

5.13 In the element of Exercise 5.12 solve Eq. (5.44) with $K_{xx} = K_{yy} = 10$, $Q = 10.0$, $T_1 = T_4 = 25$, and $q = 20.0$ on the right-hand side. The top and bottom boundaries are adiabatic. Use one bilinear element and two linear triangles; compare your results with the analytical solution.

5.14 Solve Exercise 4.14 using bilinear elements.
5.15 Solve Exercise 4.15 using bilinear elements and compare the accuracy with the solution using linear triangles.
5.16 Solve Exercise 4.27 using bilinear elements in the given mesh and compare with the solution using linear triangles.
5.17 For the element in Figure 5.6 calculate the mass matrix

$$m_{ij} = \int_{-1}^{1}\int_{-1}^{1} N_i N_j \left|J\right| d\xi d\eta$$

using
(a) One Gauss point
(b) A 2 × 2 Gauss quadrative
(c) Exact analytical integration
Discuss the results.
5.18 Assuming that the derivatives $\partial N_i/\partial x$ and $\partial N_i/\partial y$ in Eq. (5.36) have already been evaluated at the Gauss points and using the fact that the stiffness matrix is symmetric, show that it takes around 160 multiplications to evaluate the element stiffness matrix for a bilinear element using a 2 × 2 Gauss quadrature, while it takes about 1944 multiplications for the 8-node quadratic and 2436 for the biquadratic element using a 3 × 3 Gauss quadrature.
5.19 If a biquadratic element is subdivided into four bilinear elements, it still takes almost four times the number of operations to evaluate the stiffness matrix for the quadratic element. Calculate the bandwidth for a square assembly of four biquadratic elements, and compare it with the bandwidth of the same mesh when it is made using bilinear elements.
5.20 In Eq. (5.54) write the derivatives $\partial N_i/\partial x$ and $\partial N_i/\partial y$ explicitly in terms of $\partial N_i/\partial \xi$ and $\partial N_i/\partial \eta$.
5.21 Using the shape functions as in Eq. (5.,60), verify Eq. (5.61).
5.22 Solve Exercise 4.30 using bilinear elements and compare to the solution using linear triangles.

REFERENCES

Akin, J. E. (1981). *Applications and Implementation of Finite Element Methods*. New York: Academic Press.

Carey, G. F. (1997). *Computational Grids: Generation, Adaptation, and Solution Strategies*. Washington, D.C.: Taylor & Francis.

Carey, G. F. and Oden, J. T. (1983). *Finite Elements: A Second Course*. Englewood Cliffs, N.J.: Prentice-Hall.

Heinrich, J. C. and Pepper, D. W. (1999). *Intermediate Finite Element Method. Fluid Flow and Heat Transfer Applications*. New York: Taylor & Francis.

Huebner, K. H. (1975). *Finite Element Method for Engineers*: New York: John Wiley & Sons.

Oden, J. T. and Carey, G. F. (1983). *Finite Elements*. Volume IV: *Mathematical Aspects*. Englewood Cliffs, N.J.: Prentice-Hall.

Oden, J. T. and Reddy, J. N. (1976). *An Introduction to the Mathematical Theory of Finite Elements.* New York: John Wiley & Sons.
Segerlind, L. J. (1976). *Applied Finite Element Analysis.* New York: John Wiley & Sons.
Segerlind, L. J. (1984). *Applied Finite Element Analysis.* New York: John Wiley & Sons.
Zienkiewicz, O. C. (1977). *The Finite Element Method*, 3rd ed. London: McGraw-Hill.
Zienkiewicz, O. C. and Taylor, R. L. (1989). *The Finite Element Method,* 4th ed. Maidenhead, U.K.: McGraw-Hill.

CHAPTER

6

ISOPARAMETRIC TWO-DIMENSIONAL ELEMENTS

6.1 BACKGROUND

In Chapters 4 and 5, we assumed that the element sides were straight, i.e., linear, and that the side of rectangular elements were parallel to the coordinate axis. Fortunately, we are not restricted to using only straight-sided elements or rectangular quadrilateral elements; elements with arbitrarily oriented curved sides can be used to represent bodies with curved boundaries. The transformation from straight to curved sides is achieved through the use of isoparametric elements The transformation is simple and straightforward; in fact, the procedure was utilized in Chapter 5, but was constrained to straight-sided elements, with sides parallel to the coordinate axis. Remember that there are two key sets of relations that must be defined when using the finite element method: (1) a set of relations that determine the shape of the element, and (2) the order of the interpolation function. It is not necessary to use the same shape functions for the interpolation equation and the coordinate transformation, i.e., two sets of "global" nodes can exist. In the case of the isoparametric element, both sets of global nodes are identical.

Nearly all current finite element programs used in industry and academic research use isoparametric elements. Isoparametric elements are based on a local coordinate system with independent variables ξ and η, which usually vary from –1 to 1 for the spatial coordinates. The same local coordinate relations are used to define a variable within an element and to define the global coordinates of a point within an element in terms of the global coordinates of the nodal points.

6.2 NATURAL COORDINATE SYSTEM

We begin by expressing x and y as functions of ξ and η, i.e.,

$$\left.\begin{aligned} x &= x(\xi,\eta) \\ y &= y(\xi,\eta) \end{aligned}\right\} \tag{6.1}$$

The interpolation functions are also expressed in terms of ξ and η:

$$N_i = N_i(\xi,\eta) \tag{6.2}$$

Because we are dealing with Cartesian derivatives in the integral equations, we transform the derivatives of N_i using the chain rule.

$$\left.\begin{aligned} \frac{\partial N_i}{\partial \xi} &= \frac{\partial N_i}{\partial x}\frac{\partial x}{\partial \xi} + \frac{\partial N_i}{\partial y}\frac{\partial y}{\partial \xi} \\ \frac{\partial N_i}{\partial \eta} &= \frac{\partial N_i}{\partial x}\frac{\partial x}{\partial \eta} + \frac{\partial N_i}{\partial y}\frac{\partial y}{\partial \eta} \end{aligned}\right\} \tag{6.3}$$

Equation (6.3) can be rewritten using the Jacobian matrix expression we used in the previous chapters, i.e.,

$$\begin{bmatrix} \dfrac{\partial N_i}{\partial \xi} \\ \dfrac{\partial N_i}{\partial \eta} \end{bmatrix} = \begin{bmatrix} \dfrac{\partial x}{\partial \xi} & \dfrac{\partial y}{\partial \xi} \\ \dfrac{\partial x}{\partial \eta} & \dfrac{\partial x}{\partial \eta} \end{bmatrix} \begin{bmatrix} \dfrac{\partial N_i}{\partial x} \\ \dfrac{\partial N_i}{\partial y} \end{bmatrix} = \mathbf{J} \begin{bmatrix} \dfrac{\partial N_i}{\partial x} \\ \dfrac{\partial N_i}{\partial y} \end{bmatrix} \tag{6.4}$$

To find the Cartesian derivatives of N_i, we simply invert Eq. (6.4),

$$\begin{bmatrix} \dfrac{\partial N_i}{\partial x} \\ \dfrac{\partial N_i}{\partial y} \end{bmatrix} = \mathbf{J}^{-1} \begin{bmatrix} \dfrac{\partial N_i}{\partial \xi} \\ \dfrac{\partial N_i}{\partial \eta} \end{bmatrix} = \frac{1}{|\mathbf{J}|} \begin{bmatrix} \dfrac{\partial y}{\partial \eta} & -\dfrac{\partial y}{\partial \xi} \\ -\dfrac{\partial x}{\partial \eta} & \dfrac{\partial x}{\partial \xi} \end{bmatrix} \begin{bmatrix} \dfrac{\partial N_i}{\partial \xi} \\ \dfrac{\partial N_i}{\partial \eta} \end{bmatrix} \tag{6.5}$$

where $|\mathbf{J}|$ denotes the determinant of $\mathbf{J}$. From Eq. (6.5), we get

$$\frac{\partial N_i}{\partial x} = \frac{1}{|\mathbf{J}|}\left(\frac{\partial y}{\partial \eta}\frac{\partial N_i}{\partial \xi} - \frac{\partial y}{\partial \xi}\frac{\partial N_i}{\partial \eta} \right) \tag{6.6}$$

$$\frac{\partial N_i}{\partial y} = \frac{1}{|\mathrm{J}|}\left(\frac{\partial x}{\partial \xi}\frac{\partial N_i}{\partial \eta} - \frac{\partial x}{\partial \eta}\frac{\partial N_i}{\partial \xi}\right) \tag{6.7}$$

The determinant of **J** is obtained, as before:

$$|\mathrm{J}| = \frac{\partial x}{\partial \xi}\frac{\partial y}{\partial \eta} - \frac{\partial x}{\partial \eta}\frac{\partial y}{\partial \xi} \tag{6.8}$$

The relationships among the derivatives of the coordinate systems are

$$\left.\begin{aligned} \frac{\partial \xi}{\partial x} &= \frac{1}{|\mathrm{J}|}\frac{\partial y}{\partial \eta} \\ \frac{\partial \xi}{\partial y} &= -\frac{1}{|\mathrm{J}|}\frac{\partial x}{\partial \eta} \\ \frac{\partial \eta}{\partial x} &= -\frac{1}{|\mathrm{J}|}\frac{\partial y}{\partial \xi} \\ \frac{\partial \eta}{\partial y} &= \frac{1}{|\mathrm{J}|}\frac{\partial x}{\partial \xi} \end{aligned}\right\} \tag{6.9}$$

As before, the differential area transforms to the relation

$$dA = dxdy = |\mathrm{J}|\,d\xi d\eta \tag{6.10}$$

The x,y coordinates are defined in terms of their nodal point values as

$$\left.\begin{aligned} x &= \sum_{i=1}^{n} N_i\left[\xi,\eta\right]x_i \\ y &= \sum_{i=1}^{n} N_i\left[\xi,\eta\right]y_i \end{aligned}\right\} \tag{6.11}$$

when $n = 4$ for a bilinear four-node element and $n = 8$ or 9 for quadratic or biquadratic quadrilaterals. Equation (6.11) is easily established as follows: Assume that the variable ϕ can be approximated as

$$\phi = \alpha_1 + \alpha_2 x + \alpha_3 y \tag{6.12}$$

For node i,

$$\phi_i = \alpha_1 + \alpha_2 x_i + \alpha_3 y_i \tag{6.13}$$

Likewise, if we let

$$\phi = \sum_{i=1}^{n} N_i \phi_i \tag{6.14}$$

then substituting Eq. (6.14) into Eq. (6.13) gives

$$\begin{aligned}\phi &= \sum_{i=1}^{n} (\alpha_1 + \alpha_2 x_i + \alpha_3 y_i) N_i(\xi, \eta) \\ &= \alpha_1 \sum_{i=1}^{n} N_i(\xi, \eta) + \alpha_2 \sum_{i=1}^{n} x_i N_i(\xi, \eta) + \alpha_3 \sum_{i=1}^{n} y_i N_i(\xi, \eta)\end{aligned} \tag{6.15}$$

Thus,

$$\sum_{i=1}^{n} N_i(\xi, \eta) = 1, \quad \sum_{i=1}^{n} x_i N_i(\xi, \eta) = x, \quad \sum_{i=1}^{n} y_i N_i(\xi, \eta) = y.$$

6.3 SHAPE FUNCTIONS

The shape functions are identical to those used in Chapters 4 and 5 for two-dimensional triangles and quadrilaterals.

6.3.1 Bilinear Quadrilateral

The shape functions for the four-noded bilinear quadrilateral can be written as

$$N_i = \frac{1}{4}(1 + \xi_i \xi)(1 + \eta_i \eta) \qquad i = 1, \ldots, 4 \tag{6.16}$$

where ξ_i and η_i represent the element sides as defined in Figure 6.1a, and i.e., $\xi_1 = -1$, $\eta_1 = -1$ at node 1; $\xi_2 = -1$, $\eta_2 = -1$ at node; etc. Equation (6.16) in expanded form becomes Eq. (5.17):

$$N_1 = \frac{1}{4}(1 - \xi)(1 - \eta)$$

$$N_2 = \frac{1}{4}(1 + \xi)(1 - \eta)$$

$$N_3 = \frac{1}{4}(1 + \xi)(1 + \eta)$$

$$N_4 = \frac{1}{4}(1 - \xi)(1 + \eta)$$

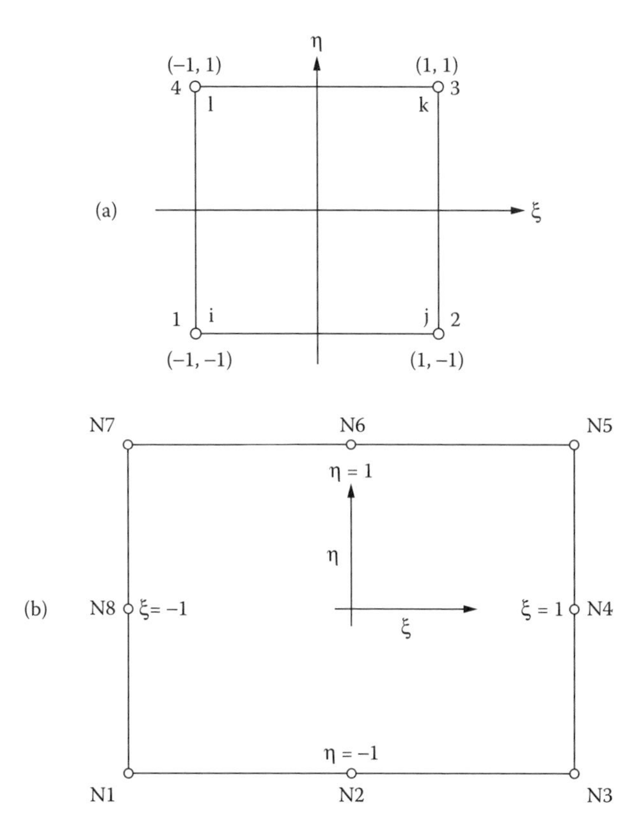

Figure 6.1 Bilinear and quadratic quadrilateral elements: (a) linear and (b) quadratic

In this case, $-1 \leq \xi \leq 1$ and $-1 \leq \eta \leq 1$. If the, §, η coordinate system is located at the lower left corner, the shape functions become Eq. (5.55),

$$N_1 = (1-\xi)(1-\eta)$$

$$N_2 = \xi(1-\eta)$$

$$N_3 = \varsigma\eta$$

$$N_4 = (1-\xi)\eta$$

where $0 \leq § \leq 1$ and $0 \leq \eta \leq 1$.

6.3.2 Eight-Noded Quadratic Quadrilateral

For the eight-noded quadratic quadrilateral, the shape functions are given as (Figure 6.1b)

Corner Nodes

$$N_i = \frac{1}{4}(1+\xi_i\xi)(1+\eta_i\eta)(\xi_i\xi+\eta_i\eta-1) \tag{6.17}$$

Mid-Side Nodes

$$\xi_i = 0: \quad \mathrm{N}_i = \frac{1}{2}(1-\xi^2)(1+\eta_i\eta) \tag{6.18}$$

$$\eta_i = 0: \quad N_i = \frac{1}{2}(1+\xi_i\xi)(1-\eta^2) \tag{6.19}$$

or in expanded form, Eq. (5.29), as

$$N_1 = -\frac{(1-\xi)(1-\eta)(1+\xi+\eta)}{4}$$

$$N_2 = \frac{(1-\xi^2)(1-\eta)}{2}$$

$$N_3 = \frac{(1+\xi)(1-\eta)(\xi-\eta-1)}{4}$$

$$N_4 = \frac{(1-\eta^2)(1+\xi)}{2}$$

$$N_5 = \frac{(1+\xi)(1+\eta)(\xi+\eta-1)}{4}$$

$$N_6 = \frac{(1-\xi^2)(1+\eta)}{2}$$

$$N_7 = -\frac{(1-\xi)(1+\eta)(1+\xi-\eta)}{4}$$

$$N_8 = \frac{(1-\eta^2)(1-\xi)}{2}$$

where $-1 \le \xi \le 1$ and $-1 \le \eta \le 1$.

6.3.3 Linear Triangle

It is a simple matter to transform the area coordinate systems for triangular elements (L_i, i = 1, 2, 3) for the ξ,η system used for quadrilateral elements (Zienkiewicz, 1977; Heinrich and Pepper, 1999).

The shape function for the three-noded linear triangle can be expressed in ξ,η coordinates as

$$\left.\begin{aligned} N_1 &= L_1 = 1-\xi-\eta \\ N_2 &= L_2 = \xi \\ N_3 &= L_3 = \eta \end{aligned}\right\} \tag{6.20}$$

with ξ,η defined in Figure 6.2a. In this instance $0 \le \xi \le 1$ and $0 \le \eta \le 1$.

6.3.4 Quadratic Triangle

For the six-noded triangle, the shape functions are defined as (Figure 6.2b),

Corner Nodes

$$N_i = (2L_i - 1)L_i \qquad i = 1, 2, 3 \tag{6.21}$$

which can be expressed in ξ,η coordinates as

$$\left.\begin{aligned} N_1 &= (2(1-\xi-\eta)-1)(1-\xi-\eta) \\ N_2 &= (2\xi-1)\xi \\ N_3 &= (2\eta-1)\eta \end{aligned}\right\} \tag{6.22}$$

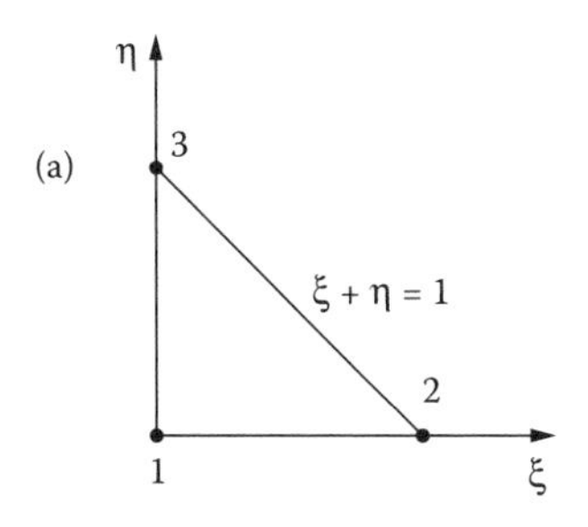

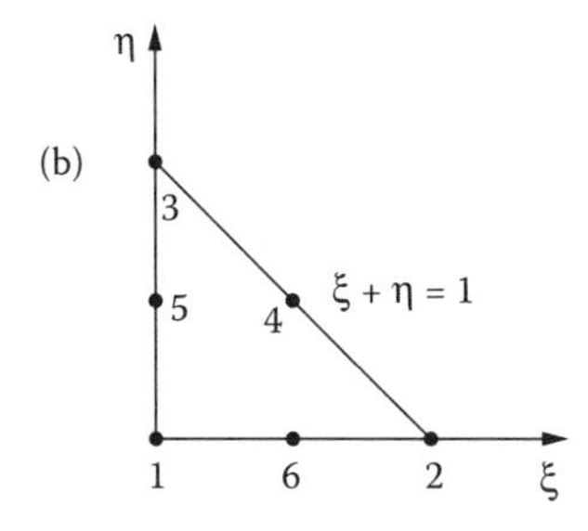

Figure 6.2 ξ,η coordinate system for the triangular element; (a) linear, and (b) quadratic

Mid-Side Nodes

$$N_i = 4L_j L_k \qquad i = 4,5,6;\ j = 2,3,1;\ k = 3,\ 1,\ 2 \tag{6.23}$$

or, in ξ,η form

$$\left.\begin{aligned} N_4 &= 4\xi\eta \\ N_5 &= 4\eta(1-\xi-\eta) \\ N_6 &= 4\xi(1-\xi-\eta) \end{aligned}\right\} \tag{6.24}$$

where $0 \le \xi \le 1$ and $0 \le \eta \le 1$.

6.3.5 Directional Cosines

The directional cosines, n_x and n_y, associated with Neumann boundary conditions can be easily obtained for an element utilizing ξ,η coordinates. For the arbitrary element side shown in Figure 6.3, the tangential vector **dr** at point **r** is defined as

$$\mathbf{dr} = dx\hat{i} + dy\hat{j} \tag{6.25a}$$

where î and are the unit vectors in the x and y directions, respectively. If s denotes the arc length coordinate along side S, we can also write

$$\mathbf{dr} = \left(\frac{dx}{ds}\hat{i} + \frac{dy}{ds} + \hat{j} \right) ds \tag{6.25b}$$

where

$$\mathrm{r} = x\hat{i} + y\hat{j} \tag{6.25c}$$

The length of the tangent is given as

$$|\mathbf{dr}| = ds\sqrt{dx^2 + dy^2} \tag{6.26}$$

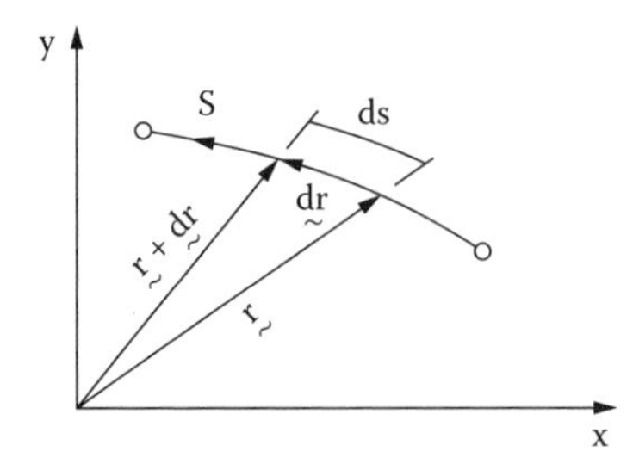

Figure 6.3 Tangential vector for an element side

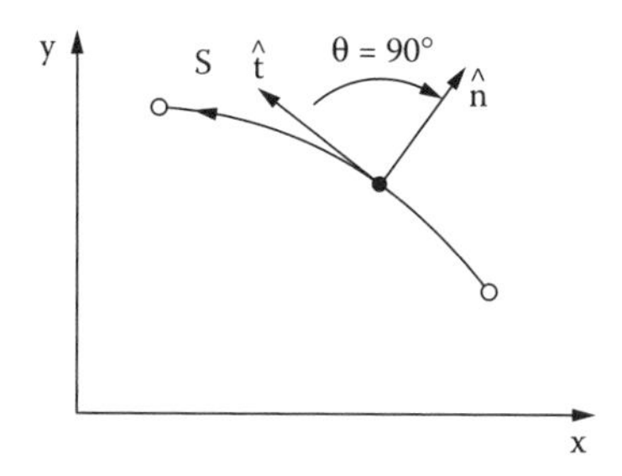

Figure 6.4 Rotation operator

The unit tangent vector $\hat{t}$ is

$$\hat{t} = \frac{d\mathbf{r}}{|d\mathbf{r}|} = \frac{d\mathbf{r}}{ds} = \left(\frac{dx}{ds}\hat{i} + \frac{dy}{ds}\hat{j} \right) \tag{6.27}$$

A unit outward normal vector can be obtained by a simple 90° rotation of the tangent vector in the clockwise direction (Figure 6.4). This is easily achieved by the rotation operator

$$Rot\theta = \begin{bmatrix} \cos\theta & -\sin\theta \\ \sin\theta & \cos\theta \end{bmatrix} \tag{6.28}$$

or

$$Rot(-90^\circ) = \begin{bmatrix} 0 & 1 \\ -1 & 0 \end{bmatrix} \tag{6.29}$$

Then the unit normal vector n, which is directed outward to the element side, is

$$\hat{n} = Rot(-90^\circ)\hat{t} = \begin{bmatrix} 0 & 1 \\ -1 & 0 \end{bmatrix} \begin{bmatrix} \frac{dx}{ds} \\ \frac{dy}{ds} \end{bmatrix} = \frac{dy}{ds}\hat{i} - \frac{dx}{ds}\hat{j} \tag{6.30}$$

The direction cosines for the quadrilateral and triangular elements, both linear and quadratic, are readily obtained using Eq. (6.30), as shown in Figure 6.5.

6.4 THE ELEMENT MATRICES

We set up the matrix relations for a local coordinate system set at local node 1 for a linear quadrilateral, as we did for the example shown in Figure 5.10. Remember that we can set up our local coordinate system wherever we like within the element.

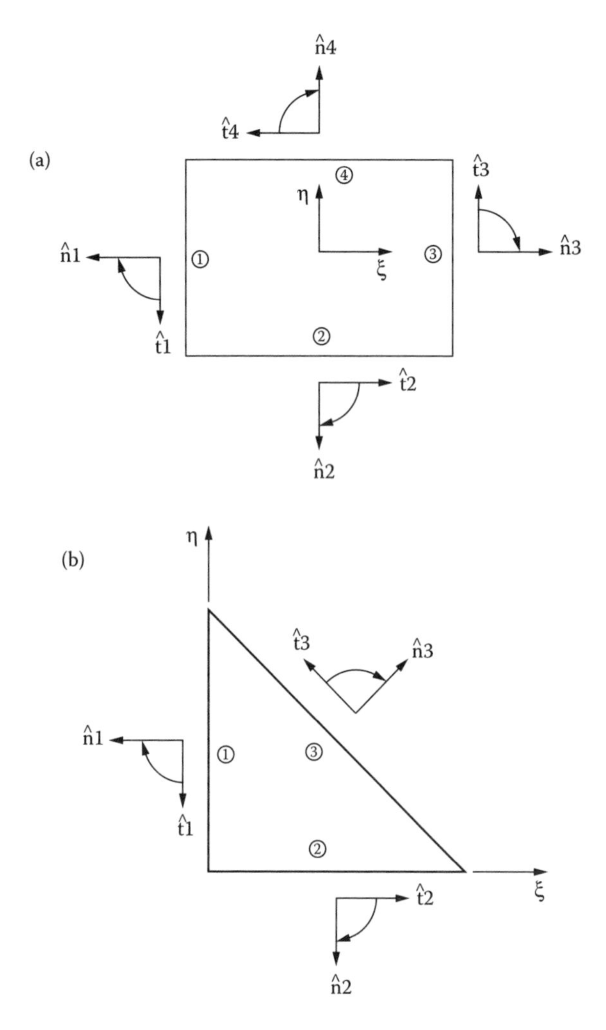

Figure 6.5 Tangential and normal vectors: (a) quadrilateral and (b) triangular

A scalar variable, ϕ, which can represent unknown temperature, concentration, or velocity, for example, is interpolated using Eq. (6.14), where the shape functions are those given by Eq. (5.55). The derivatives of ϕ, $\partial\phi/\partial\xi$, and $\partial\phi/\partial\eta$ are written using Eq. (6.4), which was used to express the derivatives for the shape functions, i.e.,

$$\begin{bmatrix} \dfrac{\partial \phi}{\partial \xi} \\ \dfrac{\partial \phi}{\partial \eta} \end{bmatrix} = \begin{bmatrix} \dfrac{\partial x}{\partial \xi} & \dfrac{\partial y}{\partial \xi} \\ \dfrac{\partial x}{\partial \eta} & \dfrac{\partial y}{\partial \eta} \end{bmatrix} \begin{Bmatrix} \dfrac{\partial \phi}{\partial x} \\ \dfrac{\partial \phi}{\partial y} \end{Bmatrix} \tag{6.31}$$

Hence, since N_i are the shape functions used to approximate ϕ, using Eqs. (6.11) we obtain

$$\begin{bmatrix} \dfrac{\partial\phi}{\partial\xi} \\ \dfrac{\partial\phi}{\partial\eta} \end{bmatrix} = \begin{bmatrix} -(1-\eta) & (1-\eta) & \eta & -\eta \\ -(1-\xi) & -\xi & \xi & (1-\xi) \end{bmatrix} \begin{bmatrix} x_1 & y_1 \\ x_2 & y_2 \\ x_3 & y_3 \\ x_4 & y_4 \end{bmatrix} \begin{bmatrix} \dfrac{\partial\phi}{\partial x} \\ \dfrac{\partial\phi}{\partial y} \end{bmatrix} \tag{6.32}$$

The first two matrices on the right-hand side of Eq. (6.32) yield the familiar 2 × 2 Jacobian matrix:

$$\mathrm{J} = \begin{bmatrix} J_{11} & J_{12} \\ J_{21} & J_{22} \end{bmatrix}$$

from which $|\mathbf{J}| = J_{11}J_{22} - J_{21}J_{12}$ and

$$\mathrm{J}^{-1} = \frac{1}{|\mathrm{J}|} \begin{bmatrix} J_{22} & -J_{12} \\ -J_{21} & J_{11} \end{bmatrix}$$

We can now find the Cartesian derivatives for ϕ from Eq. (6.14).

Evaluating the Jacobian for a bilinear rectangular element of length a and height b,

$$\begin{aligned} \mathrm{J} &= \begin{bmatrix} \sum_{i=1}^{4} \dfrac{\partial N_i}{\partial\xi} x_i & \sum_{i=1}^{4} \dfrac{\partial N_i}{\partial\xi} y_i \\ \sum_{i=1}^{4} \dfrac{\partial N_i}{\partial\eta} x_i & \sum_{i=1}^{4} \dfrac{\partial N_i}{\partial\eta} y_i \end{bmatrix} \\ &= \begin{bmatrix} \dfrac{\partial N_1}{\partial\xi} & \dfrac{\partial N_2}{\partial\xi} & \dfrac{\partial N_3}{\partial\xi} & \dfrac{\partial N_4}{\partial\xi} \\ \dfrac{\partial N_1}{\partial\eta} & \dfrac{\partial N_2}{\partial\eta} & \dfrac{\partial N_3}{\partial\eta} & \dfrac{\partial N_4}{\partial\eta} \end{bmatrix} \begin{bmatrix} x_1 & y_1 \\ x_2 & y_2 \\ x_3 & y_3 \\ x_4 & y_4 \end{bmatrix} \\ &= \frac{1}{4} \begin{bmatrix} -(1-\eta) & (1-\eta) & \eta & -\eta \\ -(1-\xi) & -\xi & \xi & (1-\xi) \end{bmatrix} \begin{bmatrix} 0 & 0 \\ a & 0 \\ a & b \\ 0 & b \end{bmatrix} \\ &= \begin{bmatrix} -(1-\eta)\cdot 0 + (1-\eta)\cdot a + \eta\cdot a - \eta\cdot 0 & -(1-\eta)\cdot 0 + (1-\eta)\cdot 0 + \eta\cdot b - \eta\cdot b \\ -(1-\xi)\cdot 0 - \xi\cdot a + \xi\cdot a + (1-\xi)\cdot 0 & -(1-\xi)\cdot 0 - \xi\cdot 0 + \xi\cdot b + (1-\xi)\cdot b \end{bmatrix} \\ &= \begin{bmatrix} a & 0 \\ 0 & b \end{bmatrix} \end{aligned} \tag{6.33}$$

Thus,

$$|J| = ab - 0 \cdot 0 = ab \tag{6.34}$$

and

$$J^{-1} = \frac{1}{ab}\begin{bmatrix} b & 0 \\ 0 & a \end{bmatrix} = \begin{bmatrix} \frac{1}{a} & 0 \\ 0 & \frac{1}{b} \end{bmatrix} \tag{6.35}$$

The derivative values become

$$\begin{bmatrix} \frac{\partial\phi}{\partial x} \\ \frac{\partial\phi}{\partial y} \end{bmatrix} = J^{-1}\begin{bmatrix} \frac{\partial\phi}{\partial\xi} \\ \frac{\partial\phi}{\partial\eta} \end{bmatrix} = \begin{bmatrix} \frac{1}{a} & 0 \\ 0 & \frac{1}{b} \end{bmatrix}\begin{bmatrix} \frac{\partial\phi}{\partial\xi} \\ \frac{\partial\phi}{\partial\eta} \end{bmatrix}$$

or

$$\left.\begin{aligned} \frac{\partial\phi}{\partial x} &= \frac{1}{a}\frac{\partial\phi}{\partial\xi} \\ \frac{\partial\phi}{\partial y} &= \frac{1}{b}\frac{\partial\phi}{\partial\eta} \end{aligned}\right\} \tag{6.36}$$

which is identical to the derivative values obtained for the shape functions as given in Eq. (5.75).

The general relations discussed in Section 6.3.5 can be used to express the differential $d\Gamma$ in terms of x, y, ξ, and η to account for boundary conditions acting on an element face. The relations are

$$\eta = \text{constant:} \qquad d\Gamma = \left[\left(\frac{dx}{d\xi}\right)^2 + \left(\frac{dy}{d\xi}\right)^2\right]^{1/2} d\xi \tag{6.37}$$

$$\xi = \text{constant:} \qquad d\Gamma = \left[\left(\frac{dx}{d\eta}\right)^2 + \left(\frac{dy}{d\eta}\right)^2\right]^{1/2} d\eta \tag{6.38}$$

The stiffness (or conductance) matrix was originally written in the form

$$K \equiv \left[K_{ij}\right] = \int_{\Omega^{(e)}}\left(K_{xx}\frac{\partial N_i}{\partial x}\frac{\partial N_j}{\partial x} + K_{yy}\frac{\partial N_i}{\partial x}\frac{\partial N_j}{\partial x}\right)dxdy \tag{6.39}$$

where the integration is over the quadrilateral element and the shape functions N_i and N_j are functions of x and y. When we transform to isoparametric elements with ξ and η centered at the midpoint, Eq. (6.39) becomes

$$K_{ij} = \int_{-1}^{1}\int_{-1}^{1}\left[K_{xx}\left(\frac{\partial N_i}{\partial \xi}\frac{\partial \xi}{\partial x}+\frac{\partial N_i}{\partial \eta}\frac{\partial \eta}{\partial x}\right)\left(\frac{\partial N_j}{\partial \xi}\frac{\partial \xi}{\partial x}+\frac{\partial N_j}{\partial \eta}\frac{\partial \eta}{\partial x}\right)\right. \\ \left.+K_{yy}\left(\frac{\partial N_i}{\partial \xi}\frac{\partial \xi}{\partial y}+\frac{\partial N_i}{\partial \eta}\frac{\partial \eta}{\partial y}\right)\left(\frac{\partial N_j}{\partial \xi}\frac{\partial \xi}{\partial y}+\frac{\partial N_j}{\partial \eta}\frac{\partial \eta}{\partial y}\right)\right]|\mathrm{J}|\, d\xi d\eta \tag{6.40}$$

where $\partial\xi/\partial x$, $\partial\xi/\partial y$, $\partial\eta/\partial x$, $\partial\eta/\partial y$ are given by Eq. (6.9). The mass matrix remains the same as before:

$$\mathrm{M} = \int_{\Omega^{(e)}} N_i N_j\, dxdy = \int_{-1}^{1}\int_{-1}^{1} N_i N_j\, |\mathrm{J}|\, d\xi d\eta \tag{6.41}$$

Likewise, the source term is unchanged:

$$\int_{\Omega^{(e)}} QN_i dxdy = \int_{-1}^{1}\int_{-1}^{1} QN_i\, |\mathrm{J}|\, d\xi d\eta \tag{6.42}$$

The flux boundary on an element face is given by the relation

$$\int_{\Gamma} qN_i d\Gamma$$

Using Eqs. (6.37) and (6.38),

$$\left.\begin{aligned} \eta = \text{constant: } \int_{\Gamma} qN_i d\Gamma &= \int_{-1}^{1} qN_i\left[\left(\frac{dx}{d\xi}\right)^2+\left(\frac{dy}{d\xi}\right)^2\right]^{1/2} d\xi \\ \xi = \text{constant: } \int_{\Gamma} qN_i d\Gamma &= \int_{-1}^{1} qN_i\left[\left(\frac{dx}{d\eta}\right)^2+\left(\frac{dy}{d\eta}\right)^2\right]^{1/2} d\eta \end{aligned}\right\} \tag{6.43}$$

The coordinate transformation is easily programmed. The ξ and η integrations are evaluated at four Gauss points for a bilinear element and at nine Gauss points for a quadratic or biquadratic element. To implement the procedure, we first obtain the derivative values for $(\partial N/\partial\xi)$ and $(\partial N/\partial\eta)$, then solve for the Jacobian and its inverse, followed by evaluation of the Cartesian derivatives $(\partial N/\partial x)$ and $(\partial N/\partial y)$.* Finally, Gaussian quadrature is used to sum all the contributions from the nodes into the appropriate matrices.

* Subroutine SHAPE in FEM-1D and FEM-2D illustrates the procedure.

6.5 INVISCID FLOW EXAMPLE

Assume that we wish to solve for inviscid, incompressible flow around the two-dimensional circular shown in Figure 6.6. While the equations describing this type of flow are written in Cartesian coordinates, the body geometry is described in polar coordinates. Because of symmetry, only a quarter plane is considered.

The equations in terms of the velocity components u,v are written as (after Fletcher, 1984)

$$\frac{\partial u}{\partial x} + \frac{\partial v}{\partial y} = 0 \tag{6.44}$$

$$\frac{\partial u}{\partial y} - \frac{\partial v}{\partial x} = 0 \tag{6.45}$$

with

$$\left.\begin{array}{ll} \dfrac{\partial u}{\partial y} = v = 0 & \text{on AB} \\ v_{normal} = 0 & \text{on BC} \\ \dfrac{\partial u}{\partial x} = v = 0 & \text{on DC} \\ u = U_{\infty}, v = 0 & \text{on AD} \end{array}\right\} \tag{6.46}$$

We introduce the trial functions for u and v as

$$u = \sum_{i=1}^{n} N_i u_i \tag{6.47}$$

$$v = \sum_{i=1}^{n} N_i v_i \tag{6.48}$$

Applying the Galerkin procedure to Eqs. (6.44) and (6.45) produces the expressions

$$\int_{\Omega} \left(\frac{\partial u}{\partial x} + \frac{\partial v}{\partial y} \right) N_i dxdy = 0 \tag{6.49}$$

$$\int_{\Omega} \left(\frac{\partial u}{\partial y} - \frac{\partial v}{\partial x} \right) N_i dxdy = 0 \tag{6.50}$$

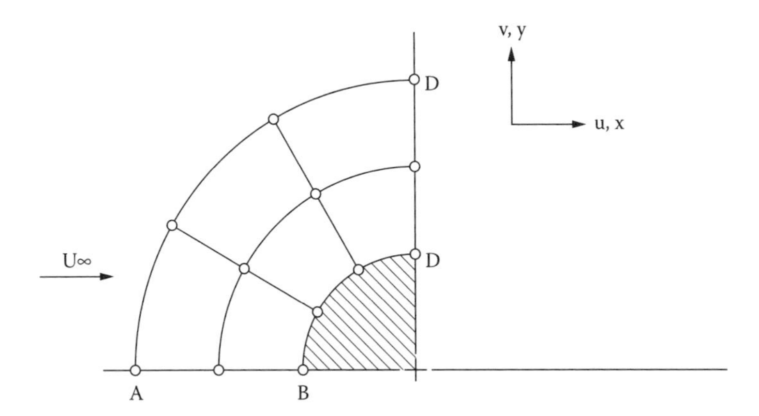

Figure 6.6 Element mesh for flow over a circular cylinder

The integrals in Eqs. (6.38) and (6.39) are rewritten so that the derivatives are moved to the weighting function using the general formula:

$$\frac{\partial f}{\partial \alpha} g = \frac{\partial}{\partial \alpha}(fg) - f \frac{\partial g}{\partial \alpha}$$

and Green's theorem is applied to the derivative of the product. Equations (6.49) and (6.50) then yield a system of equations of the form:

$$a_{ij} u_j + b_{ij} v_j = 0 \tag{6.51}$$

$$b_{ij} u_j - a_{ij} v_j = 0 \tag{6.52}$$

where

$$a_{ij} \int_{\Omega} \frac{\partial N_i}{\partial x} N_j \, dxdy - \int_{\Gamma} N_i N_j n_x d\Gamma \tag{6.53}$$

$$b_{ij} \int_{\Omega} \frac{\partial N_i}{\partial y} N_j \, dxdy - \int_{\Gamma} N_i N_j n_y d\Gamma \tag{6.54}$$

where n_x and n_y are the direction cosines.

The integrals in Eqs. (6.53) and (6.54) can be conveniently calculated using the isoparametric transformation. The integrations are performed in ξ,η space, rather than in physical space, using irregularly shaped elements. A contribution from an interior element to a_{ij} and b_{ij} can be written as

$$a_{ij} = \sum_{k=1}^{n} C_{ijk} y_k \tag{6.55}$$

and

$$b_{ij} = -\sum_{k=1}^{n} C_{ijk} x_k \tag{6.56}$$

where n is the number of local nodes within an element, x_k and y_k are the x,y coordinates of the nodes in physical space, and

$$C_{ijk} = \int_{-1}^{1}\int_{-1}^{1} N_j \left[\frac{\partial N_i}{\partial \eta}\frac{\partial N_k}{\partial \xi} - \frac{\partial N_i}{\partial \xi}\frac{\partial N_k}{\partial \eta} \right] d\xi d\eta \tag{6.57}$$

The location of the element in physical space enters through Eqs. (6.55) and (6.56). The determinant of the Jacobian associated with the transformation of the derivative in ξ,η space cancels with the derivative associated with transforming the integral, i.e.,

$$\int_{\Omega} \frac{\partial N_i}{\partial x} N_j dxdy = \int_{-1}^{1}\int_{-1}^{1} \frac{1}{|\mathrm{J}|}\left[\frac{\partial N_i}{\partial \eta}\frac{\partial y}{\partial \xi} - \frac{\partial N_i}{\partial \xi}\frac{\partial y}{\partial \eta} \right] N_j |\mathrm{J}| d\xi d\eta$$
$$= \int_{-1}^{1}\int_{-1}^{1} N_j \left[\frac{\partial N_i}{\partial \eta}\frac{\partial N_k}{\partial \xi} - \frac{\partial N_i}{\partial \xi}\frac{\partial N_k}{\partial \eta} \right] d\xi d\eta \tag{6.58}$$

The use of isoparametric elements permits the grid to be refined in regions where rapid variation is expected. In this example, it is best to cluster elements close to the cylinder surface. From Eq. (6.35), only one relation is required on boundaries

Table 6.1 Comparison of Element Type for Inviscid Incompressible Flow with a Coarse Grid

Element Type	Shape Function	No. of Unknowns	No. Elements	CPU[a] (sec)	σ^b
Rectangular	Linear	199	100	5.1	0.043
Rectangular	Quadratic	199	25	4.5	0.049
Triangular	Linear	199	200	5.1	0.061
Triangular	Quadratic	199	50	7.1	0.120

[a] IBM 370-168.

[b] $\sigma = \left[\sum_{i=1}^{e} \frac{(q_t - q_e)^2}{N} \right]^{1/2}$ where q_t = tangential velocity at surface, q_e = exact solution, and N = number of nodes

Source: Fletcher, C. A. J. (1984). *Computational Galerkin Methods*. Berlin: Springer-Verlag. With permission.

AB, BC, and CD; boundary AD requires no equations. Because the surface is assumed to permit no normal flow (it is solid), the normal velocity to BC can be expressed as

$$v_{normal} = u\cos\,\theta - v\sin\,\theta = 0 \tag{6.59}$$

A comparison of results using various elements and shape functions is shown in Table 6.1 (Fletcher, 1984).

6.6 CLOSURE

In this chapter, we have examined the use of the isoparametric transformation between physical space (x,y) and computational space (ξ,η). Although the element is distorted in physical space, isoparametric transformation produces an element with a uniform local coordinate system that can treat boundaries that do not coincide with coordinate lines.

In an isoparametric element, the same interpolating functions are used as trial functions; i.e., the polynomial expansions for x and y are identical to those for the unknown variable. Just as Eq. (6.14) leads to continuity of ϕ between elements, Eq. (6.11) produces a continuous mapping among element boundaries. This is particularly advantageous when dealing with curved surfaces.

EXERCISES

6.1 Derive Eq. (6.32) from Eq. (6.31).

6.2 Find the derivatives $\partial N_i/\partial x$ and $\partial N_i/\partial y$ for the four-node element defined by (0, 0), (0, 1), (1.25, 1.25), and (1, 0).

6.3 For the element in the figure below, find the derivatives $\partial N_i/\partial x$ and $\partial N_i/\partial y$ for i = 1, 2.

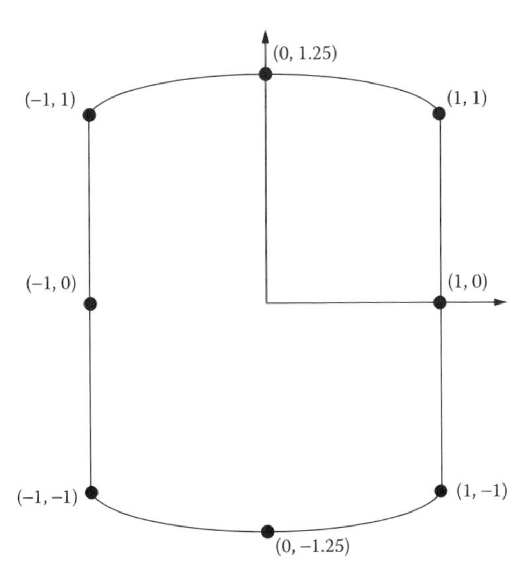

6.4 For the element in Exercise 6.3, add a node at (0, 0) and calculate the derivatives. Compare them with the previous ones.

6.5 Show that the isoparametric transformation (6.11) using bilinear shape functions transforms the sides of the parent element into the corresponding sides in an arbitrary quadrilateral.

6.6 Similarly, show that in a quadratic element the quadratic shape functions transform to arbitrary parabolic sides.

6.7 For the elements in Figure 6.2, evaluate **K** for one element using linear interpolation for temperature and for the elemental area. Discuss.

6.8 A temperature field is defined over a quadrilateral four-noded element by

Node i	T_i (C)	X_i (cm)	Y_i (cm)
1	10	2	0
2	45	4	1
3	50	3	4
4	–5	0	3

Sketch the element and calculate the temperature and temperature gradients at point (2,2).

6.9 A nine-noded element is used to model a circular segment and a temperature distribution given in polar coordinates:

Node i	T (C)	r (cm)	(degrees)
1	100	3	30
2	80	4	30
3	53	5	30
4	60	5	45
5	50	5	60
6	77	4	60
7	100	3	60
8	100	3	45
9	85	4	45

6.10 Calculate the stiffness matrix (6.39) for the element in Exercise 6.2.

6.11 Use Eqs. (6.47), (6.48), (6.53), (6.54), and (6.55) to obtain (6.57).

6.12 Integrate Eqs. (6.49) and (6.50) to obtain (6.51) and (6.52).

6.13 Use the boundary conditions to show that the line integrals in Eqs. (6.53) and (6.54) vanish except for (6.53) over *CD*.

6.14 Calculate the potential flow within the nozzle region show below. The equation for potential flow is given by the Laplacian

$$\frac{\partial^2 \phi}{\partial x^2} + \frac{\partial^2 \phi}{\partial y^2} = 0$$

Use both linear and quadratic elements and compare results at $x = 0$, $y = 0$.

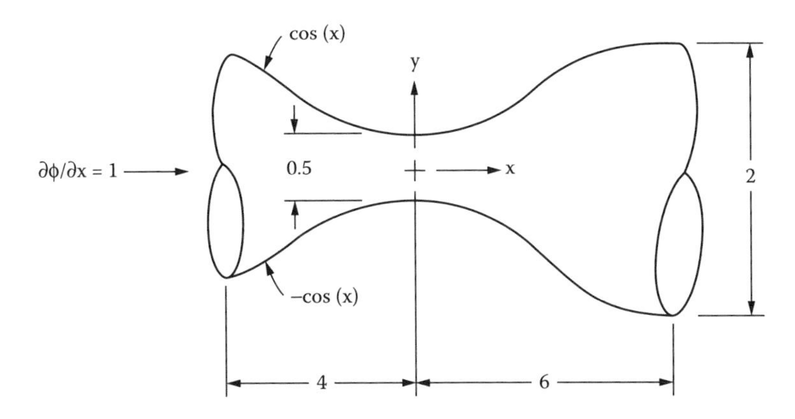

6.15 Solve for steady-state heat conduction within a cylinder if the top and bottom radial quadrants are insulated with the left quadrant at $T = 100°C$ and the right quadrant at $T = 0°C$. Let $K_{xx} = K_{yy} = 15$ W/mC. Discretize the physical domain first with linear rectangular elements; then redo the mesh using isoparametric quadrilateral elements and compare results.

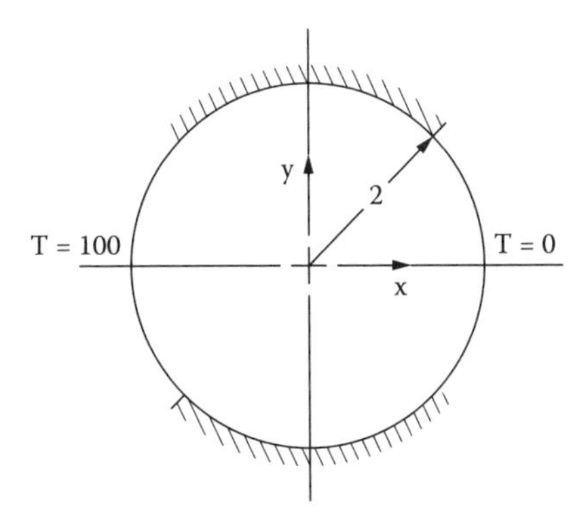

6.16 Refer to Section 6.5 to calculate inviscid flow around a circular cylinder if the radius $OC = 3$, $OD = 9$, and $U_\infty = 1$. Utilize quadratic elements.

REFERENCES

Fletcher, C. A. J. (1984). *Computational Galerkin Methods*. Berlin: Springer-Verlag.

Heinrich, J. C. and Pepper, D. W. (1999). *Intermediate Finite Element Method. Fluid Flow and Heat Transfer Applications*. New York: Taylor & Francis.

Zienkiewicz, O. C. (1977). *The Finite Element Method*, 3rd ed. London: McGraw-Hill.

CHAPTER

7

THE THREE-DIMENSIONAL ELEMENT

7.1 BACKGROUND

By now the reader should have a good understanding of the finite element method for solving one- and two-dimensional problems. The principles formulated in Chapters 2 and 3 have been seen to apply equally well to two-dimensional elements as to one-dimensional elements. In Chapters 4 through 6, we saw the finite element method utilizing either two-dimensional triangular or quadrilateral shaped elements with varying degrees of accuracy. Extension of the method to three dimensions follows logically from two dimensions. The amount of data required to establish the computational domain and boundary conditions becomes significantly greater over two-dimensional problems; likewise, the amount of computational work increases dramatically. Because of the additional number of nodes associated with the three-dimensional element (and the associated bandwidth), the computational mesh is generally limited.

In this chapter, the tetrahedron and the brick-shaped hexahedron element are developed and applied to the familiar heat conduction equation. A time-dependent heat conduction problem is solved with boundaries subjected to radiation heat transfer and convection.

7.2 ELEMENT MESH

Discretization of the problem domain into a computational domain consisting of elements and nodal points does not have a firm or exact theoretical basis. Experience and engineering judgment must be applied to produce a good mesh; poor or improper judgment will produce inaccurate results even if the remaining steps of the finite element procedure are rigorously followed.

In a one-dimensional domain, discretization into an array of elements is not difficult to grasp—small element size is generally needed where variables change rapidly. In two-dimensional regions, generation of an acceptable mesh becomes more troublesome; several meshes may need to be made before a suitable mesh is created that yields reasonable results. In a three-dimensional domain, the mesh is not only more cumbersome to make, but also difficult to visualize—many trial meshes may be required if the problem domain is complex. It should be recognized that the generation of a good mesh is an art. Recent advances in computational geometry from the field of computer science enable one to produce an optimal mesh based

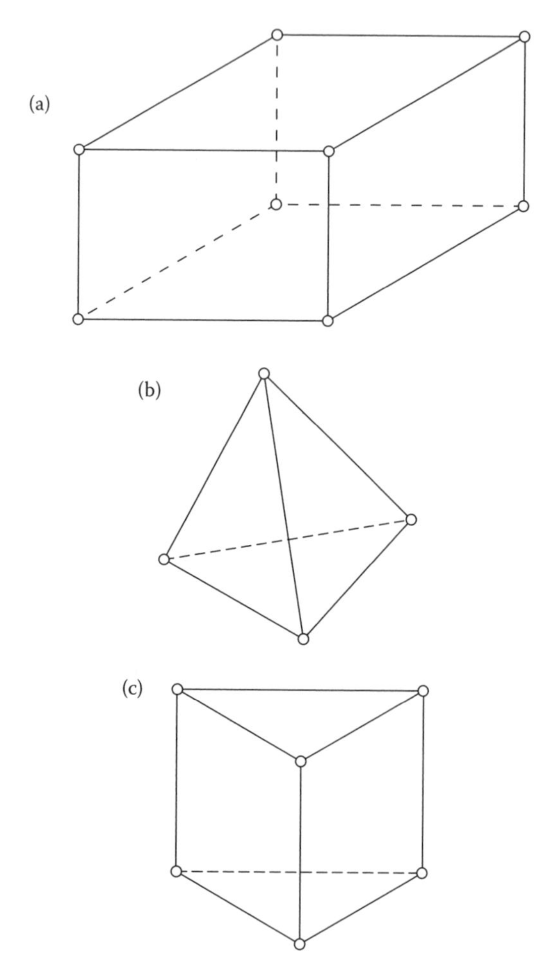

Figure 7.1 Some common three-dimensional elements; (a) hexahedron; (b) tetrahedron; and (c) cylindrical (pentahedron)

on the required number of nodes and elements (see Pepper and Gewali, 2002). A considerable amount of time can be spent in generating a mesh; however, once the mesh is made, the finite element method permits a great deal of flexibility in changing boundary conditions without any additional effort.

The most common types of three-dimensional elements are variations of the two-dimensional elements, i.e., the tetrahedron and the hexahedron, or brick, as shown in Figure 7.1. Linear elements are restricted to straight sides (planes); quadratic and

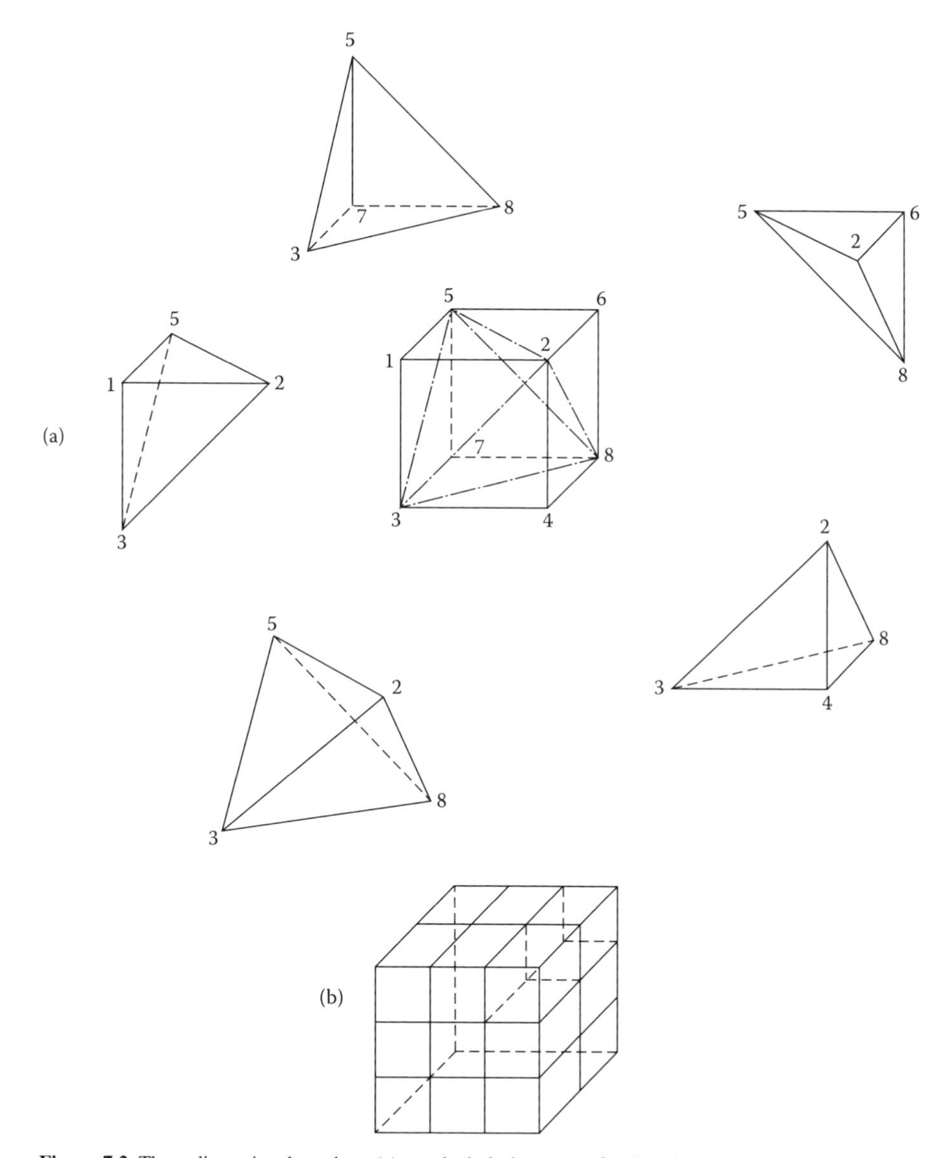

Figure 7.2 Three-dimensional meshes: (a) tetrahedral elements (after Desai and Abel, 1972), and (b) hexahedral elements

higher-order elements can have curved surfaces. A simple three-dimensional mesh made of tetrahedral elements is shown in Figure 7.2a; a mesh consisting of rectangular (brick) elements is shown in Figure 7.2b.

7.3 SHAPE FUNCTIONS

7.3.1 Tetrahedron

Four nodes are used to define the linear, three-dimensional tetrahedron. The tetrahedron is shown in Figure 7.3. The interpolation function is written as

$$\phi = \alpha_1 + \alpha_2 x + \alpha_3 y + \alpha_4 z \tag{7.1}$$

For the four nodes located at the vertices of the tetrahedron, a set of four equations is produced:

$$\left.\begin{aligned} \phi_1 &= \alpha_1 + \alpha_2 x_1 + \alpha_3 y_1 + \alpha_4 z_1 \\ \phi_2 &= \alpha_1 + \alpha_2 x_2 + \alpha_3 y_2 + \alpha_4 z_2 \\ \phi_3 &= \alpha_1 + \alpha_2 x_3 + \alpha_3 y_3 + \alpha_4 z_3 \\ \phi_4 &= \alpha_1 + \alpha_2 x_4 + \alpha_3 y_4 + \alpha_4 z_4 \end{aligned}\right\} \tag{7.2}$$

which can be written in matrix form as

$$\phi = \mathrm{C}\alpha \tag{7.3}$$

or

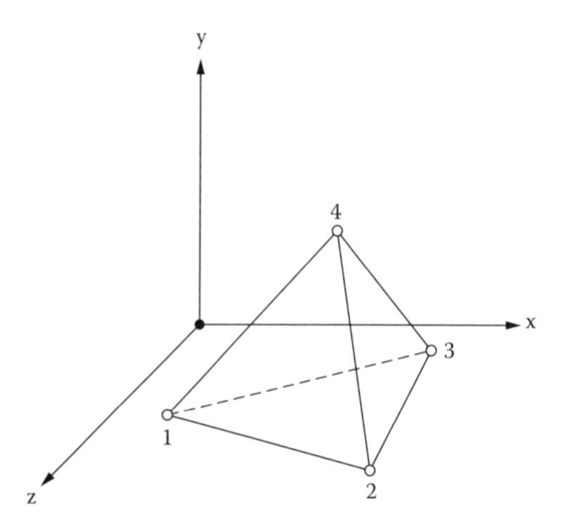

Figure 7.3 The tetrahedron element

$$\begin{bmatrix} \phi_1 \\ \phi_2 \\ \phi_3 \\ \phi_4 \end{bmatrix} = \begin{bmatrix} 1 & x_1 & y_1 & z_1 \\ 1 & x_2 & y_2 & z_2 \\ 1 & x_3 & y_3 & z_3 \\ 1 & x_4 & y_4 & z_4 \end{bmatrix} \begin{bmatrix} \alpha_1 \\ \alpha_2 \\ \alpha_3 \\ \alpha_4 \end{bmatrix} \tag{7.4}$$

Taking the inverse of **C**, the α column vector is obtained as

$$\alpha = C^{-1}\phi \tag{7.5}$$

$$\phi = \begin{bmatrix} 1 & x & y & z \end{bmatrix} \begin{bmatrix} \alpha_1 \\ \alpha_2 \\ \alpha_3 \\ \alpha_4 \end{bmatrix} = \begin{bmatrix} 1 & x & y & z \end{bmatrix} C^{-1}\phi \tag{7.6}$$

The quantity $[1\ x\ y\ z]\mathbf{C}^{-1}$ is actually the shape functions matrix, i.e.,

$$\phi = N\phi \tag{7.7}$$

where

$$N = \begin{bmatrix} 1 & x & y & z \end{bmatrix} C^{-1} \tag{7.8}$$

If we wish to use a quadratic tetrahedron, as shown in Figure 7.4, ten nodes are require to define the element. The element nodal locations directly relate to Pascal's triangle for two-dimensional space, and in three dimensions they exhibit similar properties in pyramid form (see Zienkiewicz, 1977).

The expression for ϕ becomes

$$\begin{aligned} \phi = \alpha_1 + \alpha_2 x + \alpha_3 y + \alpha_4 z + \alpha_5 x^2 + \alpha_6 xy + \alpha_7 y^2 \\ + \alpha_8 yz + \alpha_9 z^2 + \alpha_{10} xz \end{aligned} \tag{7.9}$$

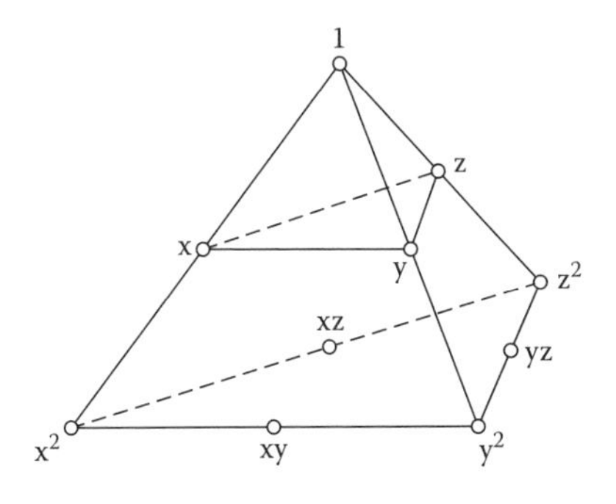

Figure 7.4 Representation of quadratic tetrahedron based on Pascal's triangle

or

$$\phi = \begin{bmatrix} 1 & x & y & z & x^2 & xy & y^2 & yz & z^2 & xz \end{bmatrix} C^{-1} \phi \tag{7.10}$$

where

$$C = \begin{bmatrix} 1 & x_1 & y_1 & z_1 & x_1^2 & x_1y_1 & y_1^2 & y_1z_1 & z_1^2 & x_1z_1 \\ \cdot & \cdot & \cdot & \cdot & \cdot & \cdot & \cdot & \cdot & \cdot & \cdot \\ \cdot & \cdot & \cdot & \cdot & \cdot & \cdot & \cdot & \cdot & \cdot & \cdot \\ \cdot & \cdot & \cdot & \cdot & \cdot & \cdot & \cdot & \cdot & \cdot & \cdot \\ 1 & x_{10} & y_{10} & z_{10} & x_{10}^2 & x_{10}y_{10} & y_{10}^2 & y_{10}z_{10} & z_{10}^2 & x_{10}z_{10} \end{bmatrix} \tag{7.11}$$

A volume coordinate system for the tetrahedron can be established in the same manner as the area coordinate system for the two-dimensional triangle. In the three-dimensional case, four distance ratios are utilized, each normal to a side: L_1, L_2, L_3, and L_4. The volume V_1 is shown in Figure 7.5.

The four coordinates are related by

$$L_1 + L_2 + L_3 + L_4 = 1 \tag{7.12}$$

from which the linear shape functions are defined:

$$\left.\begin{aligned} N_1 &= L_1 \\ N_2 &= L_2 \\ N_3 &= L_3 \\ N_4 &= L_4 \end{aligned}\right\} \tag{7.13}$$

Employing area coordinates, the volume integrals can be easily evaluated from the relation:

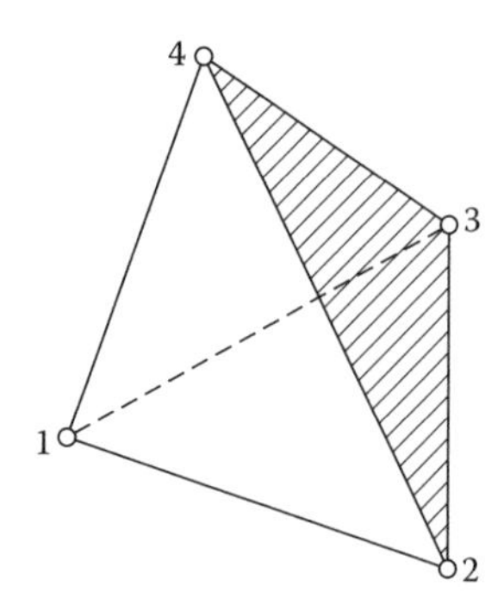

Figure 7.5 Volume coordinate system [$L_i = V_i/V$, $i = 1,\ldots, 4$]

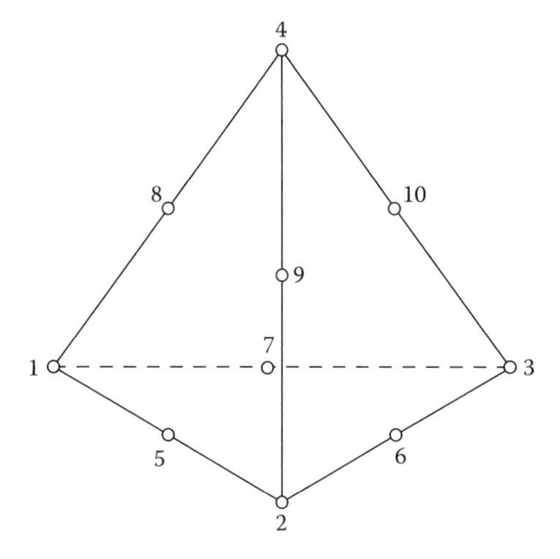

Figure 7.6 Node location for quadratic tetrahedron

$$\int_V L_1^a L_2^b L_3^c L_4^d dV = \frac{a!b!c!d!}{(3+a+b+c+d)!} 6V \tag{7.14}$$

which is an extension of Eq. (4.44b), which was used for two-dimensional triangular elements.

The shape functions for the higher-order elements follow similarly as for the linear tetrahedron; in this case, planes are passed through nodes instead of straight lines. For example, the shape function at node 1 for the quadratic tetrahedron, shown in Figure 7.6, is given as

$$N_1 = (2L_1 - 1)L_1 \tag{7.15}$$

where $L_1 = 0$ passes through nodes 2, 3, 4, 6, 9, and 10. For $L_1 = ½$, the plane passes through nodes 5, 7, and 8. At node 5,

$$N_5 = 4L_1L_2 \tag{7.16}$$

where $L_2 = 0$ passes through nodes 1, 3, 4, 7, 8, and 10. The additional shape function relations can be readily evaluated by referring to Eq. (4.26), which gives the shape functions for the two-dimensional quadratic triangular element.

Example 7.1

Determine the shape functions for the linear tetrahedron element shown in Figure 7.7 From Eq. (7.4), we first evaluate **C**:

$$C = \begin{bmatrix} 1 & x_1 & y_1 & z_1 \\ 1 & x_2 & y_2 & z_2 \\ 1 & x_3 & y_3 & z_3 \\ 1 & x_4 & y_4 & z_4 \end{bmatrix} = \begin{bmatrix} 1 & 3 & 2 & 0 \\ 1 & 0 & 4 & 0 \\ 1 & 0 & 0 & 0 \\ 1 & 1 & 2 & 5 \end{bmatrix} \tag{7.17}$$

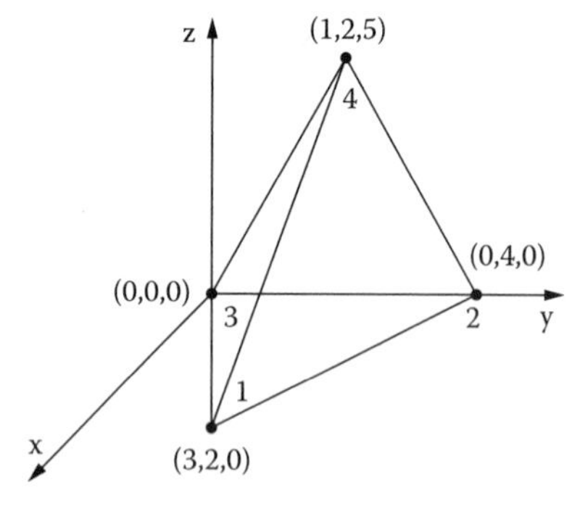

Figure 7.7 Linear tetrahedron for Example 7.1

We next need to find the inverse $\mathbf{C}^{-1}$ to obtain the shape functions as defined by Eq. (7.8). From matrix algebra (see Appendix A), the inverse is obtained as

$$C = \frac{adj\ C}{|C|} = \frac{\begin{bmatrix} 0 & 0 & 60 & 0 \\ 20 & -10 & -10 & 0 \\ 0 & 15 & -15 & 0 \\ -4 & -4 & -4 & 12 \end{bmatrix}}{60} \tag{7.18}$$

The shape functions are given by the product

$$N = \begin{bmatrix} 1 & x & y & z \end{bmatrix} C^{-1} = \frac{\begin{bmatrix} 1 & x & y & z \end{bmatrix}}{60} \begin{bmatrix} 0 & 0 & 60 & 0 \\ 20 & -10 & -10 & 0 \\ 0 & 15 & -15 & 0 \\ -4 & -4 & -4 & 12 \end{bmatrix} \tag{7.19}$$

$$= \frac{1}{60}[(20x - 4z)\ (-10x + 15y - 4z)\ (60 - 10x - 15y - 4z)\ (12z)]$$

Thus, the individual shape functions are

$$\left. \begin{aligned} N_1 &= \frac{1}{15}(5x - z) \\ N_2 &= \frac{1}{60}(-10x + 15y - 4z) \\ N_3 &= \frac{1}{60}(60 - 10x - 15y - 4z) \\ N_4 &= \frac{z}{5} \end{aligned} \right\} \tag{7.20}$$

7.3.2 Hexahedron

The tetrahedron is a simple element, which directly extends from the triangular element. However, the tetrahedron is difficult to visualize in a complex three-dimensional mesh, and may require considerable effort to define the element connectivity. A more accurate solution with less effort, but more node points, can be achieved with the hexahedron element. The simplest element in this family is the eight-noded trilinear hexahedron, or brick, shown in Figure 7.8.

The interpolation function is

$$\phi = \alpha_1 + \alpha_2 x + \alpha_3 y + \alpha_4 z + \alpha_5 xy + \alpha_6 xz + \alpha_7 yz + \alpha_8 xyz \tag{7.21}$$

Equation (7.21) can be expressed in matrix form using Eq. (7.3), as

$$\begin{bmatrix} \phi_1 \\ \cdot \\ \cdot \\ \cdot \\ \phi_8 \end{bmatrix} = \begin{bmatrix} 1 & x_1 & y_1 & z_1 & x_1y_1 & x_1z_1 & y_1z_1 & x_1y_1z_1 \\ \cdot & \cdot & \cdot & \cdot & \cdot & \cdot & \cdot & \cdot \\ \cdot & \cdot & \cdot & \cdot & \cdot & \cdot & \cdot & \cdot \\ \cdot & \cdot & \cdot & \cdot & \cdot & \cdot & \cdot & \cdot \\ 1 & x_8 & y_8 & z_8 & x_8y_8 & x_8z_8 & y_8z_8 & x_8y_8z_8 \end{bmatrix} \begin{bmatrix} \alpha_1 \\ \cdot \\ \cdot \\ \cdot \\ \alpha_8 \end{bmatrix} \tag{7.22}$$

Since $\alpha = \mathbf{C}^{-1}\phi$, we can find the values for ϕ from the expression

$$\phi = \begin{bmatrix} 1 & \mathrm{x} & \mathrm{y} & \mathrm{z} & \mathrm{xy} & \mathrm{xz} & \mathrm{yz} & \mathrm{xyz} \end{bmatrix} \mathrm{C}^{-1}\phi \tag{7.23}$$

from which the shape functions are obtained as before:

$$\mathrm{N} = \begin{bmatrix} 1 & \mathrm{x} & \mathrm{y} & \mathrm{z} & \mathrm{xy} & \mathrm{xz} & \mathrm{yz} & \mathrm{xyz} \end{bmatrix} \mathrm{C}^{-1} \tag{7.24}$$

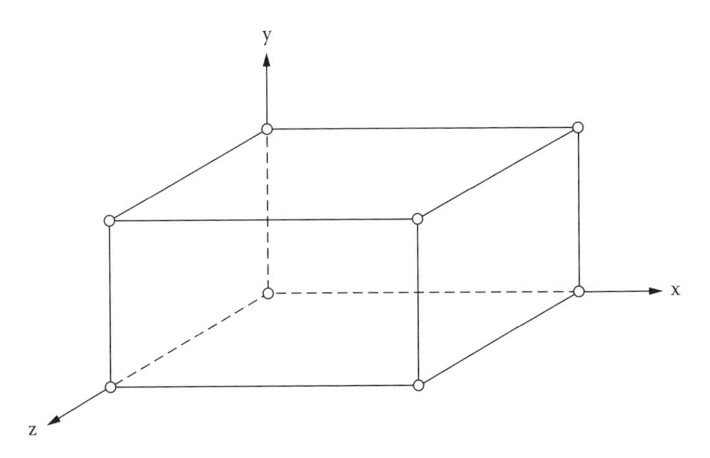

Figure 7.8 Eight-noded trilinear hexahedron element (brick)

As in the two-dimensional quadrilateral element, a natural coordinate system can also be devised for the hexahedron, as shown in Figure 7.9. The shape functions are defined (after some algebra!) as

$$\begin{bmatrix} N_1 \\ N_2 \\ N_3 \\ N_4 \\ N_5 \\ N_6 \\ N_7 \\ N_8 \end{bmatrix} = \frac{1}{8} \begin{bmatrix} (1-\xi)(1-\eta)(1-\zeta) \\ (1+\xi)(1-\eta)(1-\zeta) \\ (1+\xi)(1+\eta)(1-\zeta) \\ (1-\xi)(1+\eta)(1-\zeta) \\ (1-\xi)(1-\eta)(1+\zeta) \\ (1+\xi)(1-\eta)(1+\zeta) \\ (1+\xi)(1+\eta)(1+\zeta) \\ (1-\xi)(1+\eta)(1+\zeta) \end{bmatrix} \tag{7.25}$$

Equation (7.25) can be written in more concise form as

$$N_i = \frac{1}{8}(1+\xi\xi_i)(1+\eta\eta_i)(1+\zeta\zeta_i) \qquad i = 1, 2, \ldots, 8 \tag{7.26}$$

where ξ_i, η_i, and $\zeta_i = \pm 1$, depending on the nodal location.

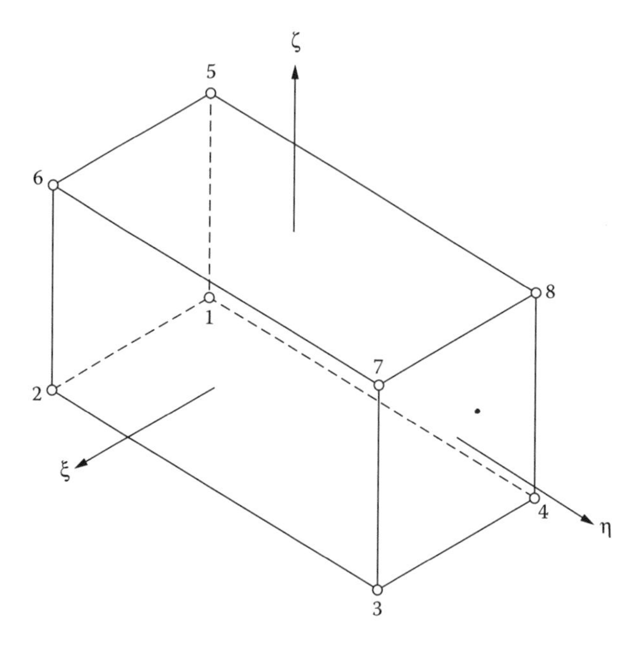

Figure 7.9 Natural coordinate system for the hexahedron element

Example 7.2

Determine the expression for N_1 in ξ, η, ζ coordinates. From Eq. (7.26),

$$N_i = \frac{1}{8}(1+\xi\xi_1)(1+\eta\eta_1)(1+\zeta\zeta_1) \tag{7.27}$$

Referring to Figure 7.9,

$$\xi_1 = -1, \quad \eta_1 = -1, \quad \zeta_1 = -1$$

Thus,

$$N^1 = \frac{1}{8}(1-\xi)(1-\eta)(1-\zeta) \tag{7.28}$$

Extension of the 8-noded quadratic element to the 20-noded quadratic hexahedron can be easily made by adding a mid-side node to each edge of the element. The three-dimensional quadratic hexahedron is shown in Figure 7.10.

The shape functions for the corner nodes are obtained from the relation

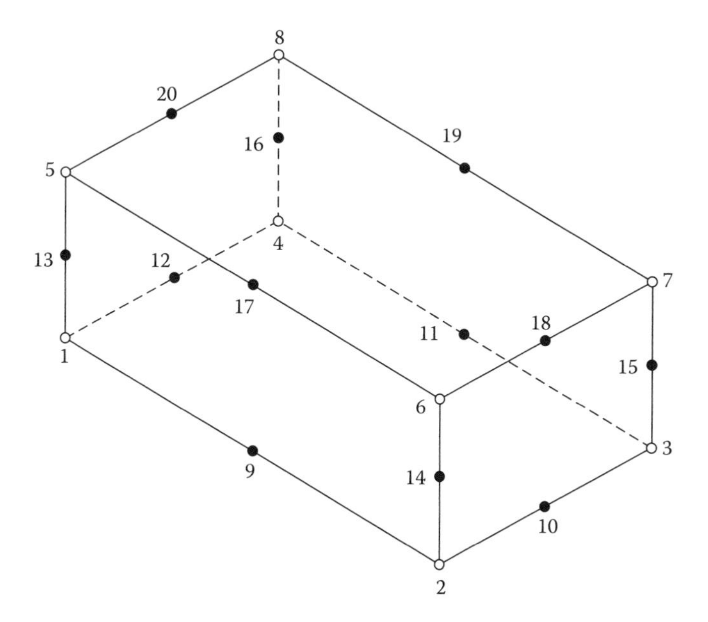

Figure 7.10 Quadratic hexahedron element

$$N_i = \frac{1}{8}(1+\xi\xi_i)(1+\eta\eta_i)(1+\zeta\zeta_i)(\xi\xi_i+\eta\eta_i+\zeta\zeta_i-2) \qquad i=1,2,...,8 \quad (7.29)$$

For the mid-side nodes,

$$N_i = \frac{1}{4}(1-\xi^2)(1+\eta\eta_i)(1+\zeta\zeta_i) \qquad \text{i =10, 12, 18, 20} \qquad (7.30)$$

$$N_i = \frac{1}{4}(1-\eta^2)(1+\xi\xi_i)(1+\zeta\zeta_i) \qquad \text{i = 9, 11, 17, 19} \qquad (7.31)$$

$$N_i = \frac{1}{4}(1-\xi^2)(1+\xi\xi_i)(1+\eta\eta_i) \qquad \text{i =13, 14, 15, 16} \qquad (7.32)$$

for a local coordinate system oriented as in Figure 7.9. The 9-noded biquadratic element extends to a 27-noded triquadratic element with a node in the center of each of the six faces and one at the center of the element.

Notice that when one of the natural coordinates is ±1, the three-dimensional elements reduce to their two-dimensional counterparts. Likewise, when two of the coordinates equal ±1, the shape function reduces to the one-dimensional shape function, permitting three-dimensional elements to be joined to the line elements. Continuity between elements exists over an elemental face (area) of a three-dimensional element; the nodal values describe the variation of the unknown variable (ϕ) identically on the element face common to adjacent elements.

7.4 NUMERICAL INTEGRATION

At this point, the difficulty in evaluating integrals with shape functions related to more than a few nodes should be obvious. Using volume coordinates, the tetrahedral elements can be evaluated exactly using Eq. (7.14); however, the amount of algebra associated with the inner product multiplications becomes quite laborious rather quickly.

The area integrals that occur in two-dimensional triangular elements are of the form (from Chapter 4):

$$I = \int_0^1 \int_0^{1-L_2} f(L_1, L_2, L_3) |\mathrm{J}| \, dL_1 dL_2$$

where the interpolation functions are expressed in terms of area coordinates. As the Jacobian is a function of the area coordinates, an explicit form of the inverse Jacobian is nearly impossible. Although some integrals can be evaluated using Eq. (4.44b), which is the two-dimensional form of Eq. (7.14), the chance for error becomes great as the number of terms increases dramatically. It is best to let the computer do the integration, i.e., Eq. (4.47),

$$\int_0^1 \int_0^{1-L_2} f(L_1, L_2, L_3) |\mathrm{J}| \, dL_1 dL_2 = \frac{1}{2} \sum_{i=1}^{n} w_i g\left[(L_1)_i, (L_2)_i, (L_3)_i \right]$$

where the order of the integration (n) is determined by summing the powers of the coordinates L_1, L_2, and L_3.

In a similar fashion, the tetrahedron elements are evaluated as

$$\begin{aligned} I &= \int_0^1 \int_0^{1-L_1} \int^{1-L_1-L_2} f(L_1, L_2, L_3, L_4) |\mathrm{J}| \, dL_1 dL_2 dL_3 \\ &= \frac{1}{6} \sum_{i=1}^{n} w_i g\left[(L_1)_i, (L_2)_i, (L_3)_i, (L_4)_i \right] \end{aligned} \tag{7.33}$$

The numerical integration sampling points for tetrahedrons are shown in Table 7.1. If we wish to integrate the product $L_1 L_2 L_4$, the sum of the exponentials is 3, and a quartic integration scheme (n = 4 points) would be required.

The procedure for evaluating the element matrices for hexahedrons is nearly identical to the procedure used for the two-dimensional quadrilateral. The Jacobian matrix now becomes 3 × 3, as opposed to 2 × 2 for the quadrilateral. The general form of the integrals is

$$\begin{aligned} &\int_{-1}^1 \int_{-1}^1 \int_{-1}^1 f(\xi, \eta, \zeta) |\mathrm{J}| \, d\xi d\eta d\zeta \\ &= \sum_{i=1}^{n} \sum_{j=1}^{n} \sum_{k=1}^{n} w_i w_j w_k f(\xi_i, \eta_j, \zeta_k) |\mathrm{J}(\xi_i, \eta_j, \zeta_k)| \end{aligned} \tag{7.34}$$

Table 7.1 Numerical Integration Formulae for Tetrahedrons

m	L_1	L_2	L_3	L_4	Weight	Order
1	1/4	1/4	1/4	1/4	1	2
1	0.5854102	0.1381966	0.1381966	0.1381966	1/4	
2	0.1381966	0.5854102	0.1381966	0.1381966	1/4	
3	0.1381966	0.1381966	0.5854102	0.1381966	1/4	
4	0.1381966	0.1381966	0.1381966	0.5854102	1/4	3
1	1/4	1/4	1/4	1/4	−16/20	
2	1/3	1/6	1/6	1/6	9/20	
3	1/6	1/3	1/6	1/6	9/20	
4	1/6	1/6	1/3	1/6	9/20	4
5	1/6	1/6	1/6	1/3	9/20	

where w_i, w_j, w_k, are the weight factors associated with the respective Gauss points. Notice that Eq. (7.34) is obtained by combining one-dimensional formulae originally established for the one-dimensional element (Chapter 3). The weighting factors and integration point coordinates are listed in Table 7.2.

Figure 7.11 shows the Gauss point locations for the cubic integration scheme. Note that the number of integration points required to evaluate the stiffness matrix, **K**, now becomes 8 for a trilinear element and 27 for a triquadratic element. If we wish to use a tricubic hexahedron (which requires 64 nodes), 64 integration points would be needed. Although the amount of work necessary to evaluate the matrices by hand is enormous, the computer performs such operations easily and quickly.

Table 7.2 Gaussian Quadrature

No. of Integration Points	$i(j)(k)$	$\xi_i(\eta_j)(\zeta_k)$	$w_i(w_j)(w_k)$
1 (linear)	1	0	2
2 (cubic)	1	$1/\sqrt{3}$	1
	2	$-1/\sqrt{3}$	1
3 (quintic)	1	0	8/9
	2	$\sqrt{15}/5$	5/9
	3	$-\sqrt{15}/5$	5/9
4 (septimal)	1	0.86113631	0.34785485
	2	–0.86113631	0.34785485
	3	0.33998104	0.65214515
	4	–0.33998104	0.65214515

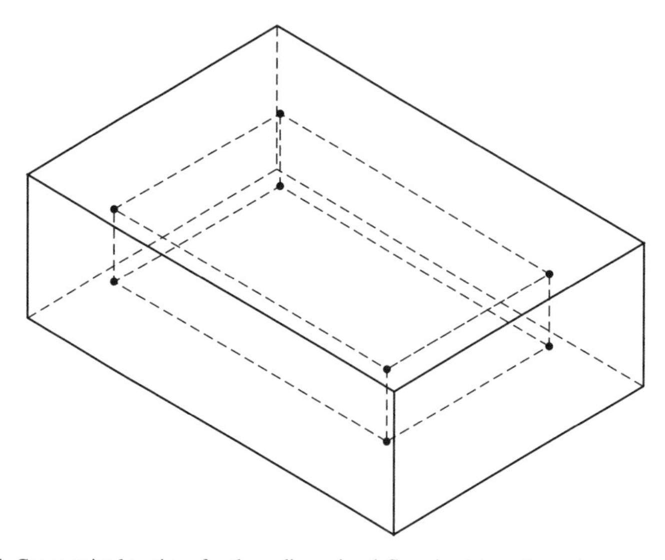

Figure 7.11 Gauss point locations for three-dimensional Gaussian integration rules

7.5 A ONE-ELEMENT HEAT CONDUCTION PROBLEM

The development of the three-dimensional equations for heat conduction directly follows from the one- and two-dimensional problems analyzed earlier. For a three-dimensional block subjected to internal heat generation and both fixed (Dirichlet) and flux (Neumann) boundary conditions, the time-dependent equation in Cartesian coordinates is

$$\rho c_p \frac{\partial T}{\partial t} = \frac{\partial}{\partial x}\left(K_{xx}\frac{\partial T}{\partial x}\right) + \frac{\partial}{\partial y}\left(K_{yy}\frac{\partial T}{\partial y}\right) + \frac{\partial}{\partial z}\left(K_{zz}\frac{\partial T}{\partial z}\right) + Q \tag{7.35}$$

valid over a volume V bounded by a surface S with

$$-\left(K_{xx}\frac{\partial T}{\partial x}n_x + K_{yy}\frac{\partial T}{\partial y}n_y + K_{zz}\frac{\partial T}{\partial z}n_z\right) = q \tag{7.36a}$$

over a portion S_B of the boundary surface S

$$T = T_S \tag{7.36b}$$

Over the rest of the boundary surface S_A, in a manner similar to the two-dimensional situation described in Eqs. (4.116) and (4.117). On that portion S_A of the boundary surface where T is specified, the gradient terms and q are zero; the quantities n_x, n_y, and n_z are the direction cosines of the unit vectors normal to the respective surfaces.

The Galerkin weak formulation of Eq. (7.35) is

$$\int_V N_i\left[\rho c_p \frac{\partial T}{\partial t} - \frac{\partial}{\partial x}\left(K_{xx}\frac{\partial T}{\partial x}\right) - \frac{\partial}{\partial y}\left(K_{yy}\frac{\partial T}{\partial y}\right) - \frac{\partial}{\partial z}\left(K_{zz}\frac{\partial T}{\partial z}\right) - Q\right]dxdydz = 0 \tag{7.37}$$

After applying Gauss's theorem, Eq. (7.37) becomes

$$\begin{aligned}&\int_V \rho c_p N_i \frac{\partial T}{\partial t} dxdydz \\ &+ \int_V\left(K_{xx}\frac{\partial N_i}{\partial x}\frac{\partial T}{\partial x} + K_{yy}\frac{\partial N_i}{\partial y}\frac{\partial T}{\partial y} + K_{zz}\frac{\partial N_i}{\partial z}\frac{\partial T}{\partial z}\right)dxdydz - \int_V N_i Q dxdydz \\ &- \int_{S_B} N_i\left(K_{xx}\frac{\partial N_i}{\partial x}n + K_{yy}\frac{\partial N_i}{\partial y}n_y + K_{zz}\frac{\partial T}{\partial z}n_z\right)dS = 0\end{aligned} \tag{7.38}$$

Equation (7.38) can be further modified to the generalized Galerkin statement:

$$\left[\int_V \rho c_p N_i N_j dxdydz\right]\dot{\mathrm{T}}$$

$$+\left[\int_V \left(K_{XX}\frac{\partial N_i}{\partial x}\frac{\partial N_j}{\partial x}+K_{yy}\frac{\partial N_i}{\partial y}\frac{\partial N_j}{\partial y}+K_{zz}\frac{\partial N_i}{\partial z}\frac{\partial N_j}{\partial z}\right)dxdydz\right]\mathrm{T} \tag{7.39}$$

$$-\int_V N_i Q dxdydz+\int_{S_B} N_i q dS=0$$

or, in matrix form as

$$\mathbf{M}\dot{\mathbf{T}}+\mathbf{K}\mathbf{T}=\mathbf{F} \tag{7.40}$$

which we saw repeatedly in Chapters 3 through 5. The matrix terms are the familiar mass matrix **M**, the stiffness matrix **K**, and the load vector **F** defined from Eq. (7.39),

$$\mathbf{M}=\left[\int_V \rho c_p N_i N_j dxdydz\right] \tag{7.41}$$

$$\mathbf{K}=\left[\int_V \left(K_{XX}\frac{\partial N_i}{\partial x}\frac{\partial N_j}{\partial x}+K_{yy}\frac{\partial N_i}{\partial y}\frac{\partial N_j}{\partial y}+K_{zz}\frac{\partial N_i}{\partial z}\frac{\partial N_j}{\partial z}\right)dxdydz\right] \tag{7.42}$$

$$\mathbf{F}=\left[\int_V N_i Q dxdydz-\int_S N_i q dS\right] \tag{7.43}$$

Because Eqs. (7.41) through (7.43) are general, we are free to choose the type of mesh and basis functions, i.e., either tetrahedron or hexahedron.

7.5.1 Tetrahedron

Assuming that the physical problem can be described geometrically by a single tetrahedron, the shape functions for the linear tetrahedron are given as [Eq. (7.13)]

$$\mathrm{N}=[N_1\ \ N_2\ \ N_3\ \ N_4]=[\mathrm{L}_1\ \ \mathrm{L}_2\ \ \mathrm{L}_3\ \ \mathrm{L}_4] \tag{7.44}$$

The mass matrix is evaluated using volume coordinates and Eq. (7.44) for constant ρc_p:

$$\mathbf{M} = \left[\rho c_p \int_V N_i N_j \, dxdydz\right] = \left[\rho c_p \int_V \begin{bmatrix} L_1 \\ L_2 \\ L_3 \\ L_4 \end{bmatrix} [L_1 \; L_2 \; L_3 \; L_4] \, dxdydz\right]$$

$$= \rho c_p \int_{\rho c_p} \begin{bmatrix} L_1^2 & L_1L_2 & L_1L_3 & L_1L_4 \\ L_2L_1 & L_2^2 & L_2L_3 & L_2L_4 \\ L_3L_1 & L_3L_2 & L_3^3 & L_3L_4 \\ L_4L_1 & L_4L_2 & L_4L_3 & L_4^2 \end{bmatrix} dxdydz$$

$$= 6V\rho c_p \begin{bmatrix} \frac{2!0!0!0!}{(2+0+0+0+3)!} & \frac{1!1!0!0!}{(1+1+0+0+3)!} & \frac{1!0!1!0!}{(1+0+0+0+3)!} & \frac{1!0!0!1!}{(1+0+0+1+3)!} \\ \frac{1!1!0!0!}{(1+1+0+0+3)!} & \frac{0!2!0!0!}{(0+2+0+0+3)!} & \frac{0!1!0!0!}{(0+1+1+0+3)!} & \frac{1!1!0!1!}{(0+1+0+1+3)!} \\ \frac{1!0!1!0!}{(1+0+1+0+3)!} & \frac{0!1!1!0!}{(0+1+1+0+3)!} & \frac{0!0!2!0!}{(0+0+2+0+3)!} & \frac{0!0!1!1!}{(0+0+1+1+3)!} \\ \frac{1!0!0!1!}{(1+0+0+1+3)!} & \frac{0!1!0!1!}{(0+1+0+1+3)!} & \frac{0!0!1!1!}{(0+0+1+1+3)!} & \frac{0!0!0!2!}{(0+0+0+2+3)!} \end{bmatrix}$$

$$= \frac{V}{20}\rho c_p \begin{bmatrix} 2 & 1 & 1 & 1 \\ 1 & 2 & 1 & 1 \\ 1 & 1 & 2 & 1 \\ 1 & 1 & 1 & 2 \end{bmatrix} \qquad (7.45)$$

The stiffness matrix is given as ($N_i = L_i$, etc.).

$$
\mathbf{K} = \left[\int_V \left(K_{xx} \frac{\partial N_i}{\partial x} \frac{\partial N_j}{\partial x} + K_{yy} \frac{\partial N_i}{\partial y} \frac{\partial N_j}{\partial y} + K_{zz} \frac{\partial N_i}{\partial z} \frac{\partial N_j}{\partial z} \right) dxdydz \right]
$$

$$
= \frac{K_{xx}}{36V} \begin{bmatrix} b_1b_1 & b_1b_2 & b_1b_3 & b_1b_4 \\ b_2b_1 & b_2b_2 & b_2b_3 & b_2b_4 \\ b_3b_1 & b_3b_2 & b_3b_3 & b_3b_4 \\ b_4b_1 & b_4b_2 & b_4b_3 & b_4b_4 \end{bmatrix}
+ \frac{K_{xx}}{36V} \begin{bmatrix} c_1c_1 & c_1c_2 & c_1c_3 & c_1c_4 \\ c_2c_1 & c_2c_2 & c_2c_3 & c_2c_4 \\ c_3c_1 & c_3c_2 & c_3c_3 & c_3c_4 \\ c_4c_1 & c_4c_2 & c_4c_3 & c_4c_4 \end{bmatrix}
+ \frac{K_{xx}}{36V} \begin{bmatrix} d_1d_1 & d_1d_2 & d_1d_3 & d_1d_4 \\ d_2d_1 & d_2d_2 & d_2d_3 & d_2d_4 \\ d_3d_1 & d_3d_2 & d_3d_3 & d_3d_4 \\ d_4d_1 & d_4d_2 & d_4d_3 & d_4d_4 \end{bmatrix} \tag{7.46}
$$

where

$$
\left.\begin{aligned}
b_1 &= (y_2 - y_4)(z_3 - z_4) - (y_3 - y_4)(z_2 - z_4) \\
b_2 &= (y_3 - y_4)(z_1 - z_4) - (y_1 - y_4)(z_3 - z_4) \\
b_3 &= (y_1 - y_4)(z_2 - z_4) - (y_2 - y_4)(z_3 - z_4) \\
b_4 &= -(b_1 + b_2 + b_3) \\
c_1 &= (x_3 - x_4)(z_2 - z_4) - (x_2 - x_4)(z_3 - z_4) \\
c_2 &= (x_1 - x_4)(z_3 - z_4) - (x_3 - x_4)(z_1 - z_4) \\
c_3 &= (x_2 - x_4)(z_1 - z_4) - (x_1 - x_4)(z_2 - z_4) \\
c_4 &= -(c_1 + c_2 + c_3) \\
d_1 &= (x_2 - x_4)(y_3 - y_4) - (x_3 - x_4)(y_2 - z_4) \\
d_2 &= (x_3 - x_4)(y_1 - y_4) - (x_1 - x_4)(y_3 - z_4) \\
d_3 &= (x_1 - x_4)(y_2 - y_4) - (x_2 - x_4)(y_1 - z_4) \\
d_4 &= -(d_1 + d_2 + d_3)
\end{aligned}\right\} \tag{7.47a}
$$

$$V = \frac{1}{6}\begin{bmatrix} 1 & x_1 & y_1 & z_1 \\ 1 & x_2 & y_2 & z_2 \\ 1 & x_3 & y_3 & z_3 \\ 1 & x_4 & y_4 & z_4 \end{bmatrix} \tag{7.47b}$$

The coefficients in Eq. (7.46) are extensions of the coefficients obtained from the two-dimensional triangular element in Chapter 4.

The load vector is evaluated as

$$\mathbf{F} = \mathbf{F}_Q + \mathbf{F}_q \tag{7.48}$$

for constant Q and q, where

$$\mathbf{F}_Q = \left[\int_V QN_i dxdydz\right] = Q\int_V \begin{bmatrix} N_1 \\ N_2 \\ N_3 \\ N_4 \end{bmatrix} dxdydz = \frac{QV}{4}\begin{bmatrix} 1 \\ 1 \\ 1 \\ 1 \end{bmatrix} \tag{7.49}$$

with (for side 2–3–4)

$$\mathbf{F}_q = \int_S qN\,dS = q\int \begin{bmatrix} 0 \\ N_2 \\ N_3 \\ N_4 \end{bmatrix} dS = q\begin{bmatrix} 0 \\ \dfrac{1!0!0!2A_{234}}{(1+0+0+2)!} \\ \dfrac{0!0!0!2A_{234}}{(0+1+0+2)!} \\ \dfrac{0!0!1!2A_{234}}{(0+0+1+2)!} \end{bmatrix} = \frac{qA_{234}}{3}\begin{bmatrix} 0 \\ 1 \\ 1 \\ 1 \end{bmatrix} \tag{7.50}$$

where A_{234} is the surface area of face 2–3–4. There are also three other forms of Eq. (7.50) which correspond to the remaining three faces of the tetrahedron. Notice that Eq. (4.44b) was used to evaluate the flux term in Eq. (7.5) since only three shape functions exist on a face, i.e., a two-dimensional triangular element.

7.5.2 Hexahedron

If the region is defined as a hexahedron, with sides of length a, b, and c, the shape functions for the trilinear element in Figure 7.9 are given by Eq. (7.26). From Eq. (7.41), the mass matrix is defined as

$$\mathbf{M} = \left[\int_V \rho c_p N_i N_j dxdydz \right] = \rho c_p \int_{-1}^{1} \int_{-1}^{1} \int_{-1}^{1} N_i N_j \left| J \right| d\xi d\eta d\zeta$$

$$= \rho c_p \int_{-1}^{1} \int_{-1}^{1} \int_{-1}^{1} \frac{1}{64} \{(1+\xi\xi_i)(1+\eta\eta_i)(1+\zeta\zeta_i)\}[(1+\xi\xi_j)(1+\eta\eta_j)(1+\zeta\zeta_j)] \left| J \right| d\xi d\eta d\zeta$$

$$= \frac{V}{27} \rho c_p \begin{bmatrix} 8 & 4 & 2 & 4 & 4 & 2 & 1 & 2 \\ 4 & 8 & 4 & 2 & 2 & 4 & 2 & 1 \\ 2 & 4 & 8 & 4 & 1 & 2 & 4 & 2 \\ 4 & 2 & 4 & 8 & 2 & 1 & 2 & 4 \\ 4 & 2 & 1 & 2 & 8 & 4 & 2 & 4 \\ 2 & 4 & 2 & 1 & 4 & 8 & 4 & 2 \\ 1 & 2 & 4 & 2 & 2 & 4 & 8 & 4 \\ 2 & 1 & 2 & 4 & 4 & 2 & 4 & 8 \end{bmatrix} \tag{7.51}$$

The stiffness matrix becomes

$$\mathbf{K} = \int_V \left(K_{xx} \frac{\partial N_i}{\partial x} \frac{\partial N_j}{\partial x} + K_{yy} \frac{\partial N_i}{\partial y} \frac{\partial N_j}{\partial y} + K_{zz} \frac{\partial N_i}{\partial z} \frac{\partial N_j}{\partial z} \right) dxdydz$$

$$= \int_{-1}^{1} \int_{-1}^{1} \int_{-1}^{1} \left(K_{xx} \frac{\partial N_i}{\partial x} \frac{\partial N_j}{\partial x} + K_{yy} \frac{\partial N_i}{\partial y} \frac{\partial N_j}{\partial y} + K_{zz} \frac{\partial N_i}{\partial z} \frac{\partial N_j}{\partial z} \right) \left| \mathrm{J} \right| d\xi d\eta d\zeta$$

$$= \frac{K_{xx}}{8V} \begin{bmatrix} 4 & -4 & -2 & 2 & 2 & -2 & -1 & 1 \\ -4 & 4 & 2 & -2 & -2 & 2 & 1 & -1 \\ -2 & 2 & 4 & -4 & -1 & 1 & 2 & -2 \\ 2 & -2 & -4 & 4 & 1 & -1 & -2 & 2 \\ 2 & -2 & 1 & 1 & 4 & -4 & -2 & 2 \\ -2 & 2 & -1 & -1 & -4 & 4 & 2 & -2 \\ -1 & 1 & -2 & -2 & -2 & 2 & 4 & -4 \\ 1 & 1 & 2 & 2 & 2 & -2 & -2 & 4 \end{bmatrix} \tag{7.52}$$

$$
+\frac{K_{yy}}{8V}\begin{bmatrix} 4 & 2 & -2 & -4 & 2 & 1 & -1 & -2 \\ 2 & 4 & -4 & -2 & 1 & 2 & -2 & -1 \\ -2 & -4 & 4 & 2 & -1 & -2 & 2 & 1 \\ -4 & -2 & 2 & 4 & -2 & -1 & 1 & 2 \\ 2 & 1 & -1 & -2 & 4 & 2 & -2 & -4 \\ 1 & 2 & -2 & -1 & 2 & 4 & -4 & -2 \\ -1 & -2 & 2 & 1 & -2 & -4 & 4 & 2 \\ 2 & 1 & 1 & 2 & -4 & -2 & 2 & 4 \end{bmatrix}
$$

$$
+\frac{K_{zz}}{8V}\begin{bmatrix} 4 & 2 & 1 & 2 & -4 & -2 & -1 & -2 \\ 2 & 4 & 2 & 1 & -2 & -4 & -2 & -1 \\ 1 & 2 & 4 & 2 & -1 & -2 & -4 & -2 \\ 2 & 1 & 2 & 4 & -2 & -1 & -2 & -4 \\ -4 & -2 & -1 & -2 & 4 & 2 & 1 & 2 \\ -2 & -4 & -2 & -1 & 2 & 4 & 2 & 1 \\ -1 & -2 & -4 & -2 & 1 & 2 & 4 & 2 \\ -2 & -1 & -2 & -4 & 2 & 1 & 2 & 4 \end{bmatrix}
$$

where

$$
\begin{bmatrix} \dfrac{\partial N_i}{\partial x} \\ \dfrac{\partial N_i}{\partial y} \\ \dfrac{\partial N_i}{\partial z} \end{bmatrix} = \mathrm{J}^{-1} \begin{bmatrix} \dfrac{\partial N_i}{\partial \xi} \\ \dfrac{\partial N_i}{\partial \eta} \\ \dfrac{\partial N_i}{\partial \zeta} \end{bmatrix} \tag{7.53}
$$

and

$$
\mathrm{J} = \begin{bmatrix} \dfrac{\partial x}{\partial \xi} & \dfrac{\partial y}{\partial \xi} & \dfrac{\partial z}{\partial \xi} \\ \dfrac{\partial x}{\partial \eta} & \dfrac{\partial y}{\partial \eta} & \dfrac{\partial z}{\partial \eta} \\ \dfrac{\partial x}{\partial \zeta} & \dfrac{\partial y}{\partial \zeta} & \dfrac{\partial z}{\partial \zeta} \end{bmatrix} \tag{7.54}
$$

The load vector is evaluated as before, using Eq. (7.48), for constant Q and q:

$$\mathbf{F}_Q = \int_V QN_i dxdydz = \int_{-1}^{1}\int_{-1}^{1}\int_{-1}^{1} Q(1+\xi\xi_i)(1+\eta\eta_i)(1+\zeta\zeta_i)|\mathrm{J}|\, d\xi d\eta d\zeta$$

$$= \frac{QV}{8}\begin{bmatrix}1\\1\\1\\1\\1\\1\\1\\1\end{bmatrix} \tag{7.55}$$

with (for side 2–3–6–7).

$$\mathbf{F}_Q = \int_{S_{2-3-6-7}} qN_i dS$$

$$= q\int_{S_{2-3-6-7}} \begin{bmatrix}0\\N_2\\N_3\\0\\0\\N_6\\N_7\\0\end{bmatrix} dS = \frac{qA_{2-3-6-7}}{4}\begin{bmatrix}0\\1\\1\\0\\0\\1\\1\\0\end{bmatrix} \tag{7.56}$$

where $A_{2-3-6-7}$ is the surface area of face 2–3–6–7. Similar relations can be written for the other faces.

Example 7.3

Evaluate the integral

$$\int_V N_i \frac{\partial N_j}{\partial x}\ \mathrm{dxdydz}$$

for the linear hexahedron.

$$\left[\int_V N_i \frac{\partial N_j}{\partial x} dxdydz\right] = \int_{-1}^{1}\int_{-1}^{1}\int_{-1}^{1}\left[N_i(\xi,\eta,\zeta)\frac{\partial N_j}{\partial §}\right]|\mathrm{J}|\,d\xi d\eta d\zeta$$

$$= \frac{V}{12}\begin{bmatrix} -4 & 4 & 2 & -2 & -2 & 2 & 1 & -1 \\ -4 & 4 & 2 & -2 & -2 & 2 & 1 & -1 \\ -2 & 2 & 4 & -4 & -1 & 1 & 2 & -2 \\ -2 & 2 & 4 & -4 & -1 & 1 & 2 & -2 \\ -2 & 2 & 1 & -1 & -4 & 4 & 2 & -2 \\ -2 & 2 & 1 & -1 & -4 & 4 & 2 & -2 \\ -1 & 1 & 2 & -2 & -2 & 2 & 4 & -4 \\ -1 & 1 & 2 & -2 & -2 & 2 & 4 & -4 \end{bmatrix} \tag{7.57}$$

where $\partial N_j/\partial x$ is obtained from Eq. (7.53).

7.6 TIME-DEPENDENT HEAT CONDUCTION WITH RADIATION AND CONVECTION

Three-dimensional time-dependent heat transfer due to conduction with radiative, flux, and convective surfaces and internal heat generation is described by Eq. (7.35) with a flux boundary condition of the form:

$$-K_{xx}\frac{\partial T}{\partial x}n_x - K_{yy}\frac{\partial T}{\partial y}n_y - K_{zz}\frac{\partial T}{\partial z}n_x + aT + q = 0 \tag{7.58}$$

where a represents the various coefficients associated with radiation and convection, which we shall describe shortly. Applying the Galerkin weak statement to Eq. (7.35) yields Eq. (7.38). After replacing

$$T = \sum_{i=1}^{n} N_i T_i$$

and applying appropriate boundary conditions, Eq. (7.38) becomes

$$\begin{aligned} &\left[\int_V \rho c_p N_i N_j dxdydz\right]\dot{\mathrm{T}} \\ &+\left[\int_V\left(K_{xx}\frac{\partial N_i}{\partial x}\frac{\partial N_j}{\partial x} + K_{yy}\frac{\partial N_i}{\partial y}\frac{\partial N_j}{\partial y} + K_{zz}\frac{\partial N_i}{\partial z}\frac{\partial N_j}{\partial z}\right)dxdydz\right]\mathrm{T} \\ &-\int_V QN_i dxdydz + \int_{S_1} qN_i dS_1 + \int_{S_2} q_r N_i dS_2 \\ &+\left[\int_{S_3} hN_iN_j dS_3\right]\mathrm{T} - \int_{S_3} hT_\infty N_i dS_3 = 0 \end{aligned} \tag{7.59}$$

where q_r denotes the flux due to radiation, q is the heat flux due to conduction, and terms involving h (h = convection coefficient) represent convection surfaces. In matrix form, Eq. (7.59) becomes

$$\mathbf{M}\dot{\mathbf{T}} + (\mathbf{K} + \mathbf{H})\mathbf{T} = \mathbf{F} \tag{7.60}$$

where

$$\mathbf{M} = \left[\int_V \rho c_p N_i N_j dxdydz \right] \tag{7.61}$$

$$\mathbf{K} = \left[\int_V \left(K_{xx} \frac{\partial N_i}{\partial x} \frac{\partial N_j}{\partial x} + K_{yy} \frac{\partial N_i}{\partial y} \frac{\partial N_j}{\partial y} + K_{zz} \frac{\partial N_i}{\partial z} \frac{\partial N_j}{\partial z} \right) dxdydz \right] \tag{7.62}$$

$$\mathbf{H} = \left[\int_S hN_i N_j S_3 \right] \tag{7.63}$$

$$\mathbf{F} = \mathbf{F}_Q + \mathbf{F}_h + \mathbf{F}_q + \mathbf{F}_{qr} \tag{7.64}$$

The load vectors are defined as

$$\mathbf{F}_Q = \int_V QN_i dxdydz \tag{7.65}$$

$$\mathbf{F}_h = \int_{S_3} hT_\infty N_i dS_3 \tag{7.66}$$

$$\mathbf{F}_q = \int_{S_1} qN_i dS_1 \tag{7.67}$$

$$\mathbf{F}_{qr} = \int_{S_2} q_r N_i dS_2 \tag{7.68}$$

We now have only to decide what type of element to use, and then evaluate the matrices defined by Eqs. (7.61) through (7.68) for the specific problem domain and solve Eq. (7.60). Before proceeding with a specific example problem, Eq. (7.68) describing the radiation boundary condition is first discussed.

7.6.1 Radiation

Radiation between two boundary (surface) segments, a and b, can be described by the relation:

$$q_r = -K\frac{\partial T}{\partial \eta} = \sigma F_{a-b}\left(\varepsilon_a T_a^4 - \alpha_b T_b^4\right) \tag{7.69}$$

where ε_a is emissivity and α_b is absorptivity; σ is the Stefan–Boltzmann constant (5.66961×10^{-8} W/m^{2}K^4) and F_{a-b} is the form (shape) factor between surfaces *a* and *b*. The form factor describes the fraction of energy leaving one surface and reaching another surface. A more thorough description of form factors is discussed in the heat transfer text by Howell (2001).

Equation (7.69) can also be written in a form similar to heat flux due to convective heat transfer,

$$q_r = h_r A_a\left(T_a - T_b\right) \tag{7.70}$$

where T_a and T_b are the temperatures of the two surfaces (bodies) exchanging heat by radiation, and h_r is defined as

$$h_r = \frac{\sigma\left(T_a^2 + T_b^2\right)\left(T_a + T_b\right)}{\dfrac{1}{\varepsilon_a} + \left(\dfrac{A_a}{A_b}\right)\left(\dfrac{1}{\varepsilon_b} - 1\right)} \tag{7.71}$$

Obviously, h_r is strongly coupled to the temperatures at the surfaces; the convection heat transfer coefficient is generally temperature dependent and, hence, nonlinear.

Substituting Eq. (7.69) for q_r, Eq. (7.68) becomes

$$\int_{S_2} q_r N_i dS = \int_{S_2}\left(\sigma\varepsilon\, T_a^4 - \sigma\alpha_a \sum_b F_{a-b}T_b^4\right)N_i dS$$
$$= A_a \int_{S_2} h_r (T_a - T_b) dS \tag{7.72}$$

where subscript *a* denotes the element and subscript *b* denotes other radiative elements (or boundaries). Hence, during a specific time step, q_r is treated as a specific constant flux. While not elegant, the approximation for q_r yields fairly good "macroscopic" values for the effects of radiation.

Example 7.4

What are the relations that describe radiation between (a) two boundaries or a boundary and a constant temperature source, and (b) two surfaces or a surface and a constant heat source?

For the two-dimensional linear triangle shown in Figure 7.12, the flux normal to a boundary is defined (from previous analysis for two-dimensional triangular elements) as

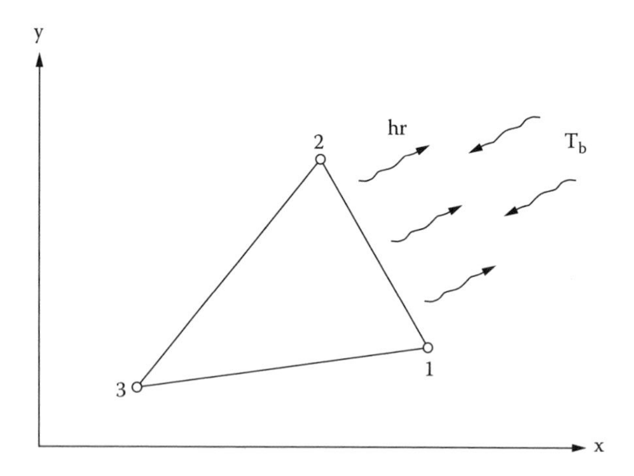

Figure 7.12 Radiation exchange for a two-dimensional linear triangular element

$$\int_{S_1} qN_i dS = \frac{q\ell_{1-2}}{2}\begin{bmatrix}1\\1\\0\end{bmatrix} \tag{7.73}$$

Assuming that q_r is constant,

$$\int_{S_2} q_r N_i dS = \frac{q_r \ell_{1-2}}{2}\begin{bmatrix}1\\1\\0\end{bmatrix} = (\sigma\varepsilon_a T_a^4 - \sigma\alpha_a \sum F_{a-b} T_b^4)\frac{\ell_{1-2}}{2}\begin{bmatrix}1\\1\\0\end{bmatrix} \tag{7.74}$$

or

$$= \frac{h_r \ell_{1-2}(T_a - T_b)}{2}\begin{bmatrix}1\\1\\0\end{bmatrix} \tag{7.75}$$

where $T_a = 1/2(T_1 + T_2)$, T_b = specified temperature (or average temperature over an element face), and h_r is defined by Eq. (7.71) with the areas replaced by element lengths, i.e., $A_a = \ell_{1-2}$.

The flux normal to one side of the tetrahedron shown in Figure 7.13 is given by Eq. (7.50), If q_r is constant,

$$\int q_r N_i dS = \frac{q_r A_{234}}{3}\begin{bmatrix}0\\1\\1\\1\end{bmatrix} = \frac{A_{234}}{3}(\sigma\varepsilon_a T_a^4 - \sigma\varepsilon_a \sum F_{a-b} T_b^4)\begin{bmatrix}0\\1\\1\\1\end{bmatrix} \tag{7.76}$$

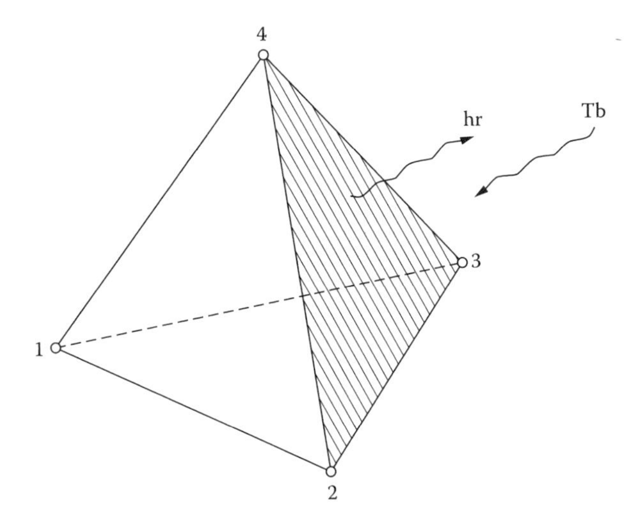

Figure 7.13 Radiation exchange for a three-dimensional linear tetrahedron element

or

$$= \frac{h_r A_{234}}{3}(T_a - T_b)\begin{bmatrix} 0 \\ 1 \\ 1 \\ 1 \end{bmatrix} \tag{7.77}$$

where $T_a = 1/3(T_2 + T_3 + T_4)$; T_b is specified temperature or average temperature of an element face), and h_r is given by Eq. (7.71).

7.6.2 Shape Factors

The amount of radiant energy leaving element a and arriving at element b is

$$\sigma T_a^4 A_a F_{a-b}$$

The amount of energy leaving element b and arriving at element a is

$$\sigma T_b^4 A_b F_{b-a}$$

where F_{a-b} and F_{b-a} represent the fraction of energy leaving each element surface. The net heat exchange is calculated as

$$A_a F_{a-b}\left(\sigma T_a^4 - \sigma T_b^4\right) = A_b F_{b-a}\left(\sigma T_a^4 - \sigma T_b^4\right) \tag{7.78}$$

which is the well-known reciprocity theorem. The problem is to determine the value of F_{a-b} and F_{b-a}. The net energy exchange between two surfaces (or element faces) can be described for any surface geometry by the relation

$$q_{a-b}^{net} = \sigma(T_a^4 - T_b^4)\int_{A_b}\int_{A_a} \cos\phi_a \ \cos\phi_b \ \frac{dA_a dA_b}{\pi r^2} \tag{7.79}$$

where dA_a and dA_b represent differential areas on surfaces a and b, ϕ_a and ϕ_b are the direction angles relative to the normal to each surface along with the radiation acts, and r is the distance between dA_a and dA_b. To evaluate Eq. (7.77), specific geometries of surfaces A_a and A_b must be known, along with the separating distance and angles. Fortunately, many of the shape factors associated with commonly known geometries have been calculated and can be found in many undergraduate textbooks on heat transfer (see Howell, 2001). Evaluation of the formal integral relations common to radiation problems can quickly become complex and time-consuming on a computer. For those readers who wish to use the more formal relations for radiation flux given by Eq. (7.69), or who wish to learn more about radiation, many textbooks on radiation are readily available in the literature.

Evaluating the matrix terms in Eq. (7.60) and combining all terms yield the general expression for a tetrahedron element experiencing convection and radiation fluxes over one face (2–3–4) with internal heat generation:

$$\rho c_p \frac{V}{20}\begin{bmatrix} 2 & 1 & 1 & 1 \\ 1 & 2 & 1 & 1 \\ 1 & 1 & 2 & 1 \\ 1 & 1 & 1 & 2 \end{bmatrix}\begin{bmatrix} \dot{T}_1 \\ \dot{T}_2 \\ \dot{T}_3 \\ \dot{T}_4 \end{bmatrix} + \left(\frac{K_{xx}}{36V}\begin{bmatrix} b_1b_1 & b_1b_2 & b_1b_3 & b_1b_4 \\ b_2b_1 & b_2b_2 & b_2b_3 & b_2b_4 \\ b_3b_1 & b_3b_2 & b_3b_3 & b_3b_4 \\ b_4b_1 & b_4b_2 & b_4b_3 & b_4b_4 \end{bmatrix} + \right.$$

$$\left. + \frac{K_{yy}}{36V}\begin{bmatrix} c_1c_1 & c_1c_2 & c_1c_3 & c_1c_4 \\ c_2c_1 & c_2c_2 & c_2c_3 & c_2c_4 \\ c_3c_1 & c_3c_2 & c_3c_3 & c_3c_4 \\ c_4c_1 & c_4c_2 & c_4c_3 & c_4c_4 \end{bmatrix} + \frac{K_{zz}}{36V}\begin{bmatrix} d_1d_1 & d_1d_2 & d_1d_3 & d_1d_4 \\ d_2d_1 & d_2d_2 & d_2d_3 & d_2d_4 \\ d_3d_1 & d_3d_2 & d_3d_3 & d_3d_4 \\ d_4d_1 & d_4d_2 & d_4d_3 & d_4d_4 \end{bmatrix}\right)\begin{bmatrix} T_1 \\ T_2 \\ T_3 \\ T_4 \end{bmatrix}$$

$$-\frac{hV}{12}\begin{bmatrix} 0 & 0 & 0 & 0 \\ 0 & 2 & 1 & 1 \\ 0 & 1 & 2 & 1 \\ 0 & 1 & 1 & 2 \end{bmatrix}\begin{bmatrix} T_1 \\ T_2 \\ T_3 \\ T_4 \end{bmatrix} = \frac{QV}{4}\begin{bmatrix} 1 \\ 1 \\ 1 \\ 1 \end{bmatrix} - \frac{qA_{234}}{4}\begin{bmatrix} 0 \\ 1 \\ 1 \\ 1 \end{bmatrix} - \frac{hT_\infty}{3}\begin{bmatrix} 0 \\ 1 \\ 1 \\ 1 \end{bmatrix}$$

$$-\frac{h_r A_{234}}{3}\left[\frac{1}{3}(T_2 + T_3 + T_4) - T_R\right]\begin{bmatrix} 0 \\ 1 \\ 1 \\ 1 \end{bmatrix} \tag{7.80}$$

where V represents the volume of the element, A_{234} is the surface area of face 2–3–4; T_∞ is the external temperature used for convection, and T_R (in this instance $T_R = T_\infty$) is the specified external heat source radiating to element surface 2–3–4. The only remaining step is to substitute the appropriate elemental values into Eq. (7.80) and solve the resulting global matrix equation (assuming assemblage is over only one element) using forward-in-time differencing for the time derivative terms. Solution of Eq. (7.80) is best left to the computer.*

7.7 CLOSURE

Construction of the three-dimensional finite element algorithm follows directly from procedures used to construct the one- and two-dimensional schemes; the calculation procedure is essentially the same for any type of element in any geometry. Careful examination of computer programs will reveal the same logic generally used in one-, two-, and three-dimensional programs—the basic differences are due to geometrical considerations dealing with lines, areas, and volumes. Obviously three-dimensional programs could be used to solve for both one- and two-dimensional problems; however, the overhead associated with generality can greatly limit the capability of a computer program. At this point, we have made enough progress that the interested reader should already have sufficient skills to develop a three-dimensional finite element program following the logic and ideas given in the two-dimensional program presented here. Once enough confidence has been gained in the solution of simple three-dimensional problems, a good exercise would be to introduce the necessary alterations to the three-dimensional program to solve one- and two-dimensional problems, as well.

It becomes quickly evident when solving three-dimensional problems that a relatively coarse mesh requires a significant number of nodes; a fine mesh will not only require a considerable number of nodes, but also a much larger computer with fast processing speed. We must also consider generation of the three-dimensional mesh and associated nodal connectivity (especially if using hexahedrons)—loading the local nodal points and their associated x, y, z locations per element quickly becomes laborious and subject to error. Commercially available multidimensional mesh generators are routinely advertised in various trade and professional journals.† With a little effort, the data set created by MESH-2D can be modified to handle three-dimensional elements.

* FEM-3D and COMSOL files are available.

† There are several popular multidimensional mesh generation programs, e.g., GAMBIT, TrueGrid, Geomesh, and LAPCAD, which runs on a PC. GAMBIT, TrueGrid, and Geomesh can generate meshes with $>10^6$ nodes, and run on PCs and workstations.

EXERCISES

7.1 Determine the shape functions for the linear tetrahedron element shown below.

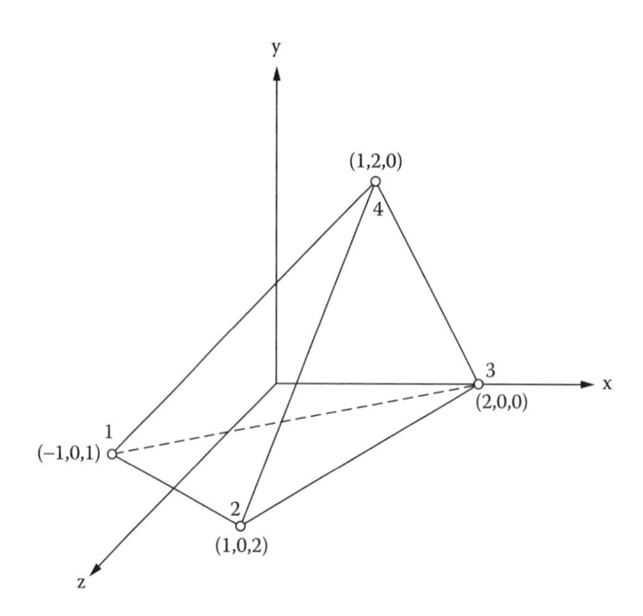

7.2 For a tetrahedral element, show that the determinant of the matrix **C** in Eq. (7.4) is good to $6V$, where V is the element volume.

7.3 Find $\mathbf{C}^{-1}$ for a general tetrahedral element.

7.4 Find the shape functions for a general linear tetrahedron in Cartesian coordinates.

7.5 With the above shape functions, verify Eq. (7.46).

7.6 Find the expression equivalent to Eq. (7.50) for the other three faces of a tetrahedral element.

7.7 Find the expressions equivalent to Eq. (7.56) for the other faces of the brick element.

7.8 Find the isoparametric transformation between the element in Figure 7.8 with sides of length a, b, and c in the x, y, and z direction, respectively, and the element in Figure 7.9.

7.9 Find the Jacobian **J** in Eq. (7.57) and its inverse.

7.10 Fill in the details needed to obtain Eq. (7.57).

7.11 Calculate the shape functions for the hexahedron shown in the figure, and assume the x, y, z coordinate system is located at (0, 0).

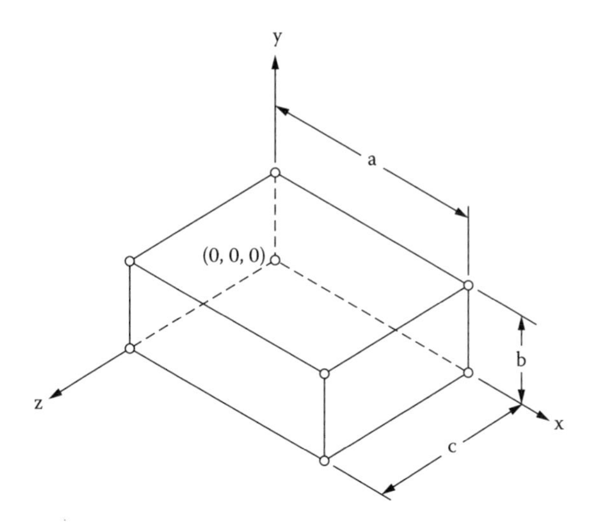

7.12 Determine the global matrix expression for the nodal temperature values within the two-element computational domain shown in the figure.

$\varepsilon_2 = 0.1$
$T_\infty = 100°C$
$\alpha_{xx} = \alpha_{yy} = \alpha_{zz} = 2\ m^2/s$
element 1: 1, 2, 4, 5
element 2: 2, 3, 5, 4

Node	x	y	z
1	0	0	0
2	3	0	2
3	5	0	0
4	2	3	0
5	4	1	1

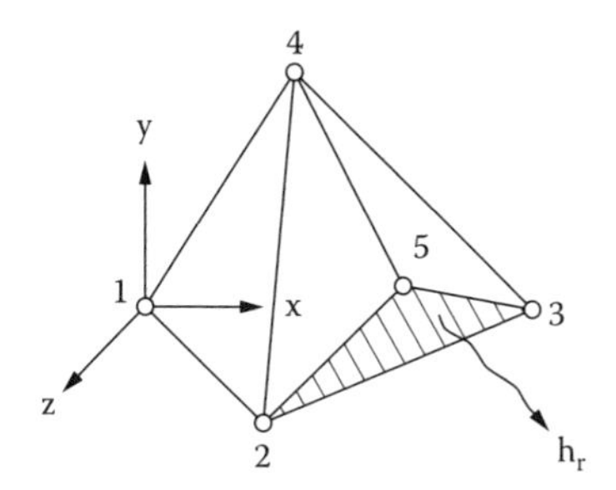

7.13 Find the shape functions for a triquadratic three-dimensional brick. Recall Exercise 5.1; the same procedure of taking products of one-dimensional functions can be used to simplify the work. Sketch the element and label the nodes carefully.

7.14 Calculate the nodal temperature values within the two-element domain shown below (values normalized by ρc_p). The sides and bottom surfaces are insulated.

$h = 10$ m/s
$T_\infty = 50°C$
$\alpha_{xx} = \alpha_{yy} = \alpha_{zz} = 15$ m²/s
$Q_1 = 10$ C/s
element 1: 1, 2, 3, 4, 5, 6, 7, 8
element 2: 4, 3, 9, 10, 8, 7, 11, 12

		(meters)	
Node	x	y	z
1	0	0	4
2	2	0	4
3	2	0	2
4	0	0	2
5	0	2	4
6	2	2	4
7	2	2	2
8	0	2	2
9	2	0	0
10	0	0	0
11	2	2	0
12	0	2	0

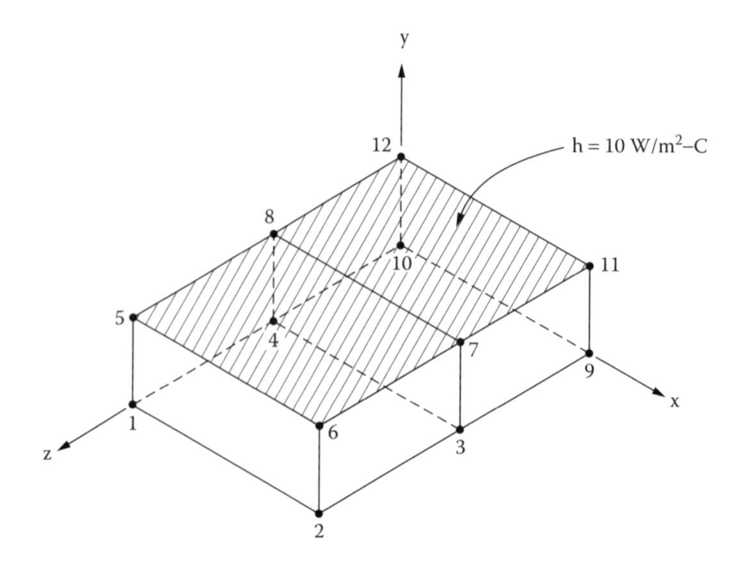

7.15 Show that

$$\begin{bmatrix} 1 \\ x \\ y \\ z \end{bmatrix} = \begin{bmatrix} 1 & 1 & 1 & 1 \\ x_1 & x_2 & x_3 & x_4 \\ y_1 & y_2 & y_3 & y_4 \\ z_1 & z_2 & z_3 & z_4 \end{bmatrix} \begin{bmatrix} L_1 \\ L_2 \\ L_3 \\ L_4 \end{bmatrix}$$

where $L_1 = V_1/V$, $L_2 = V_2/V$, etc.

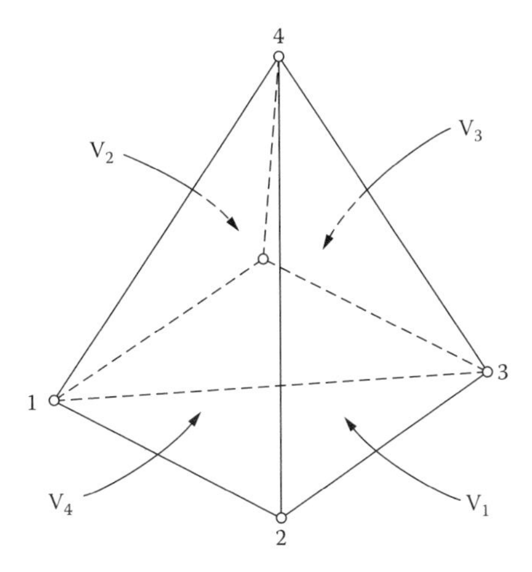

7.16 The triangular prism (wedge) is useful in cylindrical geometries. The six-noded linear element, shown in the figure, can be constructed from the tensor product of triangular and one-dimensional interpolation functions. Show that

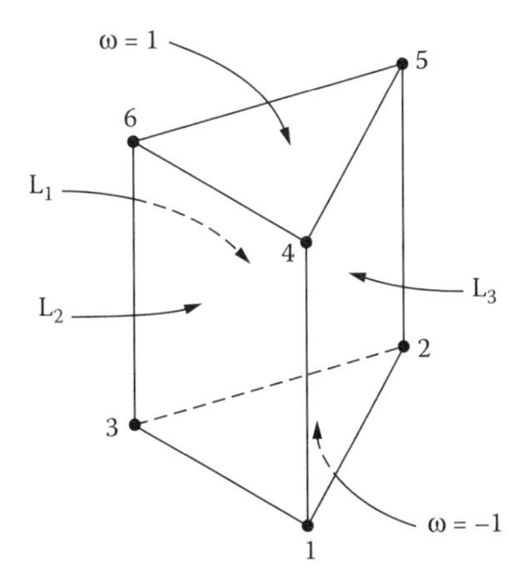

$$N_1 = \frac{1}{2}L_1(1-w) \qquad N_4 = \frac{1}{2}L_1(1+w)$$

$$N_2 = \frac{1}{2}L_2(1-w) \qquad N_5 = \frac{1}{2}L_2(1+w)$$

$$N_3 = \frac{1}{2}L_3(1-w) \qquad N_6 = \frac{1}{2}L_3(1+w)$$

7.17 Calculate the steady-state heat distribution within the block shown below if it experiences both radiation and convection on the left and right sides of the block (values normalized by ρc_p), respectively. The front and back sides are also insulated.

$T_R = 1000°C$
$T_\infty = 100°C$
$\bar{h}_r = 2$ m/s
$\bar{h} = 10$ m/s
$\bar{Q} = 50$ C/s
$\alpha_{xx} = \alpha_{yy} = \alpha_{zz} = 10$ m²/s

Node	x	y (meters)	z
1	0	0	4
2	4	0	4
3	4	4	4
4	0	4	4
5	0	0	0
6	4	0	0
7	4	4	0
8	0	4	0

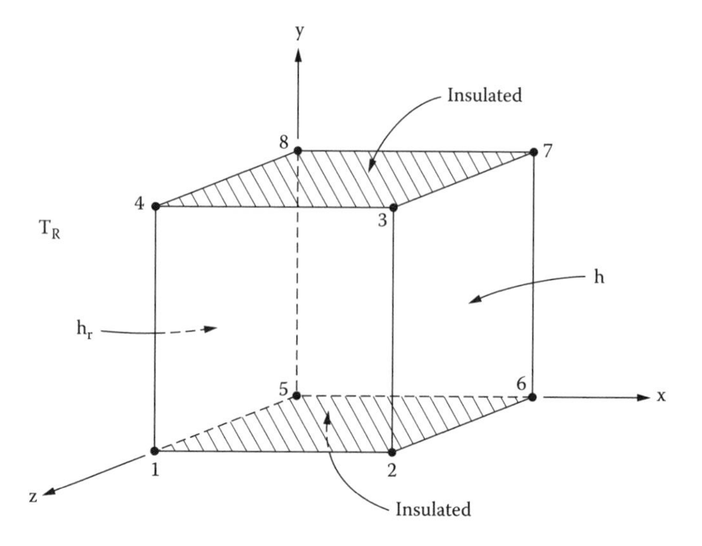

REFERENCES

Desai, C. S. and Abel, J. F. (1972). *Introduction to the Finite Element Method; A Numerical Method for Engineering Analysis.* New York: Van Nostrand-Reinhold.

GAMBIT 2.2 (2004). User's Guide, Fluent, Inc., Lebanon, N.H.

Geomesh (2004). *Users Manual.* San Diego, Calif.: ESI US R&D, Inc.

Heinrich, J. C. and Pepper, D. W. (1999). *Intermediate Finite Element Method. Fluid Flow and Heat Transfer Applications.* New York: Taylor & Francis.

Heuser, J. (1972). "Finite Element Method for Thermal Analysis." NASA Technical Note TN-D-7274. Greenbelt, Md.: Goddard Space Flight Center.

Howell, J. P. (2001). *Radiation Heat Transfer.* New York: John Wiley & Sons.

LAPCAD (2004). *User's Manual.* San Diego, Calif.: LAPCAD Engineering, Inc.

Pepper, D. W. and Gewali, L. (2002). "HTADAPT: A Web-Based, h-Adapting Finite Element Model for Heat Transfer." 8th AIAA/ASME Joint Thermophysics and Heat Transfer Conf., St. Louis, MO, June 24–26.

TrueGrid (2004). *Users Manual.* Livermore, Calif.: XYZ Scientific Applications, Inc.

Zienkiewicz, O. C. (1977). *The Finite Element Method,* 3rd ed. London: McGraw-Hill.

CHAPTER

8

FINITE ELEMENTS IN SOLID MECHANICS

8.1 BACKGROUND

As mentioned in Chapter 1, the finite element method was originally developed to help resolve design problems in aero structures and was only later extended to general field problems. In previous chapters we have dealt almost exclusively with problems involving heat transfer and, therefore, only one unknown variable at each node. In solid mechanics, which includes the analysis of stress/strain, structural design, fatigue, cracking, deformation, cyclic loading, and other areas of mechanics, we must deal with at least one unknown per spatial direction. So, for example, a problem of linear elasticity in two dimensions involves two unknowns per node.

In this chapter we establish how the finite element method is applied to the analysis of linear elastic structures and how to handle more than one degree of freedom per node. We also show how thermal conduction analysis can be utilized in structural analysis, making finite element analysis of heat transfer directly applicable in the area of structural analysis.

8.2 TWO-DIMENSIONAL ELASTICITY—STRESS STRAIN

Two-dimensional elasticity is generally categorized into two modes: plane strain and plane stress. When the thickness of a solid object is large, a state of plane strain is considered to exist. If this thickness is small compared to its overall dimensions (x,y), the condition of plane stress is assumed. Both cases are subsets of general three-dimensional elasticity problems. In this instance, body forces (or loads) cannot have components in the z-direction, nor vary in the direction of the body thickness. Figure 8.1 illustrates plane stress and plane strain.

The governing equations that describe two-dimensional elastic stress are defined as (Reddy, 1993)

$$\frac{\partial \sigma_x}{\partial x} + \frac{\partial \tau_{xy}}{\partial y} + f_x = 0 \tag{8.1a}$$

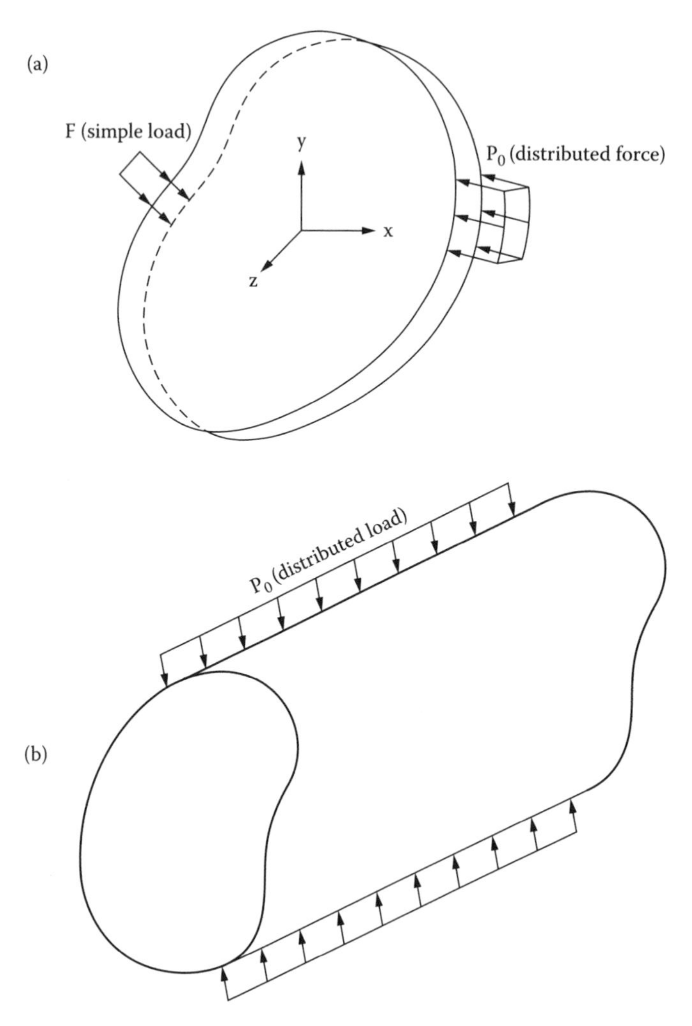

Figure 8.1 Plane stress and plane strain: (a) thin body in plane stress; and (b) thick body in plane strain

$$\frac{\partial \tau_{xy}}{\partial x} + \frac{\partial \sigma_y}{\partial y} + f_y = 0 \tag{8.1b}$$

where σ_x and σ_y are the normal stress components in the x–y directions, respectively; τ_{xy} is the shear stress which acts in the x–y plane; and f_x and f_y are the body force terms. The strain-displacement relations are defined as

$$\varepsilon_x = \frac{\partial u}{\partial x} \tag{8.2}$$

$$\varepsilon_y = \frac{\partial v}{\partial y} \tag{8.3}$$

$$\tau_{xy} = \frac{\partial u}{\partial y} + \frac{\partial v}{\partial x} \tag{8.4}$$

where u and v denote the x and y displacements (the amount of movement due to applied load). Hooke's law relates stress to strain through the relations:

$$\sigma = \mathrm{D}\varepsilon \tag{8.5}$$

where

$$\sigma_x = d_{11}\varepsilon_x + d_{12}\varepsilon_y \tag{8.6}$$

$$\sigma_y = d_{21}\varepsilon_x + d_{22}\varepsilon_y \tag{8.7}$$

$$\tau_{xy} = d_{33}\gamma_{xy} \tag{8.8}$$

with d being the "material stiffness matrix," and (for an isotropic elastic body)

Plane Stress:

$$\left.\begin{aligned} d_{11} = d_{22} &= \frac{E}{1-\mu^2} \\ d_{12} = d_{21} &= \frac{\mu E}{1-\mu^2} \\ d_{33} &= \frac{E}{2(1+\mu)} \end{aligned}\right\} \tag{8.9}$$

Plane Strain:

$$\left.\begin{aligned} d_{11} = d_{22} &= \frac{E(1-\mu)}{(1+\mu)(1-2\mu)} \\ d_{12} = d_{21} &= \frac{E\mu}{(1+\mu)(1-2\mu)} \\ d_{33} &= \frac{E}{2(1+\mu)} \end{aligned}\right\} \tag{8.10}$$

with $E \equiv$Young's modulus of elasticity and $\mu \equiv$ Poisson's ratio. Notice that

$$\mathbf{D} = \begin{bmatrix} d_{11} & d_{12} & 0 \\ d_{21} & d_{22} & 0 \\ 0 & 0 & d_{33} \end{bmatrix} \text{ and } \varepsilon = \mathbf{D}^{-1}\sigma$$

which explains the difference in relations for $\boldsymbol{D}$ in Eqs. (8.9) and (8.10). The components of $\boldsymbol{D}$ are the elasticity (or material) constants.

The boundary conditions associated with Eqs. (8.6) through (8.10) are

$$\left.\begin{aligned} \sigma_x n_x + \tau_{xy} n_y = t_x \\ \tau_{xy} n_x + \sigma_y n_y = t_y \end{aligned}\right\} on\ S_1 \tag{8.11}$$

where S_1 is part of the boundary surface, and

$$\left.\begin{aligned} u = \bar{u} \\ v = \bar{v} \end{aligned}\right\} on\ S_2 \tag{8.12}$$

with S_2 the remainder of the boundary, n_x and n_y the unit normal vectors, t_x and t_y specified (traction) boundary forces, and $\bar{u}$ and $\bar{v}$ specified displacements.

Equations (8.1) and (8.2) can be rewritten in terms of u and v displacements as

$$-\frac{\partial}{\partial x}\left(d_{11}\frac{\partial u}{\partial x} + d_{12}\frac{\partial v}{\partial y}\right) - d_{33}\frac{\partial}{\partial y}\left(\frac{\partial u}{\partial y} + \frac{\partial v}{\partial x}\right) = f_x \tag{8.13}$$

$$-d_{33}\frac{\partial}{\partial x}\left(\frac{\partial u}{\partial y} + \frac{\partial v}{\partial x}\right) - \frac{\partial}{\partial y}\left(d_{12}\frac{\partial u}{\partial x} + d_{22}\frac{\partial v}{\partial y}\right) = f_y \tag{8.14}$$

with the boundary tractions (8.11) given by

$$t_x = \left(d_{11}\frac{\partial u}{\partial x} + d_{12}\frac{\partial v}{\partial y}\right)n_x + d_{33}\left(\frac{\partial u}{\partial y} + \frac{\partial v}{\partial x}\right)n_y \tag{8.15}$$

$$t_y = d_{33}\left(\frac{\partial u}{\partial y} + \frac{\partial v}{\partial x}\right)n_x + \left(d_{12}\frac{\partial u}{\partial x} + d_{22}\frac{\partial v}{\partial y}\right)n_y \tag{8.16}$$

8.3 GALERKIN APPROXIMATION

The u and v displacements are approximated over an element e as

$$u^{(e)} = \sum_{i=1}^{n} N_i u_i = \eta u^{(e)} \tag{8.17}$$

$$v^{(e)} = \sum_{i=1}^{n} N_i v_i = \eta v^{(e)} \tag{8.18}$$

where $\eta = [N_1\ \mathrm{N}_2\ \ldots\ \mathrm{N}_n]$, $\mathrm{u}^{(e)} = [u_1\ \mathrm{u}_2\ \ldots\ \mathrm{u}_n]^T$, and $v^{(e)} = [\mathrm{v}_1\ \mathrm{v}_2\ \ldots\ \mathrm{v}_n]^T$. For the three-noded linear triangular element,

$$u^{(e)} = N_1 u_1 + N_2 u_2 + N_3 u_3 \tag{8.19}$$

$$v^{(e)} = N_1 v_1 + N_2 v_2 + N_3 v_3 \tag{8.20}$$

We define the vector **a** where **a** denotes the u and v nodal displacements in the form:

$$\mathrm{a} = [a_1\ \mathrm{a}_2\ \mathrm{a}_3\ \mathrm{a}_4\ \mathrm{a}_5\ \mathrm{a}_6]^T \equiv [u_1\ \mathrm{v}_1\ \mathrm{u}_2\ \mathrm{v}_2\ \mathrm{u}_3\ \mathrm{v}_3]^T$$

The nodal displacements are shown in Figure 8.2. Using matrix notation, u and v can be expressed as

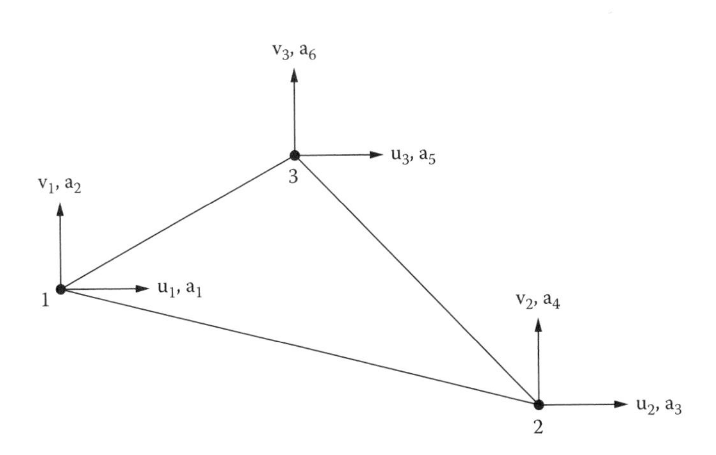

Figure 8.2 Nodal displacements for a linear triangular element

$$u = \begin{bmatrix} u^{(e)} \\ v^{(e)} \end{bmatrix} = \begin{bmatrix} N_1 & 0 & N_2 & 0 & N_3 & 0 \\ 0 & N_1 & 0 & N_2 & 0 & N_3 \end{bmatrix} \begin{bmatrix} a_1 \\ a_2 \\ a_3 \\ a_4 \\ a_5 \\ a_6 \end{bmatrix} \tag{8.21}$$

or

$$\mathbf{u} = \mathbf{Na} \tag{8.22}$$

where $\boldsymbol{N}$ is the 2 × 6 matrix consisting of the linear shape functions, and **a** is the column vector containing the discrete values of the displacements as defined above.

Applying the method of weighted residuals to Eqs. (8.13) through (8.16) and integrating using Green's theorem yield the relations:

$$\int_A \left[\frac{\partial N_i}{\partial x}\left(d_{11}\frac{\partial u}{\partial x} + d_{12}\frac{\partial v}{\partial y} \right) + d_{33}\frac{\partial N_i}{\partial y}\left(\frac{\partial u}{\partial y} + \frac{\partial v}{\partial x} \right) - N_i f_x \right] dx\,dy - \int_{S_1} N_i t_x dS = 0 \tag{8.23}$$

$$\int_A \left[d_{33}\frac{\partial N_i}{\partial x}\left(\frac{\partial u}{\partial y} + \frac{\partial v}{\partial x} \right) + \frac{\partial N_i}{\partial y}\left(d_{12}\frac{\partial u}{\partial x} + d_{22}\frac{\partial v}{\partial y} \right) - N_i f_y \right] dx\,dy - \int_{S_1} N_i t_y dS = 0 \tag{8.24}$$

where t_x and t_y are the traction boundary forces given in Eq. (8.11). Equations (8.23) and (8.24) can be written in matrix form, using Eqs. (8.17) and (8.18), as

$$\mathrm{K}_{11}\mathrm{u}^{(e)} + \mathrm{K}_{12}\mathrm{v}^{(e)} = \mathrm{F}_1 \tag{8.25}$$

$$\mathrm{K}_{21}\mathrm{u}^{(e)} + \mathrm{K}_{22}\mathrm{v}^{(e)} = \mathrm{F}_2 \tag{8.26}$$

where

$$K_{11} = \int_A \left(d_{11}\frac{\partial N_i}{\partial x}\frac{\partial N_j}{\partial x} + d_{33}\frac{\partial N_i}{\partial y}\frac{\partial N_j}{\partial y} \right) dx\,dy \tag{8.27}$$

$$K_{12} = K_{21}^T = \int_A \left(d_{12}\frac{\partial N_i}{\partial x}\frac{\partial N_j}{\partial y} + d_{33}\frac{\partial N_i}{\partial y}\frac{\partial N_j}{\partial x} \right) dxdy \tag{8.28}$$

$$K_{22} = \int_A \left(d_{33}\frac{\partial N_i}{\partial x}\frac{\partial N_j}{\partial x} + d_{22}\frac{\partial N_i}{\partial y}\frac{\partial N_j}{\partial y} \right) dxdy \tag{8.29}$$

$$F_1 = \int_A N_i f_x dxdy + \int_{S_1} N_i t_x dS \tag{8.30}$$

$$F_2 = \int_A N_i f_y dxdy + \int_{S_1} N_i t_y dS \tag{8.31}$$

Equations (8.25) through (8.31) are valid for elements of any shape. At any node on the domain boundary, there are four possible boundary conditions (Reddy, 1984): (1) u and v specified, (2) u and t_y specified, (3) t_x and v specified, and (4) t_x and t_y specified. Only one of the quantities of each pair (u, t_x) and (v, t_y) need be specified at a boundary point.

The stress–strain relations can be rewritten as

$$\begin{bmatrix} \varepsilon_x \\ \varepsilon_y \\ \gamma_{xy} \end{bmatrix} = \begin{bmatrix} \frac{\partial}{\partial x} & 0 \\ 0 & \frac{\partial}{\partial y} \\ \frac{\partial}{\partial y} & \frac{\partial}{\partial x} \end{bmatrix} \begin{bmatrix} u \\ v \end{bmatrix} \tag{8.32}$$

$$\begin{bmatrix} \sigma_x \\ \sigma_y \\ \tau_{xy} \end{bmatrix} = \begin{bmatrix} d_{11} & d_{12} & 0 \\ d_{21} & d_{22} & 0 \\ 0 & 0 & d_{33} \end{bmatrix} \begin{bmatrix} \varepsilon_x \\ \varepsilon_y \\ \gamma_{xy} \end{bmatrix} \tag{8.33}$$

and the boundary conditions involving boundary tractions as

$$\begin{bmatrix} \frac{\partial}{\partial x} & 0 & \frac{\partial}{\partial y} \\ 0 & \frac{\partial}{\partial y} & \frac{\partial}{\partial x} \end{bmatrix} \begin{bmatrix} \sigma_x \\ \sigma_y \\ \gamma_{xy} \end{bmatrix} = \begin{bmatrix} f_x \\ f_y \end{bmatrix} \tag{8.34}$$

A simple expression for the three-noded linear triangular element can then be obtained, such that

$$\mathbf{Ka} = \mathbf{F} \tag{8.35}$$

where

$$\mathbf{K} = \int_A \mathbf{B}^T \mathbf{D} \mathbf{B} dxdy \tag{8.36}$$

with $\varepsilon = \mathbf{Ba}$ and

$$\mathbf{B} = \begin{bmatrix} \frac{\partial}{\partial x} & 0 \\ 0 & \frac{\partial}{\partial y} \\ \frac{\partial}{\partial y} & \frac{\partial}{\partial x} \end{bmatrix} \mathbf{N} = \begin{bmatrix} \frac{\partial N_1}{\partial x} & 0 & \frac{\partial N_2}{\partial x} & 0 & \frac{\partial N_3}{\partial x} & 0 \\ 0 & \frac{\partial N_1}{\partial y} & 0 & \frac{\partial N_2}{\partial y} & 0 & \frac{\partial N_3}{\partial y} \\ \frac{\partial N_1}{\partial y} & \frac{\partial N_1}{\partial x} & \frac{\partial N_2}{\partial y} & \frac{\partial N_2}{\partial x} & \frac{\partial N_3}{\partial y} & \frac{\partial N_3}{\partial x} \end{bmatrix} \tag{8.37}$$

The load vector is given by

$$\mathbf{F} = \int_A \mathbf{N}^T \mathbf{f} dxdy + \int_S \mathbf{N}^T \mathbf{t} ds \tag{8.38}$$

with

$$\mathbf{f} = \begin{bmatrix} f_x \\ f_y \end{bmatrix} \text{ and } \mathbf{t} = \begin{bmatrix} t_x \\ t_y \end{bmatrix}$$

For a constant load, $f_x = f_0$ and $f_y = 0$

$$\int_A \mathbf{N}^T \mathbf{f} dxdy = \frac{A f_0}{3} \begin{bmatrix} 1 \\ 0 \\ 1 \\ 0 \\ 1 \\ 0 \end{bmatrix} \tag{8.39}$$

Notice that the strains are easily calculated from the displacements using $\varepsilon = \mathbf{Ba}$, that is

$$\varepsilon_{xx} = \frac{\partial u}{\partial x} = \frac{1}{2A}(b_1 a_1 + b_2 a_3 + b_3 a_5) \tag{8.40}$$

$$\varepsilon_{yy} = \frac{\partial v}{\partial y} = \frac{1}{2A}(c_1 a_2 + c_2 a_4 + c_3 a_6) \tag{8.41}$$

$$\gamma_{xy} = \frac{\partial u}{\partial y} + \frac{\partial v}{\partial x} = \frac{1}{2A}(c_1 a_1 + b_1 a_2 + c_2 a_3 + b_2 a_4 + c_3 a_5 + b_3 a_6) \tag{8.42}$$

where the coefficients b_i and c_i are the same as in Table 4.2 and a_i are the degrees of freedom as defined in (8.21). In matrix form:

$$\begin{bmatrix} \varepsilon_{xx} \\ \varepsilon_{yy} \\ \gamma_{xy} \end{bmatrix} = \frac{1}{2A} \begin{bmatrix} b_1 & 0 & b_2 & 0 & b_3 & 0 \\ 0 & c_1 & 0 & c_2 & 0 & c_3 \\ c_1 & b_1 & c_2 & b_2 & c_3 & b_3 \end{bmatrix} \begin{bmatrix} a_1 \\ a_2 \\ a_3 \\ a_4 \\ a_5 \\ a_6 \end{bmatrix} \tag{8.43}$$

The 3 × 6 matrix **B** containing b_i and c_i coefficients is the extension to the case of two degrees of freedom per node of the gradient matrix for a triangular element previously defined in Chapter 4, Eq. (4.85). The number of rows exceeds the dimensions of the problem because there are three unknown components of strain in a two-dimensional problem.

Example 8.1*

A thin elastic plate of thickness t is subjected to a load uniformly distributed along one edge, as shown in Figure 8.3a. Determine the nodal displacements using two triangular elements to discretize the problem domain shown in Figure 8.3b. Assume the plate is isotropic and that plane stress is valid. This problem is discussed in many structurally oriented finite element texts (Chandrupatla and Belegundu, 1991; Bickford, 1990; Stasa, 1985; Reddy, 1993).

Utilizing the plane stress assumptions defined by Eqs. (8.6) through (8.9), the element stiffness matrices associated with the two elements are computed from Eq. (8.36) as

Element 1:

$$x_1 = y_1 = y_2 = 0 \qquad x_2 = x_3 = \ell \qquad y_3 = w$$

$$b_1 = -w \qquad b_2 = w \qquad b_3 = 0$$

$$c_1 = 0 \qquad c_2 = -\ell \qquad c_3 = \ell$$

$$\mathrm{F}^{(1)} = \left[F_{1x}^{(1)} \;\; F_{1y}^{(1)} \;\; F_{2x}^{(1)} \;\; F_{2y}^{(1)} \;\; F_{3x}^{(1)} \;\; F_{3y}^{(1)} \right]^T$$

$$\mathrm{k}^{(1)} = \frac{Et}{2\ell w(1-\mu^2)} \begin{bmatrix} w^2 & & & & & (\text{symmetric}) \\ 0 & \frac{1-\mu}{2}w^2 & & & & \\ -w^2 & \frac{1-\mu}{2}\ell w & w^2 + \frac{1-\mu}{2}\ell^2 & & & \\ \mu\ell w & -\frac{1-\mu}{2}w^2 & -\frac{1+\mu}{2}\ell w & \ell^2 + \frac{1-\mu}{2}w^2 & & \\ 0 & -\frac{1-\mu}{2}\ell w & -\frac{1-\mu}{2}\ell^2 & \frac{1-\mu}{2}\ell w & \frac{1-\mu}{2}\ell^2 & \\ -\mu\ell w & 0 & \mu\ell w & -\ell^2 & 0 & \ell^2 \end{bmatrix} \tag{8.44}$$

* COMSOL file is available.

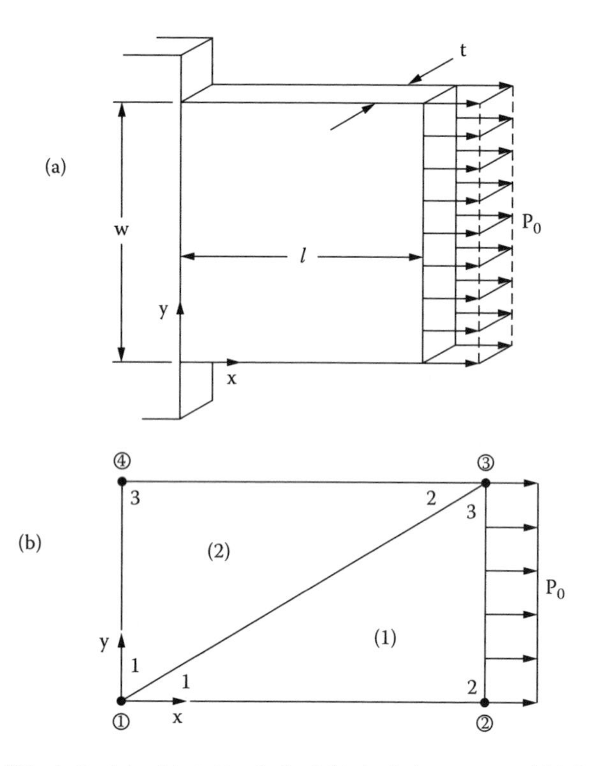

Figure 8.3 Thin elastic plate subjected to edge load: (a) physical geometry; and (b) discretization into two triangular elements

Element 2:

$$x_1 = x_3 = y_1 = 0 \qquad x_2 = \ell \qquad y_2 = y_3 = w$$

$$b_1 = 0 \qquad b_2 = w \qquad b_3 = -w$$

$$c_1 = -\ell \qquad c_2 = 0 \qquad c_3 = \ell$$

$$F^{(2)} = \left[F_{1x}^{(2)} \; F_{1y}^{(2)} \; F_{2x}^{(2)} \; F_{2y}^{(2)} \; F_{3x}^{(2)} \; F_{3y}^{(2)} \right]^T$$

$$k^{(2)} = \frac{Et}{2\ell w(1-\mu^2)} \begin{bmatrix} \frac{1-\mu}{2}\ell^2 & & & & & \text{(symmetric)} \\ 0 & \ell^2 & & & & \\ 0 & -\mu\ell w & w^2 & & & \\ -\frac{1-\mu}{2}\ell w & 0 & 0 & \frac{1-\mu}{2}w^2 & & \\ -\frac{1-\mu}{2}\ell^2 & \mu\ell w & -w^2 & \frac{1-\mu}{2}\ell w & w^2 + \frac{1-\mu}{2}\ell^2 & \\ \frac{1-\mu}{2}\ell w & \ell^2 & \mu\ell w & -\frac{1-\mu}{2}w^2 & -\frac{1-\mu}{2}\ell w & \ell^2 + \frac{1-\mu}{2}w^2 \end{bmatrix} \tag{8.45}$$

Because there are two unknowns (u, v) per node, assembly of the element matrices must account for two degrees of freedom (DOF). Thus, an 8 × 8 global matrix (DOF × 4) is formed which consists of individual coefficients of **K**, i.e., k_{11}, k_{12}, etc. Thus,

$$\mathbf{K} = \begin{array}{c} \begin{matrix} u_1 & v_1 & u_2 & v_2 & u_3 & v_3 & u_4 & v_4 \end{matrix} \\ \begin{bmatrix} k_{11}^{(1)} + k_{11}^{(2)} & k_{12}^{(1)} + k_{12}^{(2)} & k_{13}^{(1)} & k_{14}^{(1)} & k_{15}^{(1)} + k_{13}^{(2)} & k_{16}^{(1)} + k_{14}^{(2)} & k_{15}^{(2)} & k_{16}^{(2)} \\ & k_{22}^{(1)} + k_{22}^{(2)} & k_{23}^{(1)} & k_{24}^{(1)} & k_{25}^{(1)} + k_{23}^{(2)} & k_{26}^{(1)} + k_{24}^{(2)} & k_{25}^{(2)} & k_{26}^{(2)} \\ & & k_{33}^{(1)} & k_{34}^{(1)} & k_{35}^{(1)} & k_{36}^{(1)} & 0 & 0 \\ & & & k_{44}^{(1)} & k_{45}^{(1)} & k_{46}^{(1)} & 0 & 0 \\ & & & & k_{55}^{(1)} + k_{33}^{(2)} & k_{56}^{(1)} + k_{34}^{(2)} & k_{35}^{(2)} & k_{36}^{(2)} \\ & & & & & k_{66}^{(1)} + k_{44}^{(2)} & k_{45}^{(2)} & k_{46}^{(2)} \\ & & & & & & k_{55}^{(2)} & k_{56}^{(2)} \\ & & & & & & & k_{66}^{(2)} \end{bmatrix} \end{array} \tag{8.46}$$

where

$$\begin{bmatrix} u_1 \\ v_1 \\ u_2 \\ v_2 \\ u_3 \\ v_3 \\ u_4 \\ v_4 \end{bmatrix} \equiv \begin{bmatrix} a_1 \\ a_2 \\ a_3 \\ a_4 \\ a_5 \\ a_6 \\ a_7 \\ a_8 \end{bmatrix}$$

The load vector is given by

$$\mathbf{F} = \begin{bmatrix} F_{1x}^{(1)} + F_{1x}^{(2)} \\ F_{1y}^{(1)} + F_{1y}^{(2)} \\ F_{2x}^{(1)} \\ F_{2y}^{(1)} \\ F_{3x}^{(1)} + F_{2x}^{(2)} \\ F_{3y}^{(1)} + F_{2y}^{(2)} \\ F_{3x}^{(2)} \\ F_{3y}^{(2)} \end{bmatrix} \begin{matrix} \text{--------- } u_1 \\ \text{--------- } v_1 \\ \text{--------- } u_2 \\ \text{--------- } v_2 \\ \text{--------- } u_3 \\ \text{--------- } v_3 \\ \text{--------- } u_4 \\ \text{--------- } v_4 \end{matrix} \tag{8.47}$$

Since nodes 1 and 4 are fixed,

$$u_1 = v_1 = u_4 = v_4 = 0 \tag{8.48}$$

Moreover,

$$\left.\begin{aligned} F_3 &= F_{2x}^{(1)} = \frac{P_0 w t}{2} \\ F_5 &= F_{3x}^{(1)} + F_{2x}^{(2)} = \frac{P_0 w t}{2} \end{aligned}\right\} \tag{8.49}$$

Consequently, the first and last rows and columns of **K**, given by Eq. (8.46), can be deleted. Thus, **K** reduces to

$$\begin{bmatrix} k_{33}^{(1)} & k_{34}^{(1)} & k_{35}^{(1)} & k_{36}^{(1)} \\ k_{43}^{(1)} & k_{44}^{(1)} & k_{45}^{(1)} & k_{46}^{(1)} \\ k_{53}^{(1)} & k_{54}^{(1)} & k_{55}^{(1)} + k_{33}^{(2)} & k_{56}^{(1)} + k_{34}^{(2)} \\ k_{63}^{(1)} & k_{64}^{(1)} & k_{65}^{(1)} + k_{43}^{(2)} & k_{66}^{(1)} + k_{44}^{(2)} \end{bmatrix} \begin{bmatrix} u_2 \\ v_2 \\ u_3 \\ v_3 \end{bmatrix} = \begin{bmatrix} \frac{P_0 w t}{2} \\ 0 \\ \frac{P_0 w t}{2} \\ 0 \end{bmatrix} \tag{8.50}$$

Substituting $\ell = 90$ in., $t = 0.025$ in, $E = 1.08 \times 10^6$ lb/in.^{2}, $w = 120$ in., $\mu = 0.25$, and $P_0 = 20$ lb/in.^{2}, Eq. (8.50) becomes

$$\begin{bmatrix} 23250 & -9000 & -4050 & 3600 \\ -9000 & 18000 & 5400 & -10800 \\ -4050 & 5400 & 23250 & 0 \\ 3600 & -10800 & 0 & 18000 \end{bmatrix} \begin{bmatrix} u_2 \\ v_2 \\ u_3 \\ v_3 \end{bmatrix} = \begin{bmatrix} 30 \\ 0 \\ 30 \\ 0 \end{bmatrix} \tag{8.51}$$

from which the displacements are easily calculated (via a matrix inversion)

$$\begin{bmatrix} u_2 \\ v_2 \\ u_3 \\ v_3 \end{bmatrix} = \begin{bmatrix} 1.69\ x\ 10^{-3} \\ 2.95\ x\ 10^{-4} \\ 1.52\ x\ 10^{-3} \\ -1.62\ x\ 10^{-4} \end{bmatrix} in \tag{8.52}$$

Notice that the solution is not symmetric. This is due to the coarse discretization—the exact solution should be symmetric about the horizontal centerline, and can be achieved with the use of finer grids.

8.4 POTENTIAL ENERGY

In the previous section we solved the elasticity equations using the Galerkin weighted residuals formulation to approximate the governing differential equations, in the same manner as we had applied the finite element method to the heat equation in previous chapters. However, this is rarely done in solid mechanics and particularly in linear elasticity. A more standard way to formulate this is using the principle of minimum potential energy in conjunction with the Rayleigh–Ritz method.

The principle of minimum potential energy states: Of all the kinematically compatible displacement fields that a structure can attain under a specified set of external loads, the one that satisfies the governing equations—(8.13) and (8.14)—also minimizes the potential energy of the system. Vice versa, the displacement field that minimizes the potential energy is also the solution to Eqs. (8.13) and (8.14).

The advantages of formulating the problem in this fashion are that the potential energy is easy to obtain and that the Rayleigh–Ritz method guarantees that if we minimize the potential energy over the approximation functions defined by a finite element mesh, then as the mesh is refined the solution converges to the exact solution. Moreover, the minimization formulation lends itself to a detailed mathematical analysis rich in useful results. However, the analysis becomes mathematically involved, and is beyond the scope of this book. Those readers eager to explore the aspects of the finite element method in more detail are referred to Strang and Fix (1973), Oden and Reddy (1976), Oden and Carey (1983), and Dawe (1984). Here we concentrate on how the method works in practice.

For a linear elastic solid, the total potential energy is given by

$$\mathrm{U} = \frac{1}{2}\int_{v} \varepsilon^{\mathrm{T}}\sigma \mathrm{dv} - \int_{S_1} \mathbf{u}^{\mathrm{T}}\mathbf{t}\mathrm{ds} - \int_{v} \mathbf{u}^{\mathrm{T}}\mathbf{f}\mathrm{dV} - \sum_{k} \mathbf{u}_{\mathrm{k}}^{\mathrm{T}}\mathbf{P}_{\mathrm{k}} \tag{8.53}$$

where P_k denotes a point load applied at node k and u_k the displacements at that node. Substituting Eqs. (8.5), (8.22), and (8.37) the above expression becomes

$$U = \frac{1}{2}\int_{v} \mathbf{a}^{T}\mathbf{B}^{T}\mathbf{D}\mathbf{B}\mathbf{a}dV - \int_{S_1} \mathbf{a}^{T}\mathbf{N}^{T}\mathbf{t}dS_1 - \int_{v} \mathbf{a}^{T}\mathbf{N}^{T}\mathbf{f}dv - \sum_{k} \mathbf{a}_{k}^{T}\mathbf{P}_{k} \tag{8.54}$$

Here, $a_k = [u_k\ v_k]^T$ are the two components of displacement at node k and $\boldsymbol{P}_k = [P_{xk}\ P_{yk}]^T$ are the components of the applied point load.

The minimization of U given by Eq. (8.54) over the finite element mesh that defines the degrees of freedom **a** is known as the Rayleigh–Ritz approximation. To accomplish this, the conditions to find the minimum are given by

$$\frac{\partial U}{\partial a_i} = 0 \qquad i = 1, \ldots, N \tag{8.55}$$

where N is the total number of displacement degrees of freedom in the discretization. The final result is the system of equations:

$$\left[\int_{v} \mathbf{B}^{T}\mathbf{D}\mathbf{B}dV\right]\mathrm{a} = \int_{S_1} \mathbf{N}^{T}\mathbf{t}dS + \int_{v} \mathbf{N}^{T}\mathbf{f}dV + \mathbf{P} \tag{8.56}$$

Notice that this expression is exactly the same as the one we obtained before using the Galerkin method. Equation (8.56) is more general than Eq. (8.35)—which represents the discretized form of Eqs. (8.23) and (8.24)—because it has the additional vector **P** that contains the applied point loads that were not included previously. When a minimum principle exists, it yields the exact same solutions as the Galerkin formulation, but the latter is more general because it can be applied in situations where no minimum formulation is possible.

Example 8.2

Consider the one-dimensional bar of length L and cross-sectional area A shown in Figure 8.4 and an applied force **F** at the right-hand end.

The bar has an elastic modulus **E**, is clamped at the left-hand side, and is subject to a distributed load $w_0$in the x-direction and a force **F** applied at the right-hand end. We will approximate the stresses and strains in the bar using two linear one-dimensional elements.

The matrices **B** and **D** reduce to

$$\mathbf{B} = \begin{bmatrix} \frac{dN_1}{dx} & \frac{dN_2}{dx} & \frac{dN_3}{dx} \end{bmatrix}, \qquad \mathbf{D} = [E]$$

we also have $\mathbf{a}^T = [u_1 u_2 u_3]$, $\mathbf{f} = [w_0]$ and $\mathbf{t} = [F/A]$. Eq. (8.56) becomes

$$\left[\int_v \begin{bmatrix} \frac{dN_1}{dx} \\ \frac{dN_2}{dx} \\ \frac{dN_3}{dx} \end{bmatrix} E \begin{bmatrix} \frac{dN_1}{dx} & \frac{dN_2}{dx} & \frac{dN_3}{dx} \end{bmatrix} dV \right] \begin{bmatrix} u_1 \\ u_2 \\ u_3 \end{bmatrix} = \int_A \begin{bmatrix} 0 \\ 0 \\ F/A \end{bmatrix} dA + \int_V \begin{bmatrix} N_1 \\ N_1 \\ N_3 \end{bmatrix} w_0 dV$$

Integrating over the cross-sectional area A,

$$\left[AE \int_0^L \begin{bmatrix} \left(\frac{dN_1}{dx}\right)^2 & \frac{dN_1}{dx}\frac{dN_2}{dx} & \frac{dN_1}{dx}\frac{dN_3}{dx} \\ \frac{dN_2}{dx}\frac{dN_1}{dx} & \left(\frac{dN_2}{dx}\right)^2 & \frac{dN_2}{dx}\frac{dN_3}{dx} \\ \frac{dN_3}{dx}\frac{dN_1}{dx} & \frac{dN_3}{dx}\frac{dN_2}{dx} & \left(\frac{dN_3}{dx}\right)^3 \end{bmatrix} dx \right] \begin{bmatrix} u_1 \\ u_2 \\ u_3 \end{bmatrix} = \begin{bmatrix} 0 \\ 0 \\ F \end{bmatrix} + \frac{w_0 L}{2} \begin{bmatrix} 1 \\ 2 \\ 1 \end{bmatrix} \tag{8.57}$$

Notice that the integrals can be done element-wise as before, so that only the element matrices are needed and the global matrix is assembled from the elements. For the linear element with the shape functions given in Eq. (3.66) the element stiffness matrix is

$$\frac{2AE}{L} \begin{bmatrix} 1 & -1 \\ -1 & 1 \end{bmatrix}$$

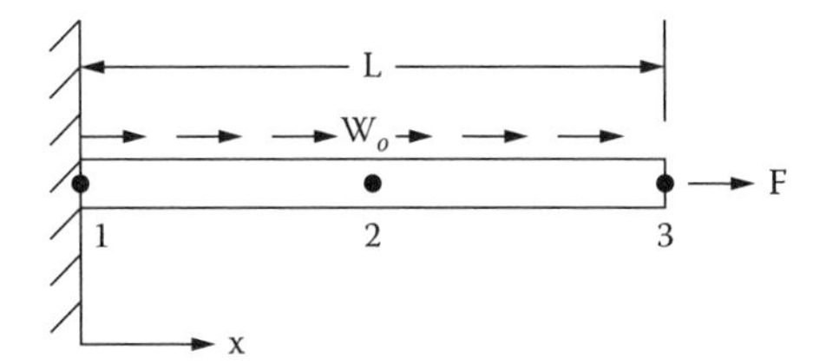

Figure 8.4 One-dimensional bar subject to a distributed sheer load w_0

and the final assembled system is

$$\frac{2AE}{L}\begin{bmatrix} 1 & -1 & 0 \\ -1 & 2 & -1 \\ 0 & -1 & 1 \end{bmatrix}\begin{bmatrix} u_1 \\ u_2 \\ u_3 \end{bmatrix} = \begin{bmatrix} \frac{w_0L}{4} \\ \frac{w_0L}{2} \\ \frac{w_0L}{4}+F \end{bmatrix}$$

As in the heat transfer problem, the system can be partitioned after applying the Dirichlet boundary condition $u_1 = 0$. We will write it in the form

$$\frac{2AE}{L}(-u_2) = R_1 + \frac{w_0L}{4} \tag{8.58}$$

$$\left.\begin{aligned} \frac{2AE}{L}(2u_2 - u_3) &= \frac{w_0L}{2} \\ \frac{2AE}{L}(-u_2 + u_3) &= F + \frac{w_0L}{4} \end{aligned}\right\} \tag{8.59}$$

The solution of (8.59) is

$$u_2 = \frac{3w_0L^2}{8AE} + \frac{FL}{2AE}$$

and

$$u_3 = \frac{FL}{AE} + \frac{w_0L^2}{2AE}$$

Replacing u_2 into (8.58) yields $R_1 = -(\mathbf{F} + w_0L)$, which is the reaction at the left wall. Just as in heat transfer the redundant equations were used to calculate the fluxes at walls with prescribed temperatures. In solid mechanics problems, these are used to find the stresses at the constraint locations.

8.5 THERMAL STRESSES

Differences in temperature are an important source of stresses and deformations in solid bodies. Now we will assume that given a solid structure, we can use the finite element method to obtain the temperature distribution at every interior point using the techniques already covered in Chapters 3 through 7. To incorporate the effect of the temperature in the structural analysis, we need to modify the constitutive relation (8.5) to become

$$\sigma = \mathbf{D}(\varepsilon - \varepsilon_T) \tag{8.60}$$

where $_T$ are the strains induced by the temperature field. $_T$ is given by

$$\begin{bmatrix} \alpha(T-T_0) \\ \alpha(T-T_0) \\ \alpha(T-T_0) \\ 0 \\ 0 \\ 0 \end{bmatrix}, \begin{bmatrix} \alpha(T-T_0) \\ \alpha(T-T_0) \\ 0 \end{bmatrix} \textit{and} \left[\alpha(T-T_0)\right] \tag{8.61}$$

in three, two, and one dimensions respectively, where T_0 is a reference constant temperature at which no thermal stresses occur, and α (mm/mm/°C or in/in/°F) is the coefficient of linear expansion of the material. Notice that only direct strains and no shear strains are produced by temperature differences.

Incorporating the potential energy due to the thermally induced strains into Eq. (8.53) and minimizing results in the system of equations

$$\mathbf{ka} = \mathbf{F} + \mathbf{F}_T \tag{8.62}$$

where **k** and **F** are exactly the same as before, and

$$\mathbf{F}_T = \int_v \mathbf{B}^T \mathbf{D} \varepsilon_T dV \tag{8.63}$$

EXAMPLE 8.3

Let us calculate $\mathbf{F}_T$ for a two-dimensional linear triangle, assuming a state of plane stress in a plate of thickness t.

$$\mathbf{D}\varepsilon_T = \frac{E}{1-\mu^2}\begin{bmatrix} 1 & \mu & 0 \\ \mu & 1 & 0 \\ 0 & 0 & \frac{1-\mu}{2} \end{bmatrix}\begin{bmatrix} \alpha(T-T_0) \\ \alpha(T-T_0) \\ 0 \end{bmatrix} = \frac{E\alpha(T-T_0)}{1-\mu}\begin{bmatrix} 1 \\ 1 \\ 0 \end{bmatrix}$$

from Eq. (8.43) and Table 4.2

$$\mathbf{B} = \frac{1}{2A}\begin{bmatrix} y_2 - y_3 & 0 & y_3 - y_1 & 0 & y_1 - y_2 & 0 \\ 0 & x_3 - x_2 & 0 & x_1 - x_3 & 0 & x_2 - x_1 \\ x_3 - x_2 & y_2 - y_3 & x_1 - x_3 & y_3 - y_1 & x_2 - x_1 & y_1 - y_2 \end{bmatrix}$$

Substituting in Eq. (8.63)

$$\mathbf{F}_T = \int_v \frac{1}{2A}\begin{bmatrix} y_2 - y_3 & 0 & x_3 - x_2 \\ 0 & x_3 - x_2 & y_2 - y_3 \\ y_3 - y_1 & 0 & x_1 - x_3 \\ 0 & x_1 - x_3 & y_3 - y_1 \\ y_1 - y_2 & 0 & x_2 - x_1 \\ 0 & x_2 - x_1 & y_1 - y_2 \end{bmatrix} \frac{E\alpha(T - T_0)}{1-\mu}\begin{bmatrix} 1 \\ 1 \\ 0 \end{bmatrix} dV$$

$$= \frac{E\alpha t}{2A(1-\mu)}\int_A (T - T_0)\,dA \begin{bmatrix} y_2 - y_3 \\ x_3 - x_2 \\ y_3 - y_1 \\ x_1 - x_3 \\ y_1 - y_2 \\ x_2 - x_1 \end{bmatrix}$$

Expressing the temperature T in terms of shape functions and its nodal values

$$T = N_1T_1 + N_2T_2 + N_3T_3$$

the integral over the area becomes

$$\int_A (T - T_0)\,dA = \frac{A}{3}(T_1 + T_2 + T_3 - 3T_0)$$

and finally

$$\mathbf{F}_T = \frac{E\alpha t(T_1 + T_2 + T_3 - 3T_0)}{6(1-\mu)}\begin{bmatrix} y_2 - y_3 \\ x_3 - x_2 \\ y_3 - y_1 \\ x_1 - x_3 \\ y_1 - y_2 \\ x_2 - x_1 \end{bmatrix} \tag{8.64}$$

We conclude this section with a simple one-dimensional example.

EXAMPLE 8.4

The bar in Figure 8.5 has cross-sectional area A, modulus E, coefficient of thermal expansion α, and has been divided into three elements of equal length L. The middle element is

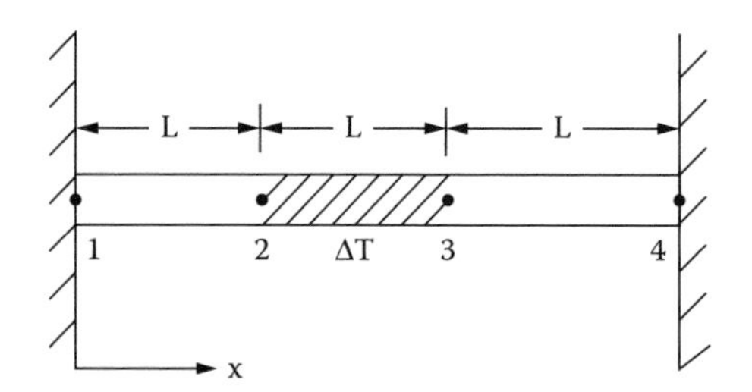

Figure 8.5 One-dimensional bar with the middle element subject to a change in temperature T

heated up to a constant temperature $T = T - T_0$ while the other two are kept at T_0. Let us find the stresses in the bar. From Eqs. (8.61) and (8.63),

$$\mathbf{F}_T = EA\alpha\Delta T\begin{bmatrix} -1 \\ 1 \end{bmatrix} \tag{8.65}$$

and this is the only load. The stiffness relation (8.62) becomes

$$\frac{EA}{L}\begin{bmatrix} 1 & -1 & 0 & 0 \\ -1 & 2 & -1 & 0 \\ 0 & -1 & 2 & -1 \\ 0 & 0 & -1 & 1 \end{bmatrix}\begin{bmatrix} u_1 \\ u_2 \\ u_3 \\ u_4 \end{bmatrix} = EA\alpha\Delta T\begin{bmatrix} 0 \\ -1 \\ 1 \\ 0 \end{bmatrix}$$

Improving the boundary conditions $u_1 = u_4 = 0$ and solving, we obtain

$$u_2 = \frac{-\alpha L\Delta T}{3} \text{ and } u_3 = \frac{+\alpha L\Delta T}{3}$$

The stresses in the bars are obtained from Eq. (8.60).

$$\sigma^{(1)} = E\left\{\begin{bmatrix} -\frac{1}{L} & \frac{1}{L} \end{bmatrix}\begin{bmatrix} 0 \\ \frac{-\alpha L\Delta T}{3} \end{bmatrix} - 0\right\} = -\frac{E\alpha\Delta T}{3}$$

$$\sigma^{(2)} = E\left\{\begin{bmatrix} -\frac{1}{L} & \frac{1}{L} \end{bmatrix}\begin{bmatrix} \frac{-\alpha L\Delta T}{3} \\ \frac{\alpha L\Delta T}{3} \end{bmatrix} - \alpha\Delta T\right\} = -\frac{E\alpha\Delta T}{3}$$

$$\sigma^{(3)} = E\left\{\begin{bmatrix} -\frac{1}{L} & \frac{1}{L} \end{bmatrix}\begin{bmatrix} \frac{\alpha L\Delta T}{3} \\ 0 \end{bmatrix} - 0\right\} = -\frac{E\alpha\Delta T}{3}$$

The elements are subject to a uniform compressive stress as expected.

8.6 THREE-DIMENSIONAL SOLID ELEMENTS

For a three-dimensional elastic solid, six components of stress and strain exist, as defined by

$$\varepsilon = \begin{bmatrix} \varepsilon_x \\ \varepsilon_y \\ \varepsilon_z \\ \gamma_{xy} \\ \gamma_{xz} \\ \gamma_{yz} \end{bmatrix} = \begin{bmatrix} \dfrac{\partial \mathrm{u}}{\partial \mathrm{x}} \\ \dfrac{\partial \mathrm{v}}{\partial \mathrm{y}} \\ \dfrac{\partial \mathrm{w}}{\partial \mathrm{z}} \\ \dfrac{\partial u}{\partial y} + \dfrac{\partial v}{\partial x} \\ \dfrac{\partial u}{\partial z} + \dfrac{\partial w}{\partial x} \\ \dfrac{\partial v}{\partial z} + \dfrac{\partial w}{\partial y} \end{bmatrix} \tag{8.66}$$

where u, v, and w are the displacement fields in the x, y, and z directions respectively. The stresses are related to the strains by Hooke's law that now take the form:

$$\varepsilon = \begin{bmatrix} \sigma_x \\ \sigma_y \\ \sigma_z \\ \tau_{xy} \\ \tau_{xz} \\ \tau_{yz} \end{bmatrix}$$

$$= \frac{E}{(1+\mu)(1-2\mu)} \begin{bmatrix} 1-\mu & \mu & \mu & 0 & 0 & 0 \\ \mu & 1-\mu & \mu & 0 & 0 & 0 \\ \mu & \mu & 1-\mu & 0 & 0 & 0 \\ 0 & 0 & 0 & \dfrac{1-2\mu}{2} & 0 & 0 \\ 0 & 0 & 0 & 0 & \dfrac{1-2\mu}{2} & 0 \\ 0 & 0 & 0 & 0 & 0 & \dfrac{1-2\mu}{2} \end{bmatrix} \begin{bmatrix} \varepsilon_x \\ \varepsilon_y \\ \varepsilon_z \\ \gamma_{xy} \\ \gamma_{xz} \\ \gamma_{yz} \end{bmatrix}$$

$$= \mathbf{D}\varepsilon \tag{8.67}$$

The surface tractions in Eq. (8.11) become

$$\left.\begin{aligned} \sigma_x n_x + \tau_{xy} n_y + \tau_{xz} n_z &= t_x \\ \tau_{xy} n_x + \sigma_y n_y + \tau_{yz} n_z &= t_y \\ \tau_{xz} n_x + \tau_{yz} n_y + \sigma_z n_z &= t_z \end{aligned}\right\} \tag{8.68}$$

and the equilibrium equations are

$$\left.\begin{aligned} \frac{\partial \sigma_x}{\partial x} + \frac{\partial \tau_{xy}}{\partial y} + \frac{\partial \tau_{xz}}{\partial z} + f_x &= 0 \\ \frac{\partial \tau_{xy}}{\partial x} + \frac{\partial \sigma_y}{\partial y} + \frac{\partial \tau_{yz}}{\partial z} + f_y &= 0 \\ \frac{\partial \tau_{xz}}{\partial x} + \frac{\partial \tau_{yz}}{\partial y} + \frac{\partial \sigma_z}{\partial z} + f_z &= 0 \end{aligned}\right\} \tag{8.69}$$

where the body force has one more component f_z. Figure 8.6 shows a tetrahedral element with the nodal degrees of freedom.

There are 12 degrees of freedom for this element, in terms of the shape functions given in Section 7.3.1. The linear displacement field and the **B** matrix are

$$\begin{bmatrix} u \\ v \\ w \end{bmatrix} = \mathbf{Na} =$$

$$\begin{bmatrix} N_1 & 0 & 0 & N_2 & 0 & 0 & N_3 & 0 & 0 & N_4 & 0 & 0 \\ 0 & N_1 & 0 & 0 & N_2 & 0 & 0 & N_3 & 0 & 0 & N_4 & 0 \\ 0 & 0 & N_1 & 0 & 0 & N_2 & 0 & 0 & N_3 & 0 & 0 & N_4 \end{bmatrix} \begin{bmatrix} a_1 \\ a_2 \\ a_3 \\ a_4 \\ a_5 \\ a_6 \\ a_7 \\ a_8 \\ a_9 \\ a_{10} \\ a_{11} \\ a_{12} \end{bmatrix} \tag{8.70}$$

$$\mathbf{B} = \begin{bmatrix} \frac{\partial N_1}{\partial x} & 0 & 0 & \frac{\partial N_2}{\partial x} & 0 & 0 & & \frac{\partial N_4}{\partial x} & 0 & 0 \\ 0 & \frac{\partial N_1}{\partial y} & 0 & 0 & \frac{\partial N_2}{\partial y} & 0 & & 0 & \frac{\partial N_4}{\partial y} & 0 \\ 0 & 0 & \frac{\partial N_1}{\partial z} & 0 & 0 & \frac{\partial N_2}{\partial z} & & 0 & 0 & \frac{\partial N_4}{\partial z} \\ \frac{\partial N_1}{\partial y} & \frac{\partial N_1}{\partial x} & 0 & \frac{\partial N_2}{\partial y} & \frac{\partial N_2}{\partial x} & 0 & \cdots & \frac{\partial N_4}{\partial y} & \frac{\partial N_4}{\partial x} & 0 \\ \frac{\partial N_1}{\partial z} & 0 & \frac{\partial N_1}{\partial x} & \frac{\partial N_2}{\partial z} & 0 & \frac{\partial N_2}{\partial x} & & \frac{\partial N_4}{\partial z} & 0 & \frac{\partial N_4}{\partial x} \\ 0 & \frac{\partial N_1}{\partial z} & \frac{\partial N_1}{\partial y} & 0 & \frac{\partial N_2}{\partial z} & \frac{\partial N_2}{\partial y} & & 0 & \frac{\partial N_4}{\partial z} & \frac{\partial N_4}{\partial y} \end{bmatrix} \tag{8.71}$$

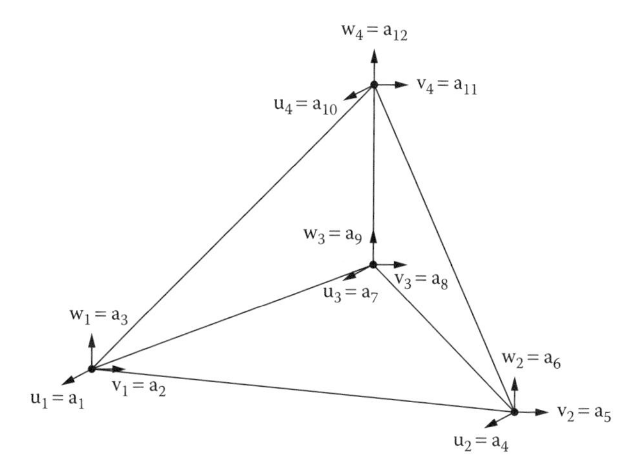

Figure 8.6 Three-dimensional tetrahedral element and displacement degrees of freedom

The element stiffness matrix can then be found from Eq. (8.36); however, this is best left to the computer.

8.7 CLOSURE

The finite element method was originally developed to solve problems of elasticity based on the principle of minimum potential energy. Only later was the method extended to field problems such as heat conduction by means of the method of weighted residuals and the Galerkin formulation. However, we have seen in this chapter that both methods have many similarities and, in fact, they give an identical answer when both formulations are possible. In the following chapters we explore further extensions of the finite element method to problems in which a minimization formulation is not possible, and make the versatility of the finite element method even more evident.

EXERCISES

8.1 Calculate the element stiffness matrix for the plane stress element shown in the figure using one linear triangular element.

$$E = 15 \times 10^6$$

$$\mu = 0.3$$

$$t = 0.1$$

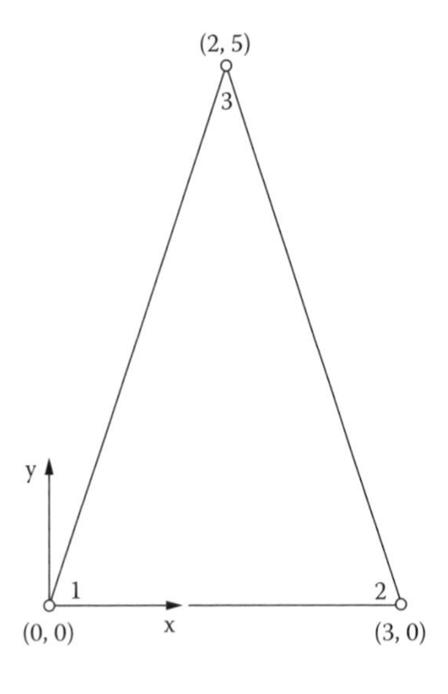

8.2 Calculate the element stiffness matrices for the two elements shown in the figure if both elements are in plane strain.

$$E = 10^6$$

$$\mu = 0.25$$

$$t = 1.0$$

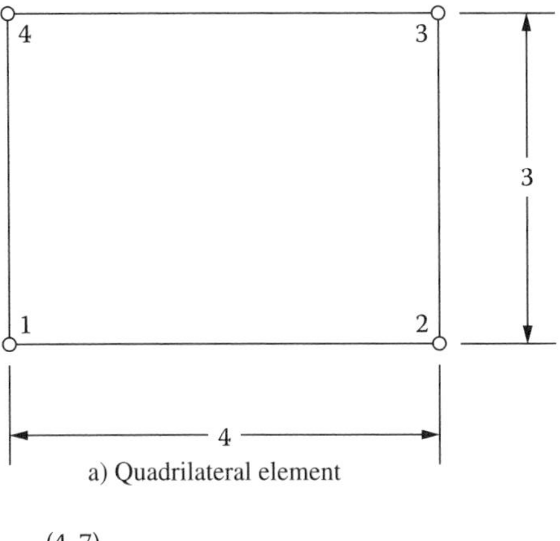

a) Quadrilateral element

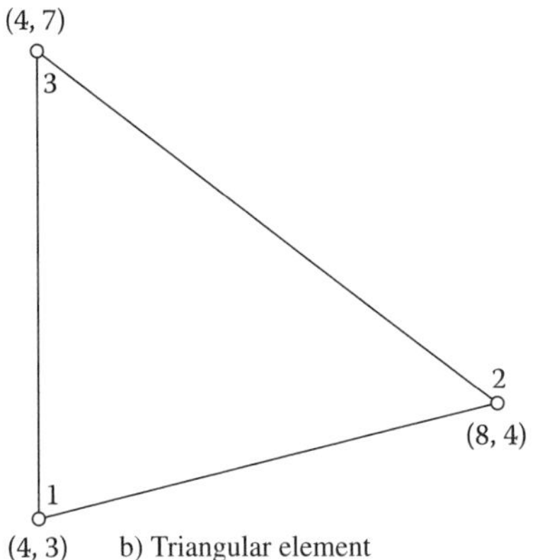

b) Triangular element

8.3 Find the element stiffness matrix given in Eq. (8.44).

8.4 Calculate the stresses in a vertical column of cross section A, elastic modulus E, height H, and density ρ under its own weight. Use a one-dimensional model with two linear elements and then with one quadratic element.

8.5 Consider a thin, rectangular panel subdivided into two triangles in two different ways as shown in the figure. Node one is displaced by a unit amount while the other three are kept fixed. Calculate the strains in the two configurations and discuss the significance of the results.

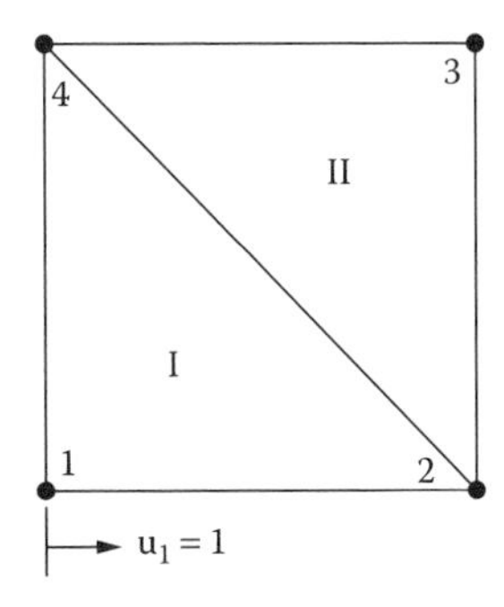

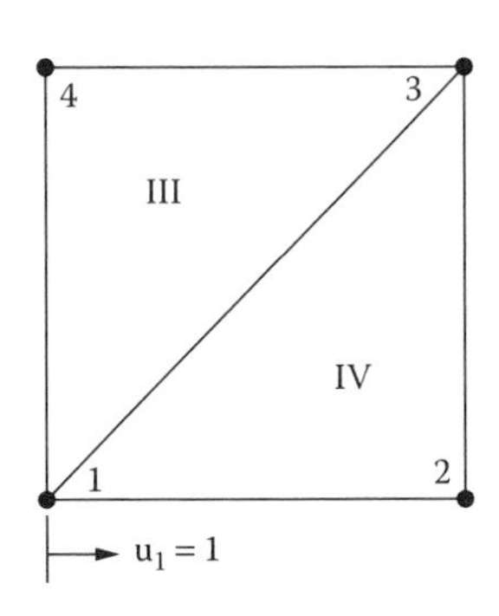

8.6 Repeat the solution to Example 8.1 using minimization of energy.

8.7 Repeat Example 8.2 using the Galerkin method.

8.8 Repeat Example 8.2 using one quadratic element.

8.9 Find the element stiffness matrix given in Eq. (8.57).

8.10 Solve the problem in Example 8.1 using one bilinear element and either the Galerkin formulation or the principle of minimum potential energy.

8.11 In the element of Exercise 8.2a, the nodes are at temperature, $T_1 = T_2 = 100°C$, $T_3 = 0°C$, $T_4 = 200°C$, and $T_0 = 0°C$. Find the load on the element due to the temperature difference if the coefficient of linear thermal expansion is $\alpha = 5.0 \times 10^{-5}$mm/mm/°C.

8.12 In Example 8.1, the load at the tip is removed and instead the thermal load defined in Exercise 8.11 above is applied. Find the displacements and the stresses.

REFERENCES

Bickford, W. B. (1990). *A First Course in the Finite Element Method*. Homewood, Ill.: Richard D. Irwin, Inc.

Chandrupatla, T. R. and Belegundu, A. D. (1991). *Introduction to Finite Elements*. Englewood Cliffs, N.J.: Prentice-Hall.

Dawe, D. J. (1984). *Matrix and Finite Element Displacement Analysis of Structures*. Oxford: Clarendon Press.

Oden, J. T. and Carey, G. F. (1983). *Mathematical Aspects*, Vol. IV. Englewood Cliffs, N.J.: Prentice-Hall.

Oden, J. T. and Reddy, J. N. (1976). *An Introduction to the Mathematical Theory of Finite Elements*. New York: John Wiley & Sons.

Reddy, J. N. (1993). *An Introduction to the Finite Element Method*. 2nd ed. New York: McGraw-Hill.

Stasa, F. L. (1985). *Applied Finite Element Analysis for Engineers*. New York: Holt, Rinehart & Winston.

Strang, G. and Fix, G. J. (1973). *An Analysis of the Finite Element Method*. Englewood Cliffs, N.J.: Prentice-Hall.

CHAPTER

9

APPLICATIONS TO CONVECTIVE TRANSPORT

9.1 BACKGROUND

Most problems in the engineering and scientific disciplines involve complex geometries and boundary conditions, for which the finite element method is ideally suited. Moreover, the method can be easily adapted to a large number of situations governed by different mathematical models. In problems involving heat and mass transfer it is common to encounter systems in which one of the mass transport mechanisms is a moving fluid. In this chapter we establish the use of the finite element method to calculate the flow field for an ideal fluid, we look at how to solve problems in which both diffusion and convection affect the transport, and we look at some practical situations including pollutant transport, groundwater flow, and lubrication.

9.2 POTENTIAL FLOW

The flow of an ideal fluid assumes that no friction exists between the fluid and a surface. Likewise, the fluid motion is irrotational. Hence, boundary layer effects are ignored and the fluid does not separate from the boundary. Velocities in the direction normal to a fixed boundary are zero. While the flow of an ideal fluid does not actually exist in nature, a great deal of insight can be gained about flow around corners or obstacles.

Two-dimensional ideal flow is normally defined in terms of either the potential function, ϕ, or the stream function, Ψ. Lines of constant ϕ are perpendicular to lines of constant Ψ. Because there is no flow, or velocity, perpendicular to a boundary, boundaries and the axis of symmetry parallel to the flow are streamlines. Both the potential function and the stream function satisfy the Laplace equation, that is,

$$\frac{\partial^2 \phi}{\partial x^2} + \frac{\partial^2 \phi}{\partial y^2} = 0 \tag{9.1}$$

and

$$\frac{\partial^2 \psi}{\partial x^2} + \frac{\partial^2 \psi}{\partial y^2} = 0 \tag{9.2}$$

While the equations appear the same, the boundary conditions differ. In Figure 9.1, we depict a two-dimensional cross section for potential flow over a cylinder in a channel, and show the differences in boundary conditions between stream function and potential function formulations.

In a stream function analysis, lines of constant Ψ are called streamlines. The difference between a pair of streamlines is equal to the volumetric flow rate Q_{ij} as defined by the two streamlines, i.e.,

$$Q_{ij} = \psi_i - \psi_j \tag{9.3}$$

The velocity components are calculated from Ψ as

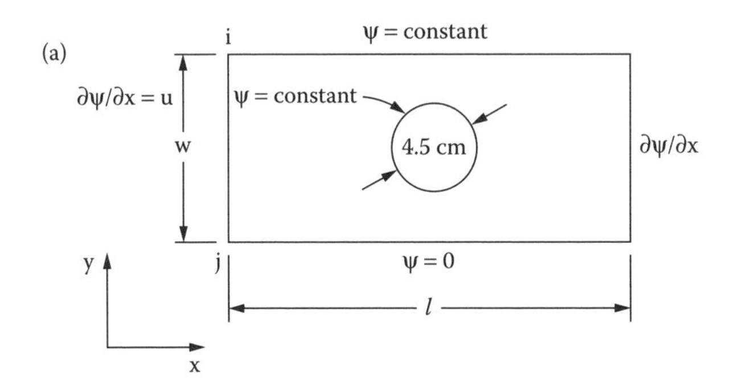

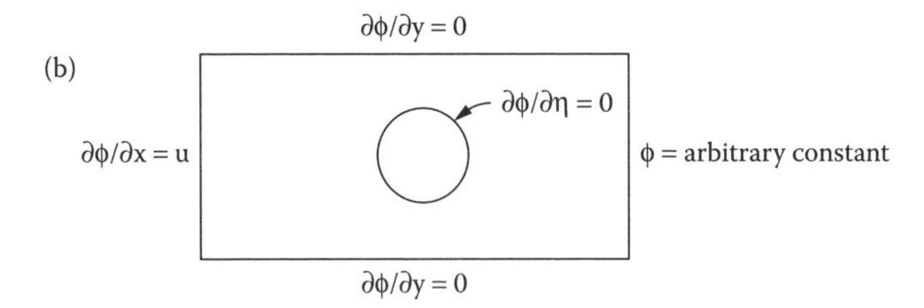

Figure 9.1 Boundary conditions for potential flow: (a) stream function; and (b) potential function

$$u = \frac{\partial \psi}{\partial y} \tag{9.4}$$

$$v = -\frac{\partial \psi}{\partial x} \tag{9.5}$$

where u and v denote the velocity components in the x- and y-directions, respectively. The slope at any point on a line of constant Ψ is always in the direction of the velocity vector. Values for Ψ can be arbitrarily specified—see Figure 9.1a. However, the flow rate must be correctly specified. Hence, for a specified inflow velocity u,

$$Q_{ji} = \int_{\psi_j}^{\psi_i} d\psi = \psi_i - \psi_j = \int_{y_j}^{y_i} u dy = u\left(y_i - y_j\right) = uw \tag{9.6}$$

In potential flow analysis, the velocity components in the x- and y-directions are calculated from the potential function as

$$u = \frac{\partial \phi}{\partial x} \tag{9.7}$$

$$v = \frac{\partial \phi}{\partial y} \tag{9.8}$$

Referring to Figure 9.1b, a constant horizontal velocity of $u = (\partial\phi/\partial x)$ is specified on the left boundary: $(\partial\phi/\partial y) = 0$ is given along the upper and lower boundaries. This condition is necessary because potential lines are perpendicular to streamlines, i.e., normal to boundaries (including the cylinder surface). The right-hand boundary is a potential line since all streamlines are perpendicular to it; the right-hand boundary acts as an outflow boundary. The value is arbitrary because the velocity components are calculated from the gradients of the potential.

Example 9.1* Ideal Flow

A grid of quadrilateral elements is used to discretize the region about a circular cylinder, as shown in Figure 9.2. The grid consists of 25 elements and 37 nodes. Because the flow is symmetric about a cylinder at low (ideal) flows, only one quadrant of the domain needs to be used. In this instance, the lower boundary represents a streamline that coincides with the axis of symmetry of the cylinder. The upper and lower surface have $(\partial\phi/\partial y) = 0$. Constant values of the potential gradient and function are specified at the left and right boundaries, respectively. The boundary conditions are defined in Figure 9.2 for the right vertical wall and the upper horizontal wall. The bottom symmetry line must be a streamline; along this streamline, the vertical velocity vanishes, and hence $(\partial\phi/\partial y)$ must be zero there. Along the right vertical

* FEM-2D and COMSOL files are available.

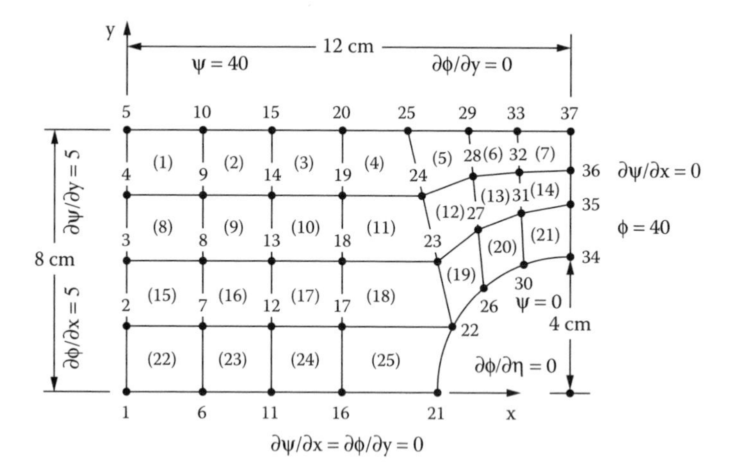

Figure 9.2 Potential flow over a cylinder

boundary, the streamline is horizontal; hence $(\partial\Psi/\partial x) = 0$. Likewise, using Eqs. (9.4) and (9.7) we see that ϕ is a constant and arbitrarily set it to 40.

Along the circular arc, the stream function must be a constant. The stream function varies linearly with y. The potential function must not produce a velocity component normal to the cylinder; hence

$$\frac{\partial\phi}{\partial x}n_x + \frac{\partial\phi}{\partial y}n_y = \frac{\partial\phi}{\partial n} = 0$$

where n_x and n_y components of the unit outward vector normal to the cylinder boundary. Nodal coordinate data and several nodal values for ψ and ϕ are given in Table 9.1. The solution of the problem using both stream function and potential function analyses is left as a computer exercise.

Equation (9.1) or (9.2) is easily solved using the finite element technique. Both formulations are governed by simple Laplacian equations, i.e., a steady-state diffusion equation with $K_{xx} = K_{yy} = 1.0$. Applying the Galerkin procedure to Eq. (9.1), we obtain

$$\int_A N_i\left(\frac{\partial^2\phi}{\partial x^2} + \frac{\partial^2\phi}{\partial y^2}\right)dx\,dy = 0 \tag{9.9}$$

By utilizing Green's theorem, Eq. (9.9) becomes

$$\int_\Omega\left(\frac{\partial N_i}{\partial x}\frac{\partial N_j}{\partial x} + \frac{\partial N_i}{\partial y}\frac{\partial N_j}{\partial y}\right)dx\,dy[\phi_i] = \int_{\Gamma_2}\left(\frac{\partial\phi}{\partial x}n_x + \frac{\partial\phi}{\partial y}n_x\right)N_i d\Gamma$$

or

Table 9.1 Nodal Coordinates, Element Connectivities, and Steady-State Solution for Flow Over a Cylinder

Nodal Coordinates			Element Connectivities					Steady-State Solution		
				Local Nodes						
Node	x	y	Element	1	2	3	4	Node	ψ	ϕ
1	0.0	0.0	1	4	9	10	5	1	0.000	32.062
2	0.0	2.0	2	9	14	15	10	3	19.413	32.119
3	0.0	4.0	3	14	19	20	15	5	40.000	32.175
4	0.0	6.0	4	19	24	25	20	18	16.396	35.130
5	0.0	8.0	5	24	28	29	25	27	13.577	37.563
6	2.0	0.0	6	28	32	33	29	36	29.246	40.000
11	4.0	0.0	7	32	36	37	33			
16	6.0	0.0	8	3	8	9	4			
21	8.0	0.0	9	8	13	14	9			
22	8.5	2.0	10	13	18	19	14			
25	8.1	8.0	.	.	.	.	.			
26	9.6	3.2	.	.	.	.	.			
29	9.3	8.0	.	.	.	.	.			
30	10.7	3.8								
33	10.6	8.0								
34	12.0	4.0								
35	12.0	5.4								
36	12.0	6.8								

$$\int_{\Omega}\left(\frac{\partial N_i}{\partial x}\frac{\partial N_j}{\partial x}+\frac{\partial N_i}{\partial y}\frac{\partial N_j}{\partial y}\right)dx\,dy[\phi_i]=-\int_0^6 5N_i dy \tag{9.10}$$

where Γ_2 comprises all but the vertical boundaries and use is made of the fact that the fluxes are zero except at the left vertical boundary. Solution of Eq. (9.10) follows in exactly the same manner as discussed in Chapters 4 and 5. Figure 9.3a shows the solution obtained for the stream function, and Figure 9.3b shows the potential function for the mesh of bilinear quadrilaterals of Figure 9.2.

9.3 CONVECTIVE TRANSPORT

A majority of the most important problems in fluid mechanics involve transport of a scalar quantity by convection, as well as diffusion. We have purposely delayed the incorporation of convective effects up to this point because they introduce special difficulties.

The study of species transport crosses many disciplines, e.g., heat transfer and combustion processes in fluid dynamics, chemical kinetics, mass transport, pollutants in lakes and oceans, aerosol dispersion in the atmosphere, contaminant transport

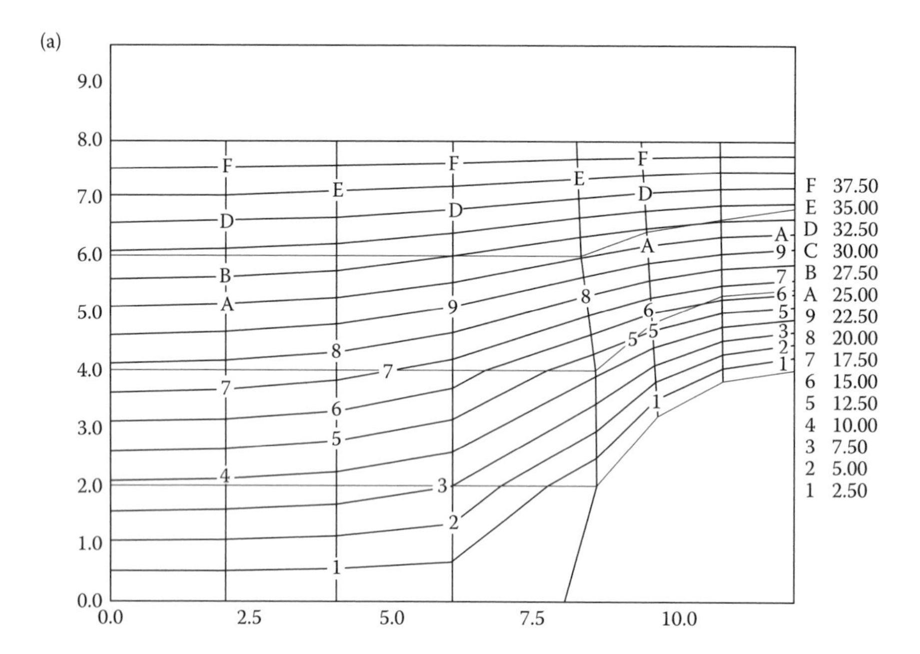

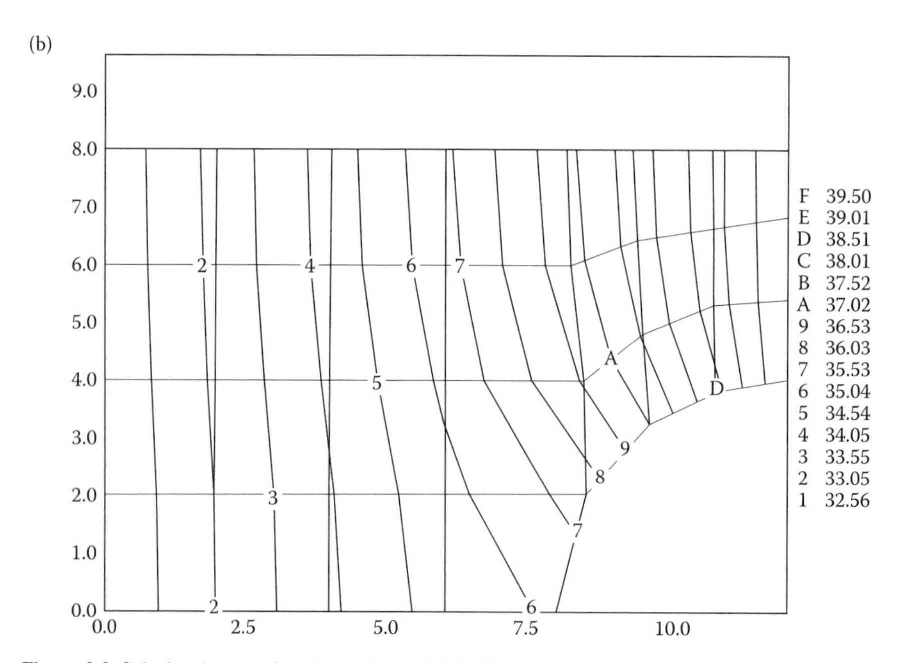

Figure 9.3 Calculated stream function and potential for flow over a circular cylinder: (a) stream function, ψ; and (b) potential function, ϕ

in groundwater flow, etc. The processes are modeled by the convection–diffusion equation, which is linear in many cases.

In a nonlinear equation, e.g., the fluid dynamic equations of motion, either an iterative or linearization procedure must normally be employed to obtain a solution. The nonlinear equations of motion (which yield the velocity components) can be quite troublesome to solve, along with the linear species transport equations, and produce numerical errors.

We start by looking at the one-dimensional, steady-state convective heat transfer equation:

$$-\frac{d}{dx}\left(K\frac{dT}{dx}\right)+\rho c_p u\frac{dT}{dx}=Q \tag{9.11}$$

with boundary conditions:

$$T(0)=T_a \tag{9.12}$$

and

$$-K\frac{dT}{dx}\bigg|_{x=L}+h(T(L)-T_\infty)=0 \tag{9.13}$$

Equation (9.11) is the same as Eq. (2.1) except that we have added the convective transport term:

$$\rho c_p u\, dT\,/\,dx$$

This equation governs the temperature distribution in a body of fluid moving with velocity $u(x)$ in the x-direction and with heat diffusion coefficient $K(x)$, assuming dimensions in the y- and z- directions are large and properties do not vary laterally.

The weak form of Eq. (9.11) is

$$\int_0^L W\left(-\frac{d}{dx}\left(K\frac{dT}{dx}\right)+\rho c_p u\frac{dT}{dx}-Q\right)dx=0 \tag{9.14}$$

After integration by parts of the diffusion term and using Eq. (9.13), Eq. (9.14) becomes

$$\int_0^L\left\{K\frac{dW}{dx}\frac{dT}{dx}+\rho c_p uW\frac{dT}{dx}-WQ\right\}dx-Wh(T-T_\infty)\big|_{x=L}=0 \tag{9.15}$$

For simplicity, assume that K and u are constant, and the interval $0\le x\le L$ has been discretized using n linear elements of equal length Δx. The Galerkin method applied to Eq. (9.15) then yields the system

$$\mathbf{KT}=\mathbf{F} \tag{9.16}$$

where the stiffness matrix **K** and forcing vector **F** are given as

$$\mathbf{K} = \begin{bmatrix} \frac{K}{\Delta x} - \frac{v}{2} & -\frac{K}{\Delta x} + \frac{v}{2} & 0 & 0 & & \\ -\frac{K}{\Delta x} - \frac{v}{2} & \frac{2K}{\Delta x} & -\frac{K}{\Delta x} + \frac{v}{2} & 0 & \cdots & \\ 0 & -\frac{K}{\Delta x} - \frac{v}{2} & \frac{2K}{\Delta x} & -\frac{K}{\Delta x} + \frac{v}{2} & & \\ & & & & \ddots & \\ & \vdots & & -\frac{K}{\Delta x} - \frac{v}{2} & \frac{2K}{\Delta x} & -\frac{K}{\Delta x} + \frac{v}{2} \\ & & & 0 & -\frac{K}{\Delta x} - \frac{v}{2} & \frac{K}{\Delta x} + \frac{v}{2} - h \end{bmatrix} \tag{9.17}$$

where $v = \rho c_p u$ and

$$F = \begin{bmatrix} \frac{Q\Delta x}{2} \\ Q\Delta x \\ Q\Delta x \\ \cdot \\ \cdot \\ \cdot \\ \cdot \\ Q\Delta x \\ \frac{Q\Delta x}{2} - hT_\infty \end{bmatrix} \tag{9.18}$$

The right-hand-side vector **F** remains the same as before. However, the stiffness matrix **K** shows important differences; namely, it is no longer symmetric and if $v/2 > K/\Delta x$, the sum of the magnitudes of the off-diagonal elements becomes larger than the magnitude of the diagonal term. We then say that the matrix is not "diagonally dominant." This is important because it means that an iterative solution procedure such as Gauss–Seidel fails to converge and a direct solver must be used.

If we introduce the nondimensional parameter:

$$\gamma = \frac{\rho c_p u \Delta x}{K} = \frac{u \Delta x}{\alpha} \tag{9.19}$$

also called the "cell Péclet number," the condition for the stiffness matrix to remain diagonally dominant can be expressed as $\gamma \leq 2$. If this condition is violated, the solution to Eq. (9.16) will exhibit spatial oscillations that render it useless. These are illustrated in Figure 9.4, where the solution to Eq. (9.16) is shown for the case $L = 1$, $Q = 0$, and = 10, and for boundary conditions $T(0) = 1$ and $T(1) = 0$.

Clearly, the Galerkin approximation shown in Figure 9.4 is useless. The problem becomes even more serious if we realize that, in most practical problems, the parameter can reach values of order of 10^4 to 10^6. A great deal of effort has been dedicated to the resolution of this kind of problem (see Heinrich and Zienkiewicz, 1979). The solution can be found by means of a Petrov–Galerkin formulation, where the weighting functions W_i are chosen to be different from the shape functions N_i in the finite element approximation.

If linear shape functions N_i are used to approximate the solution, an analysis of the numerical properties of the algorithm leads to the following Petrov–Galerkin formulation for convection-dominated problems:

1. The shape functions are linear and given by Eq. (3.66) in the natural coordinate system.
2. The weighting functions are given by

$$W_i(x) = N_i(x) + \frac{\beta \Delta x}{2}\frac{dN_i}{dx} \tag{9.20}$$

where the parameter β is obtained for each element as a function of the cell Péclet number using the relation:

$$\beta = \coth\frac{\gamma}{2} - \frac{2}{\gamma} \tag{9.21}$$

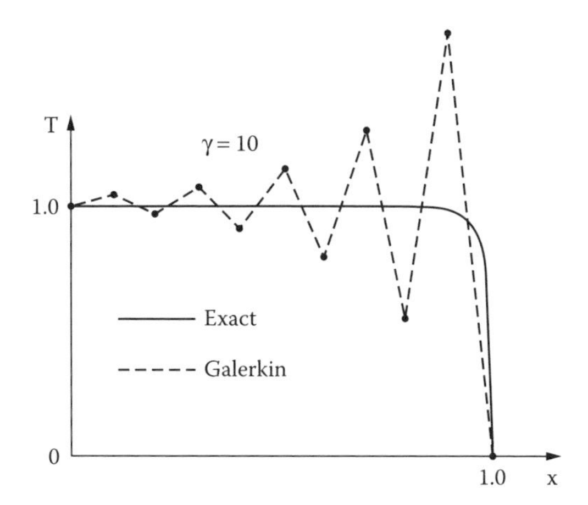

Figure 9.4 Oscillatory Galerkin solution of convection-dominated problems

The details on how this algorithm is derived are lengthy and are not given here. The interested reader should consult Hughes and Brooks (1979), Kelly et al. (1980), Yu and Heinrich (1986), and Heinrich and Pepper (1999).

The value of β given in Eq. (9.21) guarantees an oscillation-free and optimally accurate algorithm. As an exercise (Exercise 9.1), the reader should check that, for the problem of Figure 9.4, the algorithm defined by Eqs. (9.15), (9.20), and (9.21) yields the exact solution, which is given analytically by

$$T(x) = \frac{e^{uL/\alpha} - e^{(u/\alpha)x}}{e^{uL/\alpha} - 1} \tag{9.22}$$

If variable coefficients are involved, the parameter γ is calculated element-wise using average properties over the element. For time-dependent problems, the same algorithm can be used. We must be careful now, however, because the consistent mass matrix will not be symmetric (Exercise 9.3) and it may also be time dependent. For this reason, it is common to use the lumped mass matrices in these calculations.

Let us now consider two-dimensional problems. The governing equation becomes

$$\rho c_p \frac{\partial T}{\partial t} - \frac{\partial}{\partial x}\left(K_{xx}\frac{\partial T}{\partial x}\right) - \frac{\partial}{\partial y}\left(K_{yy}\frac{\partial T}{\partial y}\right) + \rho c_p\left(u\frac{\partial T}{\partial x} + v\frac{\partial T}{\partial y}\right) = Q \tag{9.23}$$

over a domain Ω with boundary conditions

$$T = T_A \qquad along\ \Gamma_A \tag{9.24}$$

and

$$-k_{xx}\frac{\partial T}{\partial x}n_x - k_{yy}\frac{\partial T}{\partial y}n_y = q\ along\ \Gamma_B \tag{9.25}$$

in a way similar to Eqs. (4.58) and (4.59) depicted in Figure 4.9. The weak form of the problem is

$$\int_\Omega \left\{ W\rho c_p \frac{\partial T}{\partial t} + K_{xx}\frac{\partial W}{\partial x}\frac{\partial T}{\partial x} + K_{yy}\frac{\partial W}{\partial y}\frac{\partial T}{\partial y} + W\left[\rho c_p\left(u\frac{\partial T}{\partial x} + v\frac{\partial T}{\partial x}\right) - Q\right]\right\} d\Omega + \int_{\Gamma_B} Wq\,d\Gamma = 0 \tag{9.26}$$

The presence of the convective terms in Eq. (9.26) causes the same kind of problems we saw in the one-dimensional case. These are illustrated in Figure 9.5, where an initial concentration profile (the spike on the left) is transported in a flow

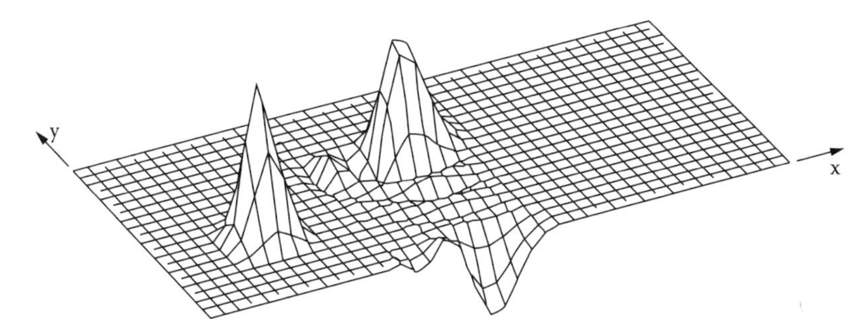

Figure 9.5 Galerkin solution to the problem of transport of a concentration spike at a 25° angle to the x-axis

field with velocity at a 25° angle to the x-axis. It is clear from Figure 9.5 that the Galerkin approximations are not adequate. The second spike in Figure 9.5 shows the calculated concentration after it has traveled a distance equal to two times the initial wavelength. A great deal of dispersion and phase shift can be observed; in fact, the calculated profile is very poor, if not useless.

A Petrov–Galerkin algorithm that alleviates most of the problems can be formulated in a way similar to the one-dimensional case if bilinear shape functions N_i are used, i.e.,

$$W_i = N_i + \frac{\beta \tilde{h}}{2\,|\,u\,|}\left(u\frac{\partial N_i}{\partial x} + v\frac{\partial N_i}{\partial y} \right) \tag{9.27}$$

where $|\,u\,| = \sqrt{u^2 + v^2}$ is the magnitude of the velocity within an element. The parameter β is calculated from Eq. (9.21) as before, with γ given by

$$\gamma = \frac{\tilde{h}}{K} \tag{9.28}$$

Here, K is the average diffusivity over an element and $\tilde{h}$ is a length scale given for a general quadrilateral by

$$\tilde{h} = |h_1| + |h_2| \tag{9.29}$$

with

$$\left.\begin{aligned} h_1 &= \left(u h_{\xi x} + v h_{\xi y} \right) \\ h_2 &= \left(u h_{\eta x} + v h_{\eta y} \right) \end{aligned}\right\} \tag{9.30}$$

and

$$\left.\begin{aligned} h_{\xi x} &= \frac{1}{2}[(x_2 + x_3) - (x_1 + x_4)] \\ h_{\xi y} &= \frac{1}{2}[(y_2 + y_3) - (y_1 + y_4)] \\ h_{\eta x} &= \frac{1}{2}[(x_3 + x_4) - (x_1 + x_2)] \\ h_{\eta y} &= \frac{1}{2}[(y_3 + y_4) - (y_1 + y_2)] \end{aligned}\right\} \tag{9.31}$$

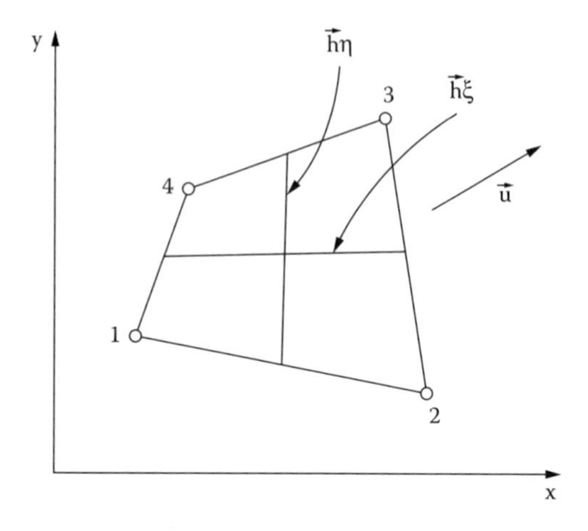

Figure 9.6 Element length scale for Petrov–Galerkin algorithm on an arbitrary quadrilateral

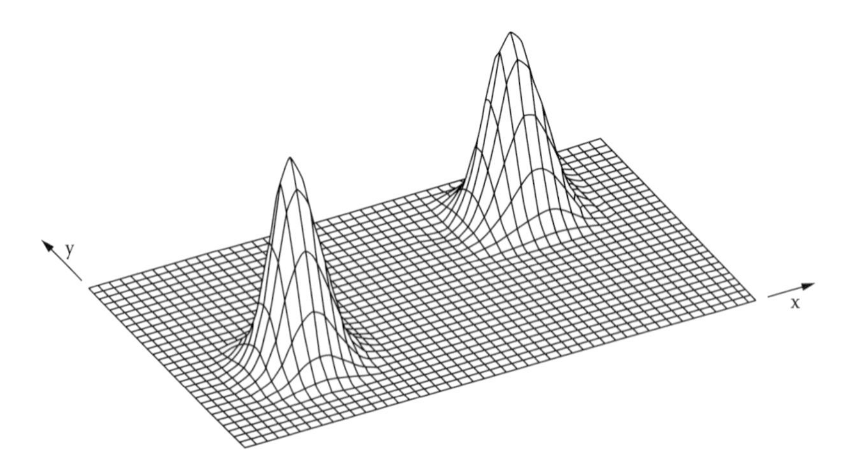

Figure 9.7 Petrov–Galerkin solution for transport of a concentration spike at 25° angle to the x-axis

as depicted in Figure 9.6. Figure 9.7 shows an improved solution for the problem of transport of a concentration spike shown in Figure 9.5. A detailed account of the algorithm, including the calculation of the fluid velocity field, can be found in Heinrich and Yu (1988) and Heinrich and Pepper (1999). There are other finite element algorithms that have been successfully used to treat convective transport. A particularly interesting account of methods for pure advection ($K = 0$) is given by Long and Pepper (1981).

Extension to three-dimensional problems follows the same idea as in two dimensions; however, the computational effort is now greater due to the lack of symmetry of the matrices. More sophisticated techniques are needed to make the computations affordable.

An algorithm that simplifies three-dimensional calculations by splitting 3-D equations into three one-dimensional calculations was developed by Pepper and Baker (1979).

Example 9.2*

Assume that we wish to calculate the advection and diffusion of a pollutant source located in the lower left corner of a rectangular domain, as shown in Figure 9.8. The governing equation is identical to Eq. (9.23), with temperature replaced by the concentration c and the thermal diffusion by the species diffusion coefficient D, which we will assume constant at $D = 10$ m^2/s, i.e.,

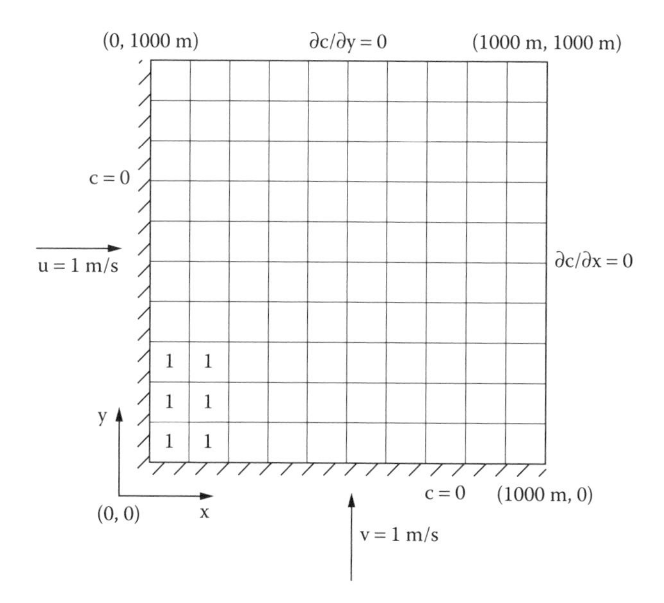

Figure 9.8 Transport of pollutant from an area source

* FEM-2D and COMSOL files are available.

$$\frac{\partial c}{\partial t} - D\left[\frac{\partial^2 c}{\partial x^2} + \frac{\partial^2 c}{\partial y^2}\right] + u\frac{\partial c}{\partial x} + v\frac{\partial c}{\partial y} = S \tag{9.32}$$

where S is the source term and $S = 1$ g/m^{3}s over the six elements in the lower left corner, as shown in Figure 9.8 and zero in the rest of the elements. The velocity components are constant $u = v = 1$ m/s, with $\Delta t = 0.1$ s. The diffusion coefficient is unrealistically large. It was chosen in order to obtain a low value of the cell Péclet number and to obtain a Galerkin solution to the problem. Remember that the condition is $\gamma \leq 2$.

The Galerkin procedure applied to Eq. (9.32) with the boundary conditions as shown in Figure 9.8 yields the familiar matrix equation:

$$\mathbf{M}\dot{\mathbf{c}} + (\mathbf{U} + \mathbf{K})\mathbf{c} = \mathbf{F} \tag{9.33}$$

where

$$\mathbf{M} = \int_\Omega N_i N_j d\Omega \tag{9.34}$$

$$\mathbf{U} = \int_\Omega N_i \left[(N_k u_k)\frac{\partial N_j}{\partial x} + (N_k v_k)\frac{\partial N_j}{\partial y}\right] d\Omega \tag{9.35}$$

$$\mathbf{K} = D\int_\Omega \left(\frac{\partial N_i}{\partial x}\frac{\partial N_j}{\partial x} + \frac{\partial N_i}{\partial y}\frac{\partial N_j}{\partial y}\right) d\Omega \tag{9.36}$$

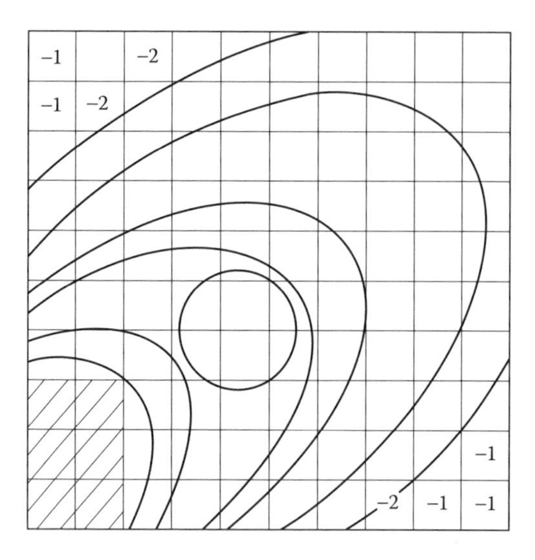

Figure 9.9 Isopleth pattern

$$\mathbf{F} = \int_\Omega N_i (N_k S_k) d\Omega \tag{9.37}$$

where the functions u, v, and S have been interpolated using the shape functions, and summation is implied in the index k.

The solution to this problem is left as a computer exercise. Results are shown in Figure 9.9 at a time of approximately 10 s. The negative values are due to numerical dispersion—there is no such thing as negative concentration. ■

9.4 NONLINEAR CONVECTIVE TRANSPORT

The fundamental equations for the two-dimensional flow of a Newtonian fluid with no body forces and constant physical properties are the two momentum (Navier–Stokes) equations (e.g., Heinrich and Pepper, 1999).

$$\rho\left(\frac{\partial u}{\partial t} + u\frac{\partial u}{\partial x} + v\frac{\partial u}{\partial y}\right) = -\frac{\partial \rho}{\partial x} + \mu\left(\frac{\partial^2 u}{\partial x^2} + \frac{\partial^2 u}{\partial y^2}\right) \tag{9.38}$$

$$\rho\left(\frac{\partial v}{\partial t} + u\frac{\partial v}{\partial x} + v\frac{\partial v}{\partial y}\right) = -\frac{\partial \rho}{\partial x} + \mu\left(\frac{\partial^2 v}{\partial x^2} + \frac{\partial^2 v}{\partial y^2}\right) \tag{9.39}$$

and the continuity equation

$$\frac{\partial \rho}{\partial t} + \frac{\partial}{\partial x}(\rho u) + \frac{\partial}{\partial y}(\rho v) = 0 \tag{9.40}$$

where u and v are the fluid velocity components in the x- and y-directions, respectively; ρ is density; p is pressure; and μ is the fluid viscosity.

If the flow is incompressible, the time derivative term in Eq. (9.40) vanishes. If the flow is compressible, an additional state equation is needed that relates density and pressure (e.g., $p = \rho RT$).

Equations (9.38) and (9.39) are nonlinear convective transport equations; the nonlinearity is due to the convective acceleration terms and, as we will see in the next chapter, they can be very difficult to solve. We introduce the basic methods of discretization for nonlinear problems through the simple one-dimensional Burgers equation (Burgers, 1948):

$$\frac{\partial u}{\partial t} + u\frac{\partial u}{\partial x} = \varepsilon\frac{\partial^2 u}{\partial x^2} \tag{9.41}$$

The most widely used methods for the implicit solution of nonlinear differential equations are all based on generalizations of Newton's method. Recall the one-dimensional problem of finding a value x^* such that $f(x^*) = 0$, where f is a real-valued function. As depicted in Figure 9.10, for a given point x_n (unless $x_n = x^*$), $f(x_n) \neq 0$. We can construct a line tangent to $f(x)$ at the point $(x_n, f(x_n))$ using the fact that the slope of this line is given by $f'(x_n)$, i.e.,

$$y = (x - x_n) f'(x_n) + f(x_n) \tag{9.42}$$

where f' denotes the derivative of f with respect to x. We can now find the intersection of the line with the x-axis by setting $y = 0$ in Eq. (9.42) and solve for x. The solution is x_{n+1}, as show in Figure 9.10, and is a better approximation to x^* than x_n. Thus,

$$x_{n+1} = x_n - \frac{f(x_n)}{f'(x_n)} \tag{9.43}$$

In finite element analysis, it is more convenient to define the algorithm in the form:

$$x_{n+1} = x_n + \Delta x \tag{9.44a}$$

$$f'(x_n)\Delta x = -f(x_n) \tag{9.44b}$$

This can then be interpreted as an implicit solution using the linearized Taylor series expansion of f about x_n, i.e.,

$$f(x_{n+1}) = f(x_n) + \Delta x f'(x_n) + O(\Delta x^2) \tag{9.45}$$

The method can be generalized to solve differential equations if x_n and Δx are vectors and f a general k-dimensional operator.

To illustrate, consider the steady-state form of Eq. (9.41), i.e.,

$$-\varepsilon \frac{d^2 u}{dx^2} + u \frac{du}{dx} = 0 \tag{9.46}$$

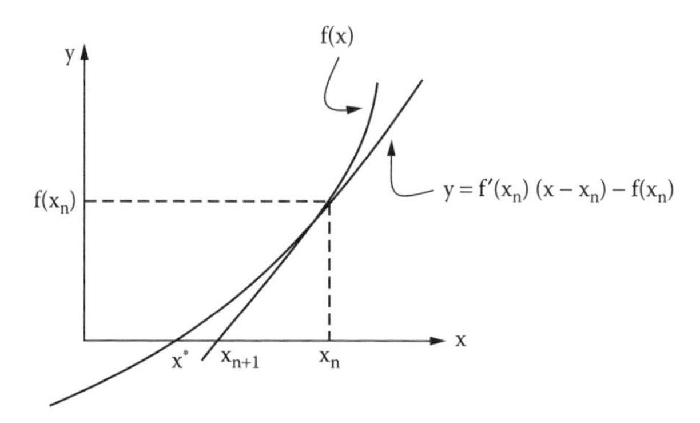

Figure 9.10 Newton's method for the one-dimensional problem, $f(x^*) = 0$

over the interval $0 < x < 1$, with boundary conditions:

$$\left.\begin{aligned} u(0) &= 1 \\ u(1) &= 0 \end{aligned}\right\} \tag{9.47}$$

We use four equal-length linear elements to discretize the interval. The weighted residuals form gives

$$\int_0^1 \left\{ \varepsilon \frac{dW}{dx}\frac{du}{dx} + Wu\frac{du}{dx} \right\} dx = 0 \tag{9.48}$$

Approximating u by

$$u = \sum_{i=1}^{5} N_i(x) u_i \tag{9.49}$$

and considering a Galerkin approximation, where $W_i = N_i$, Eq. (9.48) becomes

$$\begin{aligned} &\sum_{j=1}^{5}\left[\int_0^1 \varepsilon \frac{dN_i}{dx}\frac{dN_j}{dx}\,dx\right]u_j \\ &+\sum_{k=1}^{5}\sum_{j=1}^{5}\left[\int_0^1 N_i N_k \frac{dN_i}{dx}\,dx\right]u_k u_j = 0 \qquad \mathrm{i} = 1, \ldots, 5 \end{aligned} \tag{9.50}$$

which is a system of nonlinear algebraic equations in the unknown parameters u_i, $i = 2, 3, 4$, and the two prescribed velocities $u_1 = 1$ and $u_5 = 0$. It is left as an exercise to show that the final system of equations is

$$\left.\begin{aligned} &4\varepsilon u_1 - 4\varepsilon u_2 - u_1^2 + \frac{5}{6}u_1 u_2 + \frac{1}{6}u_2^2 = 0 \\ &-4\varepsilon u_1 + 8\varepsilon u_2 - 4\varepsilon u_3 - \frac{1}{6}u_1^2 - \frac{5}{6}u_1 u_2 + \frac{5}{6}u_2 u_3 + \frac{1}{6}u_3^2 = 0 \\ &-4\varepsilon u_2 + 8\varepsilon u_3 - 4\varepsilon u_4 - \frac{1}{6}u_2^2 - \frac{5}{6}u_2 u_3 + \frac{5}{6}u_3 u_4 + \frac{1}{6}u_4^2 = 0 \\ &-4\varepsilon u_3 + 8\varepsilon u_4 - 4\varepsilon u_5 - \frac{1}{6}u_3^2 - \frac{5}{6}u_3 u_4 + \frac{5}{6}u_4 u_5 + \frac{1}{6}u_5^2 = 0 \\ &-4\varepsilon u_4 + 4\varepsilon u_5 - \frac{1}{6}u_4^2 - \frac{5}{6}u_4 u_5 + u_5^2 = 0 \end{aligned}\right\} \tag{9.51}$$

Applying the boundary conditions, Eq. (9.51) reduces to

$$\left.\begin{aligned} &8\varepsilon u_2 - 4\varepsilon u_3 - \frac{5}{6}u_2 + \frac{5}{6}u_2u_3 + \frac{1}{6}u_3^2 = 4\varepsilon + \frac{1}{6} \\ &-4\varepsilon u_2 + 8\varepsilon u_3 - 4\varepsilon u_4 - \frac{1}{6}u_2^2 - \frac{5}{6}u_2u_3 + \frac{5}{6}u_3u_4 + \frac{1}{6}u_4^2 = 0 \\ &-4\varepsilon u_3 + 8\varepsilon u_4 - \frac{1}{6}u_3^2 - \frac{5}{6}u_3u_4 = 0 \end{aligned}\right\} \tag{9.52}$$

Equation (9.52) is of the form

$$\mathbf{F}(u) = 0 \tag{9.53}$$

and we can apply a generalized Newton method to obtain the solution u^*. To this extent, the derivative $f'(x_n)$ in Eq. (9.44b) is replaced by the matrix

$$\mathbf{F}'(u^n) = \left[\frac{\partial f_i}{\partial u_j^n}\right] \tag{9.54}$$

which is called the *tangent* matrix of $\mathbf{F}$ evaluated at u^n, where the superscript n denotes the iteration number.

In our example, $\mathbf{F}$ takes the form

$$\mathbf{F}'(u) = \begin{bmatrix} 8\varepsilon - \frac{5}{6} + \frac{5}{6}u_3^n & -4\varepsilon + \frac{5}{6}u_2^n + \frac{1}{3}u_3^n & 0 \\ -4\varepsilon - \frac{1}{3}u_2^n - \frac{5}{6}u_3^n & 8\varepsilon - \frac{5}{6}u_2^n + \frac{5}{6}u_4^n & -4\varepsilon + \frac{5}{6}u_3^n + \frac{1}{3}u_4^n \\ 0 & -4\varepsilon - \frac{1}{3}u_3^n - \frac{5}{6}u_4^n & 8\varepsilon - \frac{5}{6}u_3^n \end{bmatrix} \tag{9.55}$$

and the iterative scheme given by Eqs. (9.44a, b) becomes

$$\begin{bmatrix} u_2^{n+1} \\ u_3^{n+1} \\ u_4^{n+1} \end{bmatrix} = \begin{bmatrix} u_2^n \\ u_3^n \\ u_4^n \end{bmatrix} + \begin{bmatrix} \Delta u_2 \\ \Delta u_3 \\ \Delta u_4 \end{bmatrix} \tag{9.56}$$

where the vector Δu is the solution of the *linear* system of equations

$$\mathbf{F}'(u^n)\Delta u = -\mathbf{F}(u^n) \tag{9.57}$$

It should be very clear now that the difficulty with nonlinear problems is that the tangent matrix changes if the nonlinearity changes. To perform the calculations, we must start with an initial guess, which is usually a homogeneous function except

Table 9.2 Solution to Eqs. (9.46) and (9.47) for $\varepsilon = \Delta x = 0.1$, using linear elements

X	Galerkin Solution	Petrov–Galerkin Solution	Analytical Solution
0.0	1.000	1.000	1.000
0.2	1.000	0.999	0.999
0.4	0.997	0.995	0.995
0.6	0.970	0.964	0.964
0.8	0.770	0.762	0.762
0.9	0.464	0.462	0.462
1.0	0.000	0.000	0.000

for the boundary conditions, and iterate until the approximation is close enough to the solution. There are different criteria that may be used to decide when to stop. The most commonly used criterion is

$$\sqrt{\sum_{i=1}^{N}\left(f_i^{n+1}\right)^2} < \delta \tag{9.58}$$

where δ is a small predetermined value.

It can also be observed that the tangent matrix corresponding to the problem must be recomputed and inverted at every iteration. This makes nonlinear problems expensive. There are many variations of Newton's method. The interested reader should consult the book by Ortega and Rheinboldt (1970) for further details. Notice that the method presented here can be extended to any nonlinear problem; i.e., it provides an alternative way to treat the radiation terms encountered in Section 7.4.

As an example, we have solved Eqs. (9.46) and (9.47) for the case = 0.1 and $x = 0.1$, using the Galerkin method and the Petrov–Galerkin method of Section 9.3. The results are shown in Table 9.2 where the superior accuracy of the Petrov–Galerkin method can be observed again.

9.5 GROUNDWATER FLOW

The finite element method has been used successfully for many years to calculate the seepage of liquids and contaminants within the soil. The theory of groundwater transport is well established (Verruijt, 1982). Soil consists primarily of water, air, and solid material; water and air fill the pore space between the solid granular regions. Porosity defines the number of pores, i.e., the volume of the pores per unit volume. In sandy soil, the porosity varies from 0.20 to 0.50; in clay, the porosity may vary from 0.40 to 0.80. The amount of saturation defines the volume of water within the pores; it varies between 0 for completely dry soil to 1 for completely saturated soil. The compressibility of soil is related to the theory of elasticity, whereby Young's modulus, Poisson's ration, bulk modulus, and shear modulus can be used. Natural soils are not linearly elastic and vary with saturation and fluid pressure.

Because flow velocities within the soil are relatively slow, we can assume that the frictional resistance to the flow is proportional to the flow rate. Disregarding inertia effects (advection), and assuming the porous medium to be isotropic (i.e., the pore space geometry is independent of flow direction) with constant fluid density, the groundwater head can be defined as

$$\phi = z + \frac{p}{\rho g} \tag{9.59}$$

where ϕ is the head, z is height, p is pressure, is density, and g is the gravitational constant. Utilizing Darcy's law, the equation for equilibrium flow can be written in two dimensions as

$$\frac{\partial p}{\partial x} + \frac{\mu}{k} q_x = 0 \tag{9.60}$$

$$\frac{\partial p}{\partial y} + \frac{\mu}{k} q_y = 0 \tag{9.61}$$

where μ is the viscosity of the fluid, k is the soil permeability, and q_x and q_y represent specific discharge through the soil. We can define the discharge rates as

$$q_x = -k' \frac{\partial \phi}{\partial x} \tag{9.62}$$

$$q_y = -k' \frac{\partial \phi}{\partial y} \tag{9.63}$$

where k' is the hydraulic conductivity.

$$k' = \frac{k \rho g}{\mu} \tag{9.64}$$

which is a measurable quantity. In actuality, hydraulic conductivity is directionally dependent, $k' = k_{xx}\hat{i} + k_{yy}\hat{j}$ (in two dimensions). In this case

$$q_x = -k_{xx} \frac{\partial \phi}{\partial x} \tag{9.65}$$

$$q_y = -k_{yy} \frac{\partial \phi}{\partial y} \tag{9.66}$$

The fundamental two-dimensional equation, which describes groundwater flow in isotropic, inhomogeneous soil, is normally written as

$$S\frac{\partial h}{\partial t} = -\frac{\partial}{\partial x}\left(hq_x\right) - \frac{\partial}{\partial y}\left(hq_y\right) \tag{9.67}$$

where h is the height of the water table above an impermeable base, and S is the part of the pore space filled with water as the water table rises. In a coarse material, S is usually equal to the porosity. Because pressure can be set to zero at the upper boundary, the head is equal to the elevation of the water table, hence, equal to h. Equation (9.67) can be further transformed to the form:

$$S\frac{\partial h}{\partial t} = \frac{\partial}{\partial x}\left(kh\frac{\partial h}{\partial x}\right) + \frac{\partial}{\partial y}\left(kh\frac{\partial h}{\partial y}\right) \tag{9.68}$$

This is a nonlinear equation, and the iterative method described in the previous section can be applied. However, to simplify the presentation, an approximation to the second-derivative terms is introduced in order to linearize Eq. (9.68), i.e.,

$$\frac{\partial}{\partial x}\left(h\frac{\partial h}{\partial x}\right) = h\frac{\partial^2 h}{\partial x^2} + \left(\frac{\partial h}{\partial x}\right)^2 \cong h\frac{\partial^2 h}{\partial x^2} \tag{9.69}$$

Hence, Eq. (9.68) reduces to

$$S\frac{\partial h}{\partial t} = T'\frac{\partial^2 h}{\partial x^2} + T'\frac{\partial^2 h}{\partial y^2} \tag{9.70}$$

where $T' = kh$, which is termed the transmissivity of the aquifer. It is generally assumed that $\partial h/\partial x$ and $\partial h/\partial y$ are very small; this assumption is justified when variations in the water level are relatively small compared to their overall value (Verruijt, 1982). If we let α be the compressibility of the soil, then we can write

$$\alpha\frac{\partial p}{\partial t} \equiv \alpha\rho g\frac{\partial \phi}{\partial t} \tag{9.71}$$

which relates the volume strain to groundwater head.

Obviously, groundwater analysis has deep roots in the theory of elasticity. A separate field has risen within geology that is known as soil mechanics. Extensive research continues in this field, particularly in modeling soil dynamics.

If we define the discharge rates as velocity components,

$$u \equiv q_x = -k\frac{\partial \phi}{\partial x} \tag{9.72}$$

$$v \equiv q_y = -k\frac{\partial\phi}{\partial y} \tag{9.73}$$

the continuity of flow requires that

$$k\frac{\partial^2\phi}{\partial x^2} + k\frac{\partial^2\phi}{\partial y^2} = 0 \tag{9.74}$$

Since $\phi \equiv h$, lines in which pressure head is constant must have constant "potentials." Comparing Eqs. (9.74) and (9.71) with expressions for heat transfer, we see that groundwater flow is synonymous with heat conduction where temperature (T) $\equiv \phi$ or h; likewise, Eq. (9.74) is identical to Laplace's equation for potential flow. Hence, groundwater flow problems can be solved with finite element expressions for heat transfer, subject to proper interpretation for T, k, and Q (source term).

Once the "potential," ϕ, is obtained at each nodal point, the flow velocity in each element can be immediately obtained from the relations (from Chapter 4):

$$u = -k\sum_{i=1}^{n}\frac{\partial N_i}{\partial x}\phi_i \tag{9.75}$$

$$v = -k\sum_{i=1}^{n}\frac{\partial N_i}{\partial y}\phi_i \tag{9.76}$$

where n is the number of nodes, and N_i are the shape functions of the particular element used in the discretization.

EXAMPLE 9.3*

Assume that a regional aquifer has a single pump removing water. The physical domain is shown in Figure 9.11. Calculate the steady-state drawdown of water from the aquifer. The upper and lower boundaries are impermeable, the left and right boundaries are far enough from the pump such that a constant head of 100 m is maintained, and k_{xx} and k_{yy} are 10 m/day. The pump operates at 1000 m^3/day and is located at x = 1000 m and y = 500 m. Table 9.3 gives the nodal location data; Figure 9.12 shows the element mesh.

The governing equation is

$$k_{xx}\frac{\partial^2\phi}{\partial x^2} + k_{yy}\frac{\partial^2\phi}{\partial y^2} + S = 0 \tag{9.77}$$

* FEM-2D and COMSOL files are available.

$$k_{xx}\frac{\partial \phi}{\partial x}n_x + k_{yy}\frac{\partial \phi}{\partial y}n_y = 0 \qquad (9.78)$$

along the impermeable boundaries, where ϕ is the piezometric head measured from a reference level (usually the bottom of the aquifer), and S is the pumping rate.

Applying the Galerkin procedure to Eq. (9.77), we obtain

$$\int_{\Omega}\left(k_{xx}\frac{\partial N_i}{\partial x}\frac{\partial N_j}{\partial x} + k_{yy}\frac{\partial N_i}{\partial y}\frac{\partial N_j}{\partial y}\right)dxdy\left[\phi_i\right] = \int_{\Omega} SN_i\,dxdy \qquad (9.79)$$

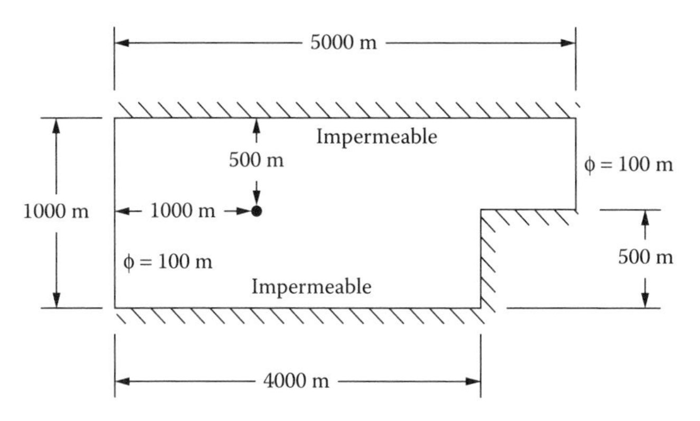

Figure 9.11 Regional aquifer with a single pump

Table 9.3 Coordinate Data for Regional Aquifer Mesh

Node No.	x	y
1	0	0
2	0	300
3	0	500
4	0	700
5	0	1000
10	500	1000
15	1000	1000
20	1500	1000
25	2300	1000
30	3300	1000
35	4000	1000
38	4500	1000
41	5000	1000

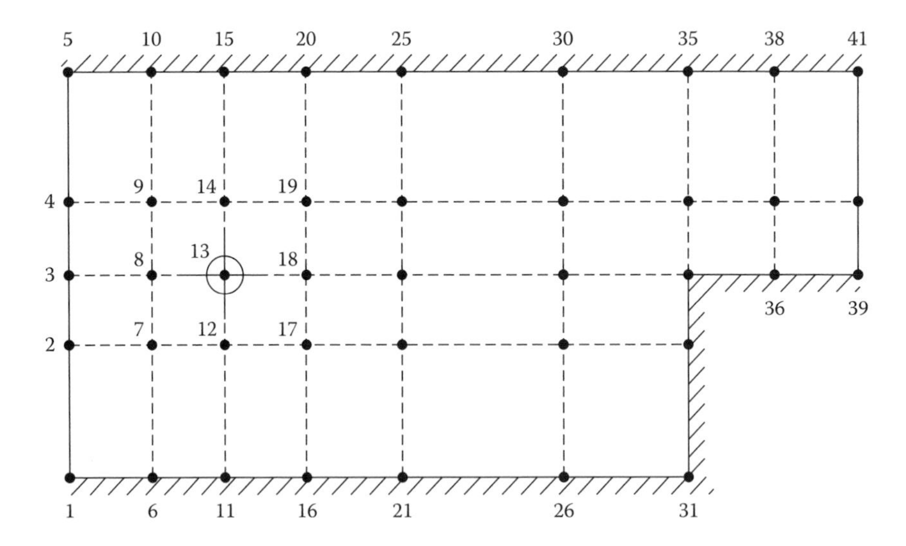

Figure 9.12 Element mesh for regional aquifer

Notice that the finite element method automatically enforces the impermeable boundary condition when no other conditions are specified.

Bilinear quadrilateral shape functions are used for N_i (N_j) in Eq. (9.79). Following similar procedures as used in Chapter 5, Eq. (9.79) can be evaluated for a generic element as

$$\left(\frac{k_{xx}^{(e)}\Delta y}{6\Delta x}\begin{bmatrix} 2 & -2 & -1 & 1 \\ -2 & 2 & 1 & -1 \\ -1 & 1 & 2 & -2 \\ 1 & -1 & -2 & 2 \end{bmatrix} + \frac{k_{yy}^{(e)}\Delta x}{6\Delta y}\begin{bmatrix} 2 & 1 & -1 & -2 \\ 1 & 2 & -2 & -1 \\ -1 & -2 & 2 & 1 \\ -2 & -1 & 1 & 2 \end{bmatrix}\right)\begin{bmatrix} \phi_1^{(e)} \\ \phi_2^{(e)} \\ \phi_3^{(e)} \\ \phi_4^{(e)} \end{bmatrix} = \frac{S^{(e)}\Delta x\Delta y}{4}\begin{bmatrix} 1 \\ 1 \\ 1 \\ 1 \end{bmatrix} \tag{9.80}$$

where Δx and Δy are the element dimensions in the x- and y-directions, respectively. Note that the source term applies only at specific locations within the mesh.

9.6 LUBRICATION

Every machine with moving parts has at least one bearing, and its successful operation depends crucially on the performance of the bearing(s). The performance of the bearing(s) can be predicted by solving the relevant Reynolds equation, which normally requires the use of a numerical procedure. Because of the complex geometrical configurations often encountered in lubrication, the finite element method is particularly well suited for the solution of such problems. Furthermore, lubrication problems often involve complicated driving functions and boundary conditions. Many of these problems are compounded by large changes in geometric scales and by abrupt changes in the field properties, such as film thickness.

The equation governing the pressure generated in a bearing is the Reynolds equation (see, e.g., Gross et al., 1980). For an incompressible fluid, the equation is usually written as

$$\frac{\partial}{\partial x}\left(\frac{h^3}{6\mu}\frac{\partial p}{\partial x}\right)+\frac{\partial}{\partial y}\left(\frac{h^3}{6\mu}\frac{\partial p}{\partial y}\right)=2\frac{\partial h}{\partial t}+U\frac{\partial h}{\partial x} \tag{9.81}$$

where h is the fluid film thickness, μ is the fluid viscosity, and U is the bearing velocity. This is illustrated in Figure 9.13 for a rigid slider bearing.

In the case of gas bearings, when the working fluid is compressible, the equation takes the form:

$$\frac{\partial}{\partial x}\left(\frac{h^3 p}{6\mu}\frac{\partial p}{\partial x}\right)+\frac{\partial}{\partial y}\left(\frac{h^3 p}{6\mu}\frac{\partial p}{\partial y}\right)=2\frac{\partial (ph)}{\partial t}+U\frac{\partial (ph)}{\partial x} \tag{9.82}$$

which is now nonlinear in the pressure and will require an iterative solution, such as the Newton iteration introduced in Section 9.3.

The solution of a lubrication problem is usually further complicated by the necessity to consider temperature variations that affect the viscosity and/or the fact that one or both surfaces may be compliant. In the first case, an energy equation of the form:

$$\frac{\partial T}{\partial t}+U\frac{\partial T}{\partial x}=\frac{\partial}{\partial x}\left(k_{xx}\frac{\partial T}{\partial x}\right)+\frac{\partial}{\partial y}\left(k_{yy}\frac{\partial T}{\partial y}\right)+\Phi \tag{9.83}$$

must be solved, where Φ is the viscous dissipation function. When the slider and/or wall are compliant, appropriate elastic equations that model the solid mechanics aspects of the problem must be incorporated; hence, the problem becomes significantly

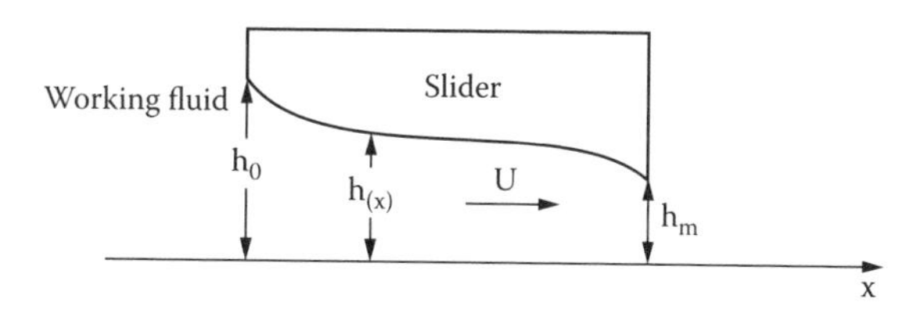

Figure 9.13 Slider bearing configuration and notation

more complex. The finite element solution of lubrication problems incorporating temperature effects is presented by Huebner (1974). For an application involving gas bearings and compliant surfaces, see Heinrich and Wadhwa (1986).

EXAMPLE 9.4

We wish to calculate the load capacity of the rigid slide bearing shown in Figure 9.14. We can neglect side leakage by assuming that the bearing is infinitely wide, thus permitting a one-dimensional analysis. In this problem we let $(\partial h/\partial t) = 0$ and $\mu = 0.6$ g/cm s. Equation (9.81) becomes

$$\frac{\partial}{\partial x}\left(\frac{h^3}{6\mu}\frac{\partial p}{\partial x}\right) = 2U\frac{\partial h}{\partial x} \tag{9.84}$$

The boundary conditions are $p(0) = p(2) = p_a$, where p_a is the ambient or reference pressure which can be conveniently prescribed. The Galerkin finite element approximation to Eq. (9.83) is

$$\left[\int_0^2 \frac{h^3}{6\mu}\frac{\partial N_i}{\partial x}\frac{\partial N_j}{\partial x}dx\right]p_j = \int_0^2 2UN_i\frac{\partial h}{\partial x}dx \tag{9.85}$$

where h is a given function. The reader should look into ways of incorporating the function h into the program in order to evaluate the integrals of Eq. (9.85). This can be done using linear elements with the linear relations:

$$p^{(e)} = N_1^{(e)}p_1^{(e)} + N_2^{(e)}p_2^{(e)} \tag{9.86}$$

$$h^{(e)} = N_1^{(e)}h_1^{(e)} + N_2^{(e)}h_2^{(e)} \tag{9.87}$$

and replacing h^3 into Eq. (9.85). A minimum of two Gaussian points will be required to evaluate the left-hand-side integral. A second possibility is to define the average value of h^3 over each element and use a piecewise constant function; notice that the coefficient $h^3/6\mu$ is equivalent to the variable diffusion coefficient in Eq. (3.45). A third possibility is to evaluate h^3 at each nodal point and use linear interpolation of those values. In the left-hand-side integral in Eq. (9.85), the

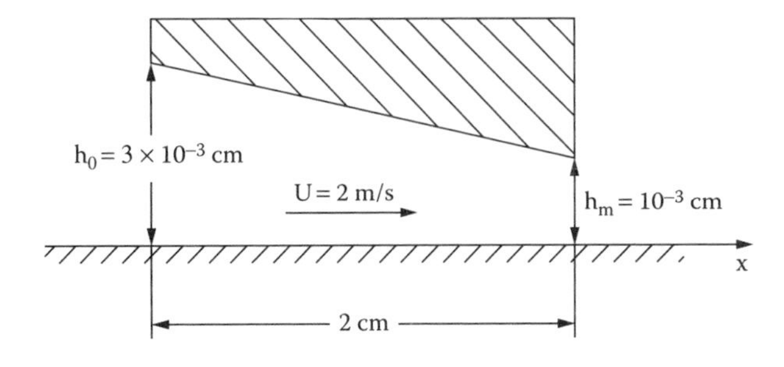

Figure 9.14 Data for slide bearing problem

quantity $2U(\partial h/\partial x)$ is a constant. Hence, this part can be treated as a source term in Eq. (3.27) with Q replaced by $2U(\partial h/\partial x)$. The corresponding terms can be readily calculated.

The mesh can consist of linear or quadratic elements. In this problem, the function $h(x)$ is linear and can be interpolated exactly by linear elements. If this were not the case, care would need to be exercised to ensure that the mesh is not too coarse due to the great difference between the horizontal and vertical scales. In this instance, the interpolation error could approach the same order of magnitude as the fluid film thickness, thereby rendering the results useless. ■

9.7 CLOSURE

The finite element method is a very versatile numerical technique, which can be applied to many types of problems. With only a few changes in the fundamental solution algorithm, a broad spectrum of problems can be addressed. Regardless of the type of problem being solved, many facets of the finite element method are generic; i.e., the same formulations and coding instructions apply. In many instances, the only changes required are boundary condition input and consistency in units. Obviously, the example problems addressed in this chapter are relatively simple; further research into the respective areas will reveal significantly more complex and difficult problems with diverging commonality. However, the underlying principles associated with the finite element method still apply.

Research and development in both the theory and application of the finite element technique are continually ongoing. Numerous alterations and improvements in solution techniques and investigations into new areas of study appear regularly in the literature. First-time users should be able to find applications of the finite element method to problems of their interest. In most instances, an overwhelming number of books, articles, and computer programs continue to surface; so many, in fact, that the newcomer can become hopelessly saturated and bewildered. In such circumstances, we should return to the basics and establish the underlying principles. A number of good textbooks on the finite element method can be found in most libraries and bookstores; the references cited in this book are recommended as a starting point. More advanced studies of the finite element technique are readily available.

EXERCISES

9.1 Calculate the Galerkin solution to the problem of Figure 9.4 using linear elements. Then obtain a solution using the Petrov-Galerkin approximation method defined by Eqs. (9.20) and (9.21), and verify that this method gives the exact solution at the nodal points.

9.2 Solve the following convective diffusion equations:

(a)

$$-\frac{d^2\phi}{dx^2} + 200\frac{d\phi}{dx} = x^2, \qquad \phi(0) = \phi(1) = 0$$

(b)

$$-\frac{d^2\phi}{dx^2}+\frac{40}{x}\frac{d\phi}{dx}=x^2, \qquad 1\le x\le 2, \quad \phi(1)=1, \quad \phi(2)=0$$

9.3 Calculate the element mass matrix for the Petrov–Galerkin finite element method. Show that even though the consistent mass matrix differs from the one obtained using a standard Galerkin approximation, the assembled lumped mass matrices are always equal, except at the end points of the interval.

9.4 Using the Galerkin method, perform a parametric study on the effect of the diffusion coefficient D in the problem of Figure 9.4 by assuming $\alpha = 100$ (diffusion dominated) and $\alpha = 0.001$ (strongly convection dominated).

9.5 Derive Eq. (9.51) from Eq. (9.50).

9.6 Calculate and plot the potential lines for the groundwater seepage problem around the dam shown in the figure. Assume the bottom layer is impermeable and $\boldsymbol{K}_{xx}$ and $\boldsymbol{K}_{yy}$ = 10 m/day.

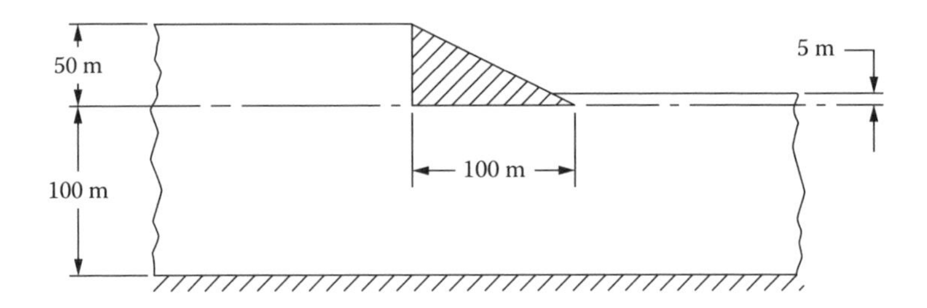

9.7 Solve the problem of Example 9.4 using 20 and 40 linear elements and compare the answers.

REFERENCES

Burgers, J. M. (1948). "A Mathematical Model Illustrating the Theory of Turbulence." *Adv. Appl. Mech.* I: 171–188.

Gross, W. A., Matsch, L. A., Castelli, V., Eshel, A., Vohr, J. H., and Wildmann, M. (1980). *Fluid Film Lubrication.* New York: John Wiley & Sons.

Heinrich, J. C. and Pepper, D. W. (1999). *Intermediate Finite Element Method. Fluid Flow and Heat Transfer Applications.* New York: Taylor & Francis.

Heinrich, J. C. and Wadhwa, S. (1986). *Analysis of Self-Acting Foil Bearings: A Finite Element Approach, Tribology and Mechanics of Magnetic Storage Systems*, Vol. III (B. Bhushan and N. Eiss, eds.). Park Ridge, Ill.: ASLE SP-21.

Heinrich, J. C. and Yu, C.-C. (1988). "Finite Element Simulations of Buoyancy-Driven Flows with Emphasis on Natural Convection in a Horizontal Circular Cylinder." *Comp. Meth. Appl, Mech. Eng*. 69:1–27.

Heinrich, J. C. and Zienkiewicz, O. C. (1970). "The Finite Element Method and 'Upwinding' Techniques in the Numerical Solution of Convection Dominated Flow Problems." In *Finite Element Methods for Convection Dominated Flows* (T. J. R. Hughes, ed.). New York: ASME AMD; 34:105–136.

Huebner, K. H. (1974). "Application of Finite Element Methods to Thermodynamic Lubrication." *Int. J. Num. Meth. Eng.* 8:139–147.

Hughes, T. J. R. and Brooks, A. (1979). "A Multidimensional Upwind Scheme with No Cross-Diffusion." In *Finite Element Methods for Convection Dominated Flows* (T. J. R. Hughes, ed.). New York: ASME AMD; 34:19–35.

Kelly, D. W., Nakazawa, S., Zienkiewicz, O. C., and Heinrich, J. C. (1980). "A Note on Upwinding and Anisotropic Balancing Dissipation in Finite Element Approximations to Convective Diffusion Problems." *Int. J. Num. Meth. Eng.* 15:1705–1711.

Long, P. E. and Pepper, D. W. (1981). "An Examination of Some Simple Numerical Schemes for Calculating Advection." *J. Appl. Meteorol.* 20:146–156.

Ortega, J. M. and Rheinboldt, W. (1970). *Iterative Solution of Nonlinear Equations in Several Variables.* New York: Academic Press.

Pepper, D. W. and Baker, A. J. (1979). "A Simple One-Dimensional Finite Element Algorithm with Multidimensional Capabilities." *Num. Heat Transfer* 2:81–95.

Verruijt, A. (1982). *Theory of Groundwater Flow.* London: Macmillian.

Yu, C.-C. and Heinrich, J. C. (1986). "Petrov-Galerkin Methods for the Time Dependent Convective Transport Equation." *Int. J. Num. Meth. Eng.* 23:883–901.

CHAPTER

10

INTRODUCTION TO VISCOUS FLUID FLOW

10.1 BACKGROUND

In Chapter 9 we introduced the Navier–Stokes equations that govern the flow of a viscous fluid, but did not attempt their solution. In fact, solutions can be very difficult to obtain depending on whether the flow is laminar or turbulent, the different scales involved in the problem, or if the flow is compressible or incompressible. Here we restrict ourselves to the case of laminar incompressible flow, and we present one finite element technique capable of solving the two-dimensional Navier–Stokes equation under several situations. This is the penalty finite element formulation, which is very robust, but is restricted mainly to two dimensions because it requires the storage of the stiffness matrix—which becomes unmanageable in three dimensions. To solve three-dimensional problems, more-sophisticated techniques will be required (Heinrich and Pepper, 1999).

10.2 VISCOUS INCOMPRESSIBLE FLOW WITH HEAT TRANSFER

The two-dimensional Navier–Stokes and energy equation for a laminar incompressible flow are the continuity, momentum, and energy equations, given by

$$\frac{\partial u}{\partial x}+\frac{\partial v}{\partial y}=0 \tag{10.1}$$

$$\rho\left(\frac{\partial u}{\partial t}+u\frac{\partial u}{\partial x}+v\frac{\partial u}{\partial y}\right)=-\frac{\partial p}{\partial x}+\mu\left(\frac{\partial^2 u}{\partial x^2}+u\frac{\partial^2 u}{\partial y^2}\right)+\rho\beta_x \tag{10.2}$$

$$\rho\left(\frac{\partial v}{\partial t}+u\frac{\partial v}{\partial x}+v\frac{\partial v}{\partial y}\right)=-\frac{\partial p}{\partial y}+\mu\left(\frac{\partial^2 v}{\partial x^2}+\frac{\partial^2 v}{\partial y^2}\right)+\rho\beta_y \tag{10.3}$$

$$\rho c_p\left(\frac{\partial T}{\partial t}+u\frac{\partial T}{\partial x}+v\frac{\partial T}{\partial y}\right)=k\left(\frac{\partial^2 T}{\partial x^2}+\frac{\partial^2 T}{\partial y^2}\right) \tag{10.4}$$

In these equations u and v are the components of velocity in the x- and y-directions respectively; p is the pressure; t is time; ρ is density; μ is the dynamic fluid viscosity; c_p is the fluid's specific heat; k is the heat conduction coefficient; and β_x, β_y are the components of body forces acting in the x- and y-directions, respectively.

We assume that the density can be considered constant except in the body force term where it can vary linearly with temperature, i.e.,

$$\rho=\rho_0\left(1+\beta\left(T-T_0\right)\right) \tag{10.5}$$

This is known as the Boussinesq approximation. Assuming that gravity acts in the negative y-direction, we nondimensionalize the equations using a characteristic length L and the convective time scale $\tau = L/U$ where U is a characteristic velocity. The equations can then be rewritten as

$$\frac{\partial u}{\partial x}+\frac{\partial v}{\partial y}=0 \tag{10.6}$$

$$\frac{\partial u}{\partial t}+u\frac{\partial u}{\partial x}+v\frac{\partial u}{\partial y}=-\frac{\partial p}{\partial x}+\frac{1}{\text{Re}}\left(\frac{\partial^2 u}{\partial x^2}+\frac{\partial^2 u}{\partial y^2}\right) \tag{10.7}$$

$$\frac{\partial v}{\partial t}+u\frac{\partial v}{\partial x}+v\frac{\partial v}{\partial y}=-\frac{\partial p}{\partial y}+\frac{1}{\text{Re}}\left(\frac{\partial^2 v}{\partial x^2}+\frac{\partial^2 v}{\partial y^2}\right)+\frac{1}{Fr}T \tag{10.8}$$

$$\frac{\partial T}{\partial t}+u\frac{\partial T}{\partial x}+v\frac{\partial T}{\partial y}=\frac{1}{Pe}\left(\frac{\partial^2 T}{\partial x^2}+\frac{\partial^2 T}{\partial y^2}\right) \tag{10.9}$$

where now all the variables are nondimensional and the Reynolds number Re, Froude number Fr, and Péclet number Pe are given by

$$Re = \frac{\rho UL}{\mu} \tag{10.10}$$

$$Fr = \frac{U^2}{\beta g \left|\Delta T\right| L} \tag{10.11}$$

$$Pe = \frac{\rho_0 c_p UL}{\kappa} \tag{10.12}$$

The discretization of Eqs. (10.6) to (10.9) is made difficult by the need to impose the incompressibility condition (10.6). The penalty formulation resolves this problem in a simple way by accepting a very small compressibility and introducing a pseudo-constitutive relation of the form:

$$p - p_s = -\lambda\left(\frac{\partial u}{\partial x} + \frac{\partial v}{\partial y}\right) \tag{10.13}$$

where p_s is the static pressure and λ, the penalty parameter, is a large number. As ∞ incompressibility is imposed exactly. We do not examine all the theory and implications of the penalty method here; these are explained in detail in Heinrich and Pepper (1999).

10.3 THE PENALTY FUNCTION ALGORITHM

We discretize equations (10.7) to (10.9) using bilinear isoparametric elements. The pressure and the continuity equation (10.6) are eliminated from Eqs. (10.7) and (10.8) using the penalty relation (10.13). This results in the modified system of equations:

$$\frac{\partial u}{\partial t} - \lambda\frac{\partial}{\partial x}\left(\frac{\partial u}{\partial x} + \frac{\partial v}{\partial y}\right) - \frac{1}{Re}\left(\frac{\partial^2 u}{\partial x^2} + \frac{\partial^2 u}{\partial y^2}\right) = -\left(u\frac{\partial u}{\partial x} + v\frac{\partial u}{\partial y}\right) \tag{10.14}$$

$$\frac{\partial v}{\partial t} - \lambda\frac{\partial}{\partial y}\left(\frac{\partial u}{\partial x} + \frac{\partial v}{\partial y}\right) - \frac{1}{Re}\left(\frac{\partial^2 v}{\partial x^2} + \frac{\partial^2 v}{\partial y^2}\right) = -\left(u\frac{\partial v}{\partial x} + v\frac{\partial v}{\partial y}\right) + \frac{1}{Fr}T \tag{10.15}$$

Equation (10.9) remains the same as before. The equations have been written in such a way that terms to be solved implicitly in the algorithm are on the left-hand side and terms that will be evaluated explicitly are on the right-hand side. The weighted residuals formulations of Eqs. (10.9), (10.14), and (10.15) are written as

$$\int_{\Omega}\left\{N_i\frac{\partial u}{\partial t}+\lambda\frac{\partial N_i}{\partial x}\left(\frac{\partial u}{\partial x}+\frac{\partial v}{\partial y}\right)+\frac{1}{\text{Re}}\left(\frac{\partial N_i}{\partial x}\frac{\partial u}{\partial x}+\frac{\partial N_i}{\partial y}\frac{\partial u}{\partial y}\right)\right\}d\Omega$$
$$=-\int_{\Omega}U_i\left(u\frac{\partial u}{\partial x}+v\frac{\partial u}{\partial y}\right)d\Omega \tag{10.16}$$

$$\int_{\Omega}\left\{N_i\frac{\partial v}{\partial t}+\lambda\frac{\partial N_i}{\partial y}\left(\frac{\partial u}{\partial x}+\frac{\partial v}{\partial y}\right)+\frac{1}{\text{Re}}\left(\frac{\partial N_i}{\partial x}\frac{\partial v}{\partial x}+\frac{\partial N_i}{\partial y}\frac{\partial v}{\partial y}\right)\right\}d\Omega$$
$$=-\int_{\Omega}\left\{U_i\left(u\frac{\partial v}{\partial x}+v\frac{\partial v}{\partial y}\right)+N_i\frac{1}{Fr}T\right\}d\Omega \tag{10.17}$$

$$\int_{\Omega}\left\{N_i\frac{\partial T}{\partial t}+\frac{1}{Pe}\left(\frac{\partial N_i}{\partial x}\frac{\partial T}{\partial x}+\frac{\partial N_i}{\partial y}\frac{\partial T}{\partial y}\right)\right\}d\Omega$$
$$=-\int_{\Omega}V_i\left(u\frac{\partial T}{\partial x}+v\frac{\partial T}{\partial y}\right)d\Omega \tag{10.18}$$

In the above equation, N_i denotes the bilinear isoparametric shape function and U_i, V_i, are the Petrov–Galerkin weights defined in Eq. (9.27) with calculated using $\gamma = \text{Re}\,\tilde{h}$ for U_i and $\gamma = \text{Pe}\,\tilde{h}$ for V_i. After expressing u, v, and T in terms of the shape functions, and using the backward implicit method— $\theta = 1$ in Eq. (3.99)— the discrete form of the equations becomes

$$\left[\int_{\Omega}\left\{\frac{1}{\Delta t}N_iN_j+\frac{1}{\text{Re}}\left(\frac{\partial N_i}{\partial x}\frac{\partial N_j}{\partial x}+\frac{\partial N_i}{\partial y}\frac{\partial N_j}{\partial y}\right)\right\}d\Omega+\lambda\int_{\Omega}\frac{\partial N_i}{\partial x}\frac{\partial N_j}{\partial x}d\Omega\right]u_j^{n+1}+$$
$$\left[\lambda\int_{\Omega}\frac{\partial N_i}{\partial x}\frac{\partial N_j}{\partial y}d\Omega\right]v_j^{n+1}=\int_{\Omega}\left\{\frac{1}{\Delta t}N_iu^n-U_i\left(u^n\frac{\partial u^n}{\partial x}+v^n\frac{\partial u^n}{\partial y}\right)\right\}d\Omega$$
$$\tag{10.19}$$

$$\left[\int_{\Omega}\left\{\frac{1}{\Delta t}N_iN_j+\frac{1}{\text{Re}}\left(\frac{\partial N_i}{\partial x}\frac{\partial N_j}{\partial x}+\frac{\partial N_i}{\partial y}\frac{\partial N_j}{\partial y}\right)\right\}d\Omega+\lambda\int_{\Omega}\frac{\partial N_i}{\partial y}\frac{\partial N_j}{\partial y}d\Omega\right]v_j^{n+1}+$$
$$\left[\lambda\int_{\Omega}\frac{\partial N_i}{\partial y}\frac{\partial N_j}{\partial x}d\Omega\right]u_j^{n+1}=\int_{\Omega}\left\{N_i\left(\frac{1}{\Delta t}v^n+\frac{1}{Fr}T^n\right)-U_i\left(u^n\frac{\partial v^n}{\partial x}+v^n\frac{\partial v^n}{\partial y}\right)\right\}d\Omega$$
$$\tag{10.20}$$

$$\left[\int_{\Omega}\left\{\frac{1}{\Delta t}N_iN_j+\frac{1}{Pe}\left(\frac{\partial N_i}{\partial x}\frac{\partial N_j}{\partial x}+\frac{\partial N_i}{\partial y}\frac{\partial N_j}{\partial y}\right)\right\}d\Omega\right]T_j^{n+1}$$
$$=\int_{\Gamma}N_i\frac{\partial T}{\partial n}d\Gamma+\int_{\Omega}\left\{\frac{1}{\Delta t}N_iT^n-V_i\left(u^n\frac{\partial T^n}{\partial x}+v^n\frac{\partial T^n}{\partial y}\right)\right\}d\Omega \tag{10.21}$$

In Eqs. (10.19) and (10.20) the symbol $\int\!\!\!\!\mathrm{R}$ indicates that this term must be reduce integrated using a single Gauss point. Also notice that the line integrals that arise from applying the Green–Gauss Theorem (see Section 4.6) have been omitted in Eqs. (10.19) and (10.20). The reasons behind these choices are discussed in greater detail in Heinrich and Pepper (1999). Here we restrict ourselves to presenting the algorithms.

Evaluation of the integrals in Eqs. (10.19) to (10.21) leads to a system of coupled equations for the velocity components and a second system of equations for the temperature, of the form:

$$\mathrm{AV}^{n+1}=\mathrm{F}\left(\mathrm{V}^n,\mathrm{T}^{n+1}\right) \tag{10.22}$$

$$\mathrm{BT}^{n+1}=\mathrm{G}\left(\mathrm{V}^n,\mathrm{T}^n\right) \tag{10.23}$$

The systems are solved alternately to advance in time. Using the latest known values $\mathbf{V}^n$ and $\mathbf{T}^n$, first we solve for $\mathbf{T}^{n+1}$ from Eq. (10.23). The same velocity $\mathbf{V}^n$ and the new temperature $\mathbf{T}^{n+1}$ are then used in the right-hand side of Eq. (10.22) to find $\mathbf{V}^{n+1}$. This formulation has the advantage that if the time step does not change, the matrices **A** and **B** remain unchanged. They can then be decomposed once at the beginning of the calculation using an **LU** algorithm (recall Section 3.8) and the solution can be efficiently advanced in time by simply updating the right-hand side and performing a forward and backward substitution.

The disadvantages of this algorithm are twofold. The fact that the convective and body force terms are evaluated explicitly introduces a stability limitation in the time step. For the convective term the stability limitation can be expressed as

$$\Delta t_1 \le \frac{1}{\dfrac{|u|}{\Delta x}+\dfrac{|v|}{\Delta y}} \tag{10.24}$$

for every element in the mesh. For the body-force term it takes the form:

$$\Delta t_2 \le Fr \tag{10.25}$$

and we must have $\Delta t < \min\,(\Delta t_1, \Delta t_2)$ or the calculation becomes unstable. This can be restrictive if steady-state solutions are sought that could be found faster with a large time step. If the time evolution of the flow is important, or if steady states do not exist, it is always recommended that Eq. (10.24) be satisfied to preserve accuracy.

The second disadvantage of this formulation comes from the fact that we are using a penalty formulation—a direct method of solution must be used to solve the system of equations (10.22). This requires the storage of the matrices in skyline form and makes the method somewhat limited in three dimensions.

A final word is necessary about the penalty parameter. The parameter has to be chosen according to the value of the Reynolds number and the Froud number. However, for most practical problems, assuming that double precision arithmetic is being used (64 bit words), we can set $\lambda = 10^9$. For more complicated situations, the criteria to choose λ are explained in Heinrich and Vionnet (1995).

10.4 APPLICATION TO NATURAL CONVECTION

We now use the above algorithm to simulate natural convection in a square enclosure. Figure 10.1 shows the domain and boundary conditions. For problems dominated by natural convection such as the one we are discussing, a more appropriate nondimensionalization is obtained using the thermal diffusion timescale $\tau = L^2/\rho c_p k$. A characteristic velocity is very difficult to determine *a priori* in these problems. The latter is chosen in terms of the buoyancy as

$$U = \sqrt{\beta g L |\Delta T|}$$

This results in a nondimensional form that depends only on two parameters, i.e.,

$$\frac{\partial u}{\partial t} + \sqrt{Ra\,\mathrm{Pr}}\left(u\frac{\partial u}{\partial x} + v\frac{\partial u}{\partial y}\right) = \frac{-1}{\sqrt{Ra\,\mathrm{Pr}}}\frac{\partial p}{\partial x} + \frac{1}{\mathrm{Pr}}\left(\frac{\partial^2 u}{\partial x^2} + \frac{\partial^2 u}{\partial y^2}\right) \qquad (10.26)$$

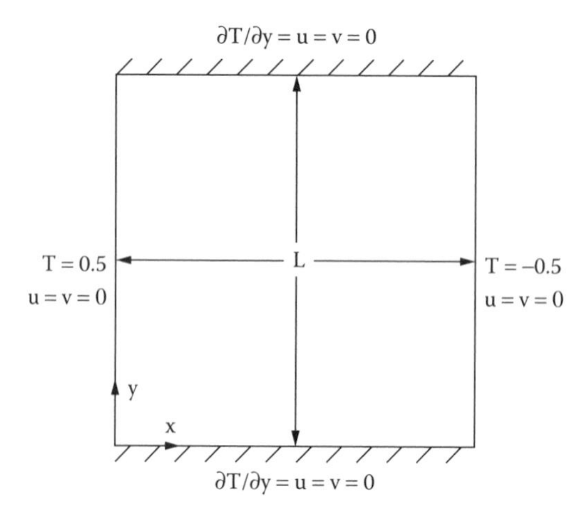

Figure 10.1 Domain and boundary conditions for natural convection in a square enclosure

$$\frac{\partial v}{\partial t} + \sqrt{Ra\,\mathrm{Pr}}\left(u\frac{\partial v}{\partial x} + v\frac{\partial v}{\partial y}\right) = \frac{-1}{\sqrt{Ra\,\mathrm{Pr}}}\frac{\partial p}{\partial y} + \frac{1}{\mathrm{Pr}}\left(\frac{\partial^2 v}{\partial x^2} + \frac{\partial^2 v}{\partial y^2}\right) + \sqrt{Ra\,\mathrm{Pr}}\,T \tag{10.27}$$

$$\frac{\partial T}{\partial t} + \sqrt{Ra\,\mathrm{Pr}}\left(u\frac{\partial T}{\partial x} + v\frac{\partial T}{\partial y}\right) = \frac{\partial^2 T}{\partial x^2} + \frac{\partial^2 T}{\partial y^2} \tag{10.28}$$

where the Rayleigh number $Ra = \left(\rho_0^2 c_p \beta g \left|\Delta T\right| L^3\right) / \mu k$ and the Prandtl number Pr = $\mu/\rho_o k$ are related to Re, Fr, and Pe by the relations:

$$Fr = \frac{\mathrm{Re}^2\,\mathrm{Pr}}{Ra} \tag{10.29}$$

and

$$Pe = \mathrm{Re}\,\mathrm{Pr} \tag{10.30}$$

Heinrich and Pepper (1999) include software in their text that offers the choice between these two nondimensional forms, the one using Re, Fr, and Pe should be used when forced convection is present. The one using Ra, Pr should be used when natural convection provides the driving force in the system.

Example 10.1*

We discretize the domain in Figure 10.2 using a uniform mesh of 40 × 40 square bilinear elements. The parameters are chosen as Ra = 10^5, Pr = 1.0, and the time step $\Delta t = 10^{-3}$. Starting from uniform initial conditions $T(x,y,0) = u(x,y,0) = v(x,y,0) = 0.0$, the steady-state solution is obtained in approximately 400 time steps. Figure 10.2 shows the calculated steady-state solution. In Figure 10.3a only every other velocity vector has been plotted. In Figure 10.3b the isotherms are shown at intervals $\Delta T = 0.1$. We observe a boundary layer flow with two recirculation cells. These were first observed experimentally by Elder (1965) who called them "cat's eyes." The reader is encouraged to run this simulation using the enclosed COMSOL example file. This will underscore the importance of identifying the correct time and length scales to successfully obtain accurate numerical approximations.

Example 10.2*

In this second example, we examine flow over a backward-facing step. The Reynolds number is 800, and the flow enters from the left and exits from the right-hand boundary. We use a 20 × 40 mesh consisting of bilinear elements. Figure 10.4 shows the initial mesh for the problem domain, and Figure 10.5 shows the streamlines overlaid on the velocity vectors. Notice the recirculation bubble on the top wall, and the recirculation length behind the step on the bottom wall.

* COMSOL file is available.

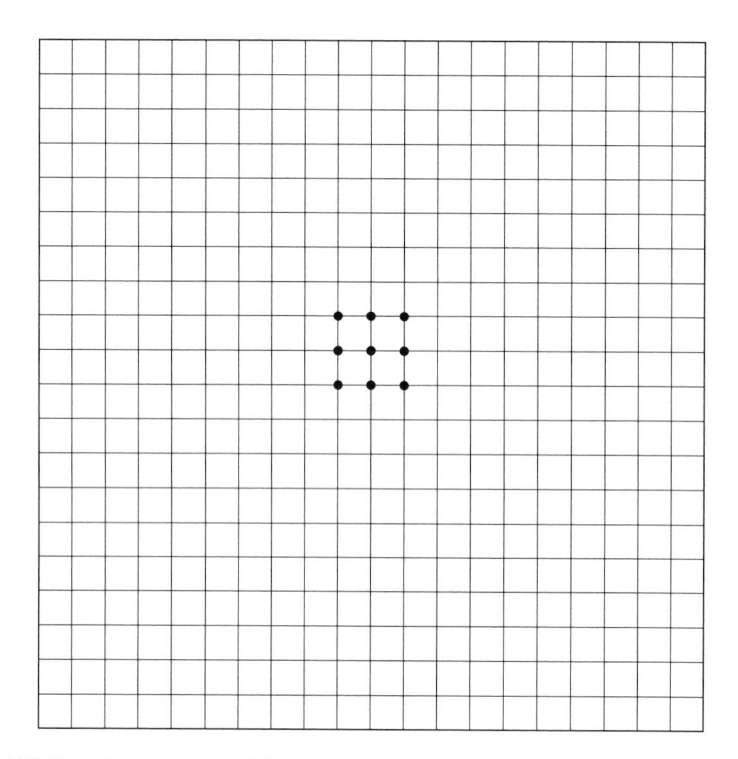

Figure 10.2 Natural convection mesh for a square enclosure, Ra = 10^5

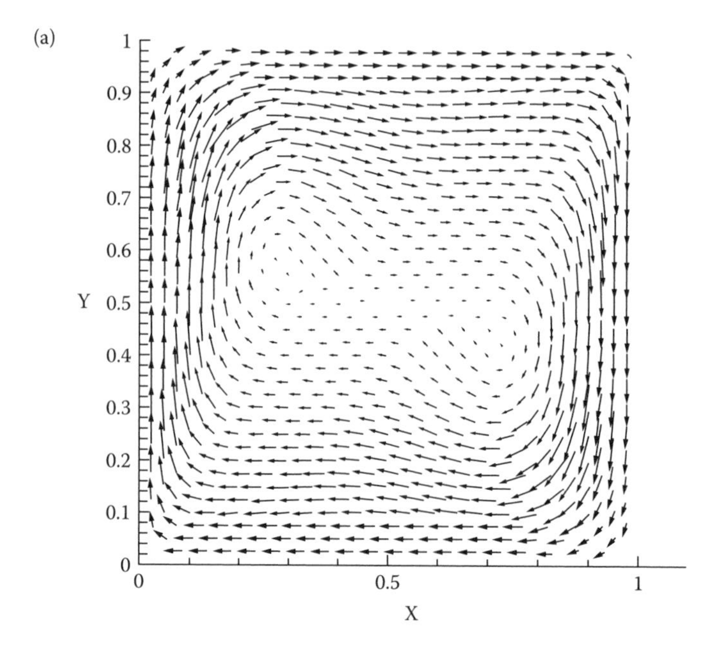

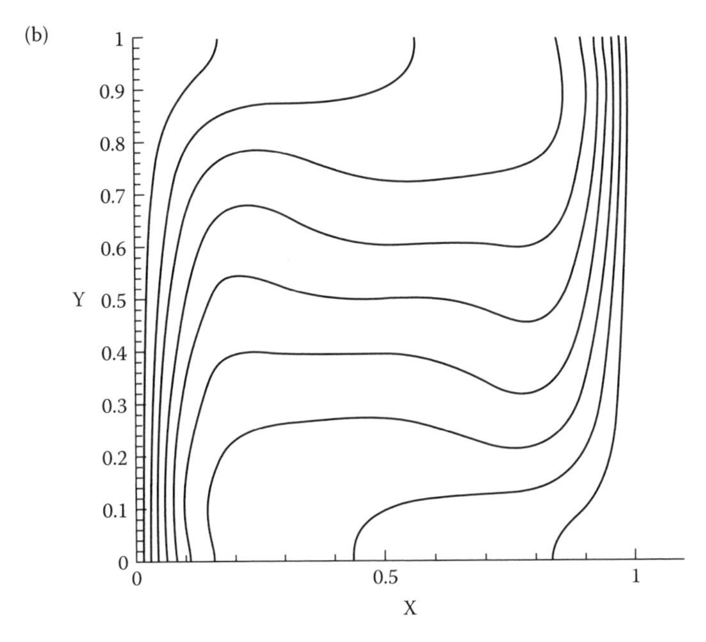

Figure 10.3 Steady-state solution for natural convection within a square enclosure, Ra = 10^5: (a) velocity vectors; and (b) isotherms

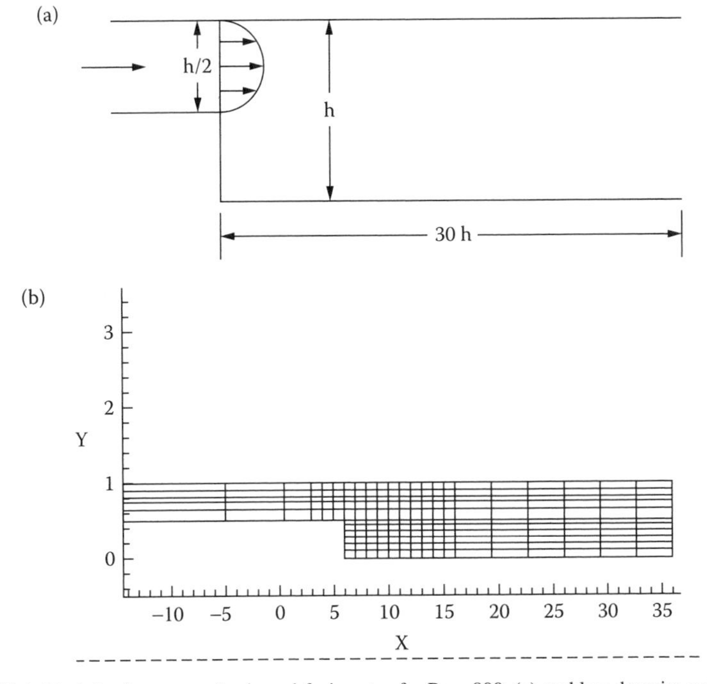

Figure 10.4 Mesh for flow over a backward-facing step for Re = 800: (a) problem domain; and (b) mesh

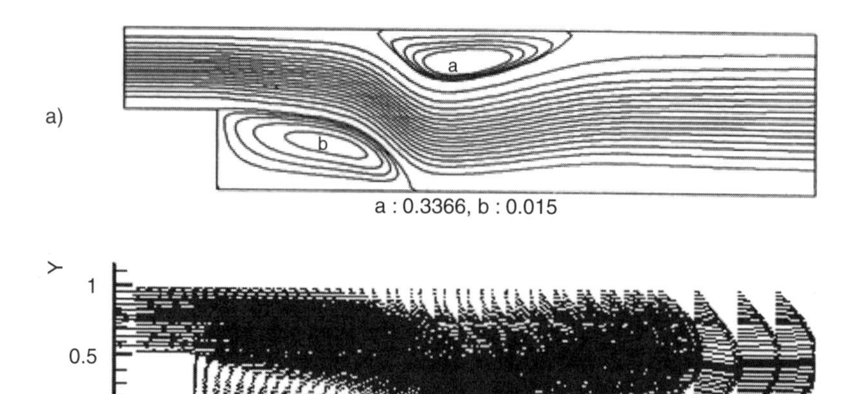

Figure 10.5 Streamlines and velocity vectors for flow over a backward-facing step at Re = 800

The recirculation length compares with results obtained from others in a benchmark exercise conducted by ASME in 1992 (Blackwell and Pepper, 1992). Both experimental data (Armaly et al., 1983) and numerical simulation using a finite element model (Gartling, 1990) were used in the benchmark study as target guidelines.

10.5 CLOSURE

Throughout this book we have covered a very large range of numerical approximations using the finite element method. We started with the simplest one-dimensional, linear, steady-state conduction problem, expanded the method to two- and three-dimensional elements, and ended with the time-dependent, nonlinear incompressible, Navier–Stokes and energy equations. All of this was accomplished using the basic concepts contained in finite element approximations, which are equally applicable to heat transfer, wave propagation, structural analysis, and fluid mechanics. The reader should by now have an appreciation for the power and versatility of the finite element technique and should be well prepared to move beyond the contents of this book into more complex problems.

EXERCISES

10.1 Calculate the flow over the square cavity shown below for Re = 100 using COMSOL.* Assume a nondimensional horizontal velocity of $u = 1$ for the top

* COMSOL mesh file is available.

moving from left to right; the remaining walls are no-slip surfaces. Plot the streamlines, pressure contours, and velocity vectors.

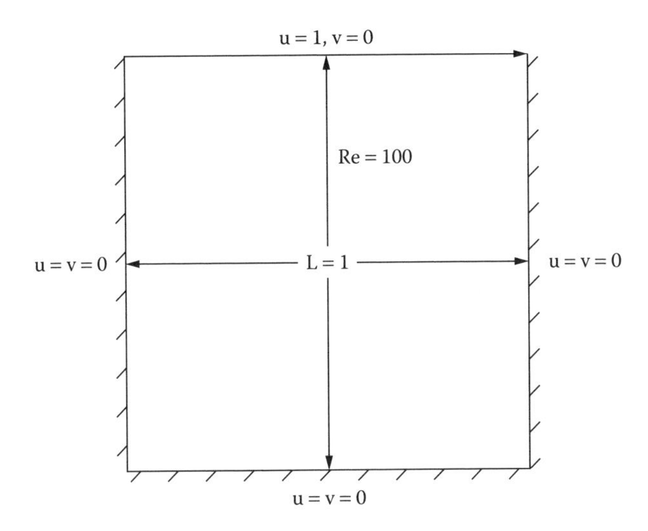

10.2 Calculate the flow over a circular cylinder shown in the figure for Re = 10 and Re = 100 using COMSOL.* What happens to the wake behind the cylinder as the flow increases in speed? Is the flow at Re = 100 steady or transitory?

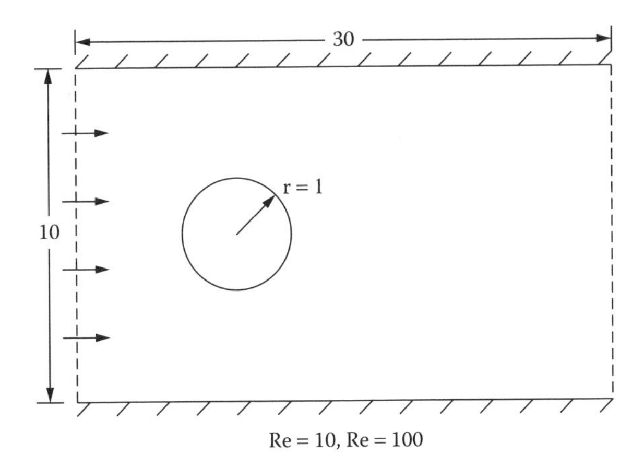

10.3 Calculate the flow and temperature distribution within the eccentric annulus shown below for Ra = 10^4 using COMSOL.* Assume a nondimensional temperature of $T = 1$ for the inner cylinder and $T = 0$ for the outer cylinder. You

* COMSOL mesh file is available.

may wish to employ vertical symmetry to model the domain—be sure to incorporate the proper boundary conditions along the line of symmetry.

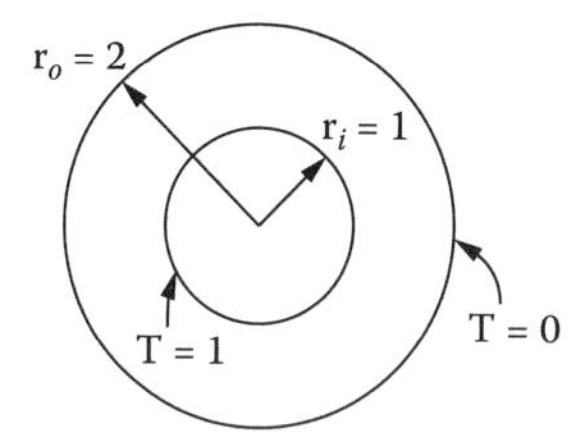

REFERENCES

Armaly, B. F., Durst, F., Pereira, J. C. F., and Schonung, B. (1983). "Experimental and Theoretical Investigation of Backward-Facing Step Flow." *J. Fluid Mech.* 172:473–496.

Baker, A. J. (1971). "A Finite Element Computational Theory for the Mechanics and Thermodynamics of a Viscous Compressible Multi-Species Fluid." Bell Aerospace Research Dept. 9500-920200.

Baker, A. J. (1983). *Finite Element Computational Fluid Mechanics.* Washington, D.C.: Hemisphere.

Blackwell, B. F. and Pepper, D. W. (1992). "Benchmark Problems for Heat Transfer Codes." HTD-Vol. 222, ASME WAM, Anaheim, Calif., Nov. 8–13.

Chung, R. W. (1960). *Finite Element Analysis in Fluid Dynamics.* New York: McGraw-Hill.

Elder, J. W. (1965). "Laminar Fall Convection in a Vertical Slot." *J. Fluid Mech.* 23:77–98.

Gartling, D. K. (1990). "A Test Problem for Outflow Boundary Conditions—Flow Over a Backward Facing Step." *Int. J. Num. Meth. Fluids.* 11:953–967.

Heinrich, J. C. and Pepper, D. W. (1999). *Intermediate Finite Element Method. Fluid Flow and Heat Transfer Applications.* New York: Taylor & Francis.

Heinrich, J. C. and Vionnet, C. A. (1995). "The Penalty Method for the Navier-Stokes Equations." *Arch. Comp. Meth. Eng.* 2:51–65.

Heinrich, J. C. and Yu, C.-C. (1988). "Finite Element Simulations of Buoyancy-Driven Flows with Emphasis on Natural Convection in a Horizontal Circular Cylinder." *Comp. Meth. Appl, Mech. Eng.* 69:1–27.

Heinrich, J. C. and Zienkiewicz, O. C. (1970). "The Finite Element Method and 'Upwinding' Techniques in the Numerical Solution of Convection Dominated Flow Problems." In *Finite Element Methods for Convection Dominated Flows* (T. J. R. Hughes, ed.). New York: ASME AMD; 34, pp. 105–136.

Hughes, T. J. R. and Brooks, A. (1979). "A Multidimensional Upwind Scheme with No Cross-Diffusion." In *Finite Element Methods for Convection Dominated Flows* (T. J. R. Hughes, ed.). New York: ASME AMD; 34, pp. 19–35.

Martin, H. C. (1968) "Finite Element Analysis of Fluid Flow." *Proc. 2nd Conf. Matrix Methods in Structural Mechanics.* Wright Patterson Air Force Bass, Dayton, Ohio.

Zienkiewicz, O. C. and Taylor, R. L. (1989). *The Finite Element Method,* 4th ed. Maidenhead, U.K.: McGraw-Hill.

A. MATRIX ALGEBRA

A basic understanding of elementary matrix algebra is necessary in using finite element methods. Assemblage of the various parts of the general equation set utilizes addition, subtraction, multiplication and division of $(n \times m)$ matrices, (m) column vectors and (n) row vectors. Integration of the integral terms by Gaussian quadrature relies on the summation of the elemental terms into a global matrix. Solution of the overall global matrix utilizes matrix reduction techniques (inversion, etc.) to solve for the column vector of unknown variables. These matrix operations are relatively simple. The relevant operations performed in the programs are discussed in the following.

A system of linear algebraic equations is usually written as the matrix statement:

$$\mathrm{Ax} = \mathrm{b} \tag{A.1}$$

Expanded in terms of the elements $\mathbf{a}_{ij}$, $\mathbf{x}_j$, and $\mathbf{b}_i$ of $\mathbf{A}$, $\mathbf{x}$, and $\mathbf{b}$, we have

$$\begin{aligned}
a_{11}x_1 + a_{12}x_2 + a_{13}x_3 + a_{14}x_4 \cdots + a_{1m} &= b_1 \\
a_{21}x_1 + a_{22}x_2 + a_{23}x_3 + a_{24}x_4 \cdots + a_{2m} &= b_1 \\
a_{11}x_1 + a_{32}x_2 + a_{33}x_3 + a_{34}x_4 \cdots + a_{3m} &= b_1 \\
\vdots \qquad\qquad\qquad & \quad \vdots \\
a_{n1}x_1 + a_n x_2 + a_n x_3 + a_n x_4 \cdots + a_{nm} &= b_n
\end{aligned} \tag{A.2}$$

For example, if $n = m = 3$, we have

$$\begin{bmatrix} a_{11} & a_{12} & a_{13} \\ a_{21} & a_{22} & a_{23} \\ a_{31} & a_{32} & a_{33} \end{bmatrix} \begin{bmatrix} x_1 \\ x_2 \\ x_3 \end{bmatrix} = \begin{bmatrix} b_1 \\ b_2 \\ b_3 \end{bmatrix} \tag{A.3}$$

In Eqs. (A.2) and (A.3), i refers to a row and j to a column location in **A**, **x**, and/or **b**. In this example, the coefficient matrix **A** is 3 × 3. The matrices **x** and **b** are called column (or vector) matrices since they are $n \times 1$ matrices. Similarly, a row vector would consist of only one row but with m columns. If $n = m$, the matrix is called a square matrix. If, furthermore, a square matrix is such that $\mathbf{a}_{ij} = \mathbf{a}_{ji}$, it is called symmetric.

A.1 Addition/Subtraction of Matrices

Two matrices can be added or subtracted if they are of the same order ($n \times m$). For example,

$$\begin{bmatrix} a_{11} & a_{12} & a_{13} \\ a_{21} & a_{22} & a_{23} \\ a_{31} & a_{32} & a_{33} \end{bmatrix} \begin{bmatrix} x_1 \\ x_2 \\ x_3 \end{bmatrix} = \begin{bmatrix} b_1 \\ b_2 \\ b_3 \end{bmatrix}$$

$$\begin{bmatrix} c_{11} & c_{12} & c_{13} \\ c_{21} & c_{22} & c_{23} \\ c_{31} & c_{32} & c_{33} \end{bmatrix} \begin{bmatrix} x_1 \\ x_2 \\ x_3 \end{bmatrix} = \begin{bmatrix} d_1 \\ d_2 \\ d_3 \end{bmatrix} \tag{A.4}$$

$$\begin{bmatrix} a_{11}+c_{11} & a_{12}+c_{12} & a_{13}+c_{13} \\ a_{21}+c_{21} & a_{22}+c_{22} & a_{23}+c_{23} \\ a_{31}+c_{31} & a_{32}+c_{32} & a_{33}+c_{33} \end{bmatrix} \begin{bmatrix} x_1 \\ x_2 \\ x_3 \end{bmatrix} = \begin{bmatrix} b_1+d_1 \\ b_2+d_3 \\ b_3+d_3 \end{bmatrix} \tag{A.5}$$

or

$$(\mathbf{A}+\mathbf{C})\mathbf{x} = \mathbf{b}+\mathbf{d} \tag{A.6a}$$

Similarly,

$$(\mathbf{A}-\mathbf{C})\mathbf{x} = \mathbf{b}-\mathbf{d} \tag{A.6b}$$

or

$$\begin{bmatrix} a_{11}-c_{11} & a_{12}-c_{12} & a_{13}-c_{13} \\ a_{21}-c_{21} & a_{22}-c_{22} & a_{23}-c_{23} \\ a_{31}-c_{31} & a_{32}-c_{32} & a_{33}-c_{33} \end{bmatrix} \begin{bmatrix} x_1 \\ x_2 \\ x_3 \end{bmatrix} = \begin{bmatrix} b_1-d_1 \\ b_2-d_3 \\ b_3-d_3 \end{bmatrix} \tag{A.7}$$

A.2 Multiplication of Matrices

A matrix can be multiplied by another matrix providing the number of columns in the first matrix equals the number of rows in the second matrix. If **A** contains n rows and m columns and **B** contains m rows and columns, then the product **C** is an $n \times \ell$ matrix:

$$\mathbf{C} = \mathbf{AB} \tag{A.8a}$$

or

$$C_{ij} = \sum_{k=1}^{\ell} a_{ik} b_{kj} \tag{A.8b}$$

It is important to remember the order in which matrices are multiplied. The product **AB** is not equal to the product of **BA**. The **A** matrix is the pre-multiplier whereas **B** is the post-multiplier. However, multiplication of matrices is associative, i.e.,

$$(\mathbf{AB})\mathbf{C} = \mathbf{A}(\mathbf{BC}) \tag{A.9}$$

If we multiply a 3 × 1 row vector matrix, **A**, times a 1 × 3 column vector, **B**, we obtain

$$\begin{bmatrix} a_1 a_2 a_3 \end{bmatrix} \begin{bmatrix} b_1 \\ b_2 \\ b_3 \end{bmatrix} = a_1 b_1 + a_2 b_2 + a_3 b_3 \tag{A.10}$$

On the other hand, **BA** yields

$$\begin{bmatrix} b_1 \\ b_2 \\ b_3 \end{bmatrix} [a_1 \; a_2 \; a_3] = \begin{bmatrix} b_1 a_1 & b_1 a_2 & b_1 a_3 \\ b_2 a_1 & b_2 a_2 & b_2 a_3 \\ b_3 a_1 & b_3 a_2 & b_3 a_3 \end{bmatrix} \tag{A.11}$$

Multiplication of row and column vectors consisting of the shape functions and their derivatives takes place repeatedly throughout the programs.

A.3 Determinant of a Matrix

The determinant of a matrix is used in the 2 × 2 Jacobian matrix for transforming the shape function derivatives from η_1, η_2 to x, y coordinates. The Jacobian is

$$\mathrm{J} = \begin{bmatrix} \dfrac{\partial x}{\partial \eta_1} & \dfrac{\partial y}{\partial \eta_1} \\ \dfrac{\partial x}{\partial \eta_2} & \dfrac{\partial y}{\partial \eta_2} \end{bmatrix} \tag{A.12}$$

which, for simplicity, can be written as

$$\mathbf{A} = \begin{bmatrix} a_{11} & a_{12} \\ a_{21} & a_{22} \end{bmatrix} \tag{A.13}$$

The determinant, det **A**, is obtained by cross-multiplying and subtracting products, i.e.,

$$|\mathbf{A}| = \det \mathbf{A} = a_{11}\, a_{22} - a_{21}\, a_{12} \tag{A.14}$$

Hence, the determinant of the Jacobian matrix can be written

$$|\mathbf{J}| = \det \mathbf{J} = \sum_{k=1}^{k} \frac{\partial \eta_k}{\partial \eta_1} x_k \cdot \sum_{k=1}^{k} \frac{\partial \eta_k}{\partial \eta_2} y_k - \sum_{k=1}^{k} \frac{\partial \eta_k}{\partial \eta_1} y_k \cdot \sum_{k=1}^{k} \frac{\partial \eta_k}{\partial \eta_2} x_k \tag{A.15}$$

where N denotes the shape function and k the number of local nodal points. If **A** = 0, the matrix is said to be singular.

A.4 Inverse of a Matrix

The inverse of a matrix is generally used to solve the column vector of unknown variables, i.e.,

$$\mathbf{x} = \mathbf{A}^{-1}\mathbf{b} \tag{A.16}$$

We define the inverse by A^{-1} such that

$$\mathbf{A}^{-1}\mathbf{A} = \mathbf{A}\,\mathbf{A}^{-1} = \mathbf{I} \tag{A.17}$$

where **I** is the identity matrix in which all elements are 0 except for the diagonal terms, which are equal to 1. The inverse is given by

$$\mathbf{A}^{-1} = \frac{1}{|\mathbf{A}|} Adj\ \mathbf{A} \tag{A.18}$$

where Adj **A** is the adjoint matrix of **A** defined as

$$Adj\ \mathbf{A} = (\mathbf{A}_{cf})^T \tag{A.19}$$

where the transport matrix, discussed in more detail later on, is obtained by interchanging columns and rows in the matrix, i.e., if $\mathbf{B} = [b_{ij}]$, $\mathbf{B}^T = [b_{ji}]$; and the cofactors matrix is determined by replacing each element in **A** by $(-1)^{i=j}$ times the determinant of a reduced matrix, in which the ith row and jth column of **A** are removed. Once this process is completed, the adjoint of **A** is obtained, which is the transpose of the co-factors in **A**. The resultant inverse of **A** is finally obtained by dividing its adjoint matrix by **A**. In elemental form,

$$\mathbf{A} = \begin{bmatrix} a_{11} & a_{12} \\ a_{21} & a_{22} \end{bmatrix}$$

Co-factor:

$$\mathbf{A}_{cf} = \begin{bmatrix} a_{22} & a_{21} \\ a_{12} & a_{11} \end{bmatrix}$$

Adjoint:

$$Adj\ \mathbf{A} = \begin{bmatrix} a_{22} & -a_{12} \\ -a_{21} & a_{11} \end{bmatrix}$$

Inverse:

$$\mathbf{A}^{-1} = \frac{1}{(a_{11}\ a_{22} - a_1\ a_{12})} \begin{bmatrix} a_{22} & a_{12} \\ a_{21} & a_{11} \end{bmatrix} \tag{A.20}$$

As an illustrative example, we wish to find the inverse of the matrix

$$\mathbf{A} = \begin{bmatrix} 4 & 2 \\ 1 & 1 \end{bmatrix}$$

Thus,

$$\det \mathbf{A} = 4 - 2 = 2$$

$$\mathbf{A}_{cf} = \begin{bmatrix} 1 & -1 \\ -2 & 4 \end{bmatrix}$$

$$Adj\ \mathbf{A} = \begin{bmatrix} 1 & -2 \\ -1 & 4 \end{bmatrix}$$

$$\mathbf{A}^{-1} = \frac{\begin{bmatrix} 1 & -2 \\ -1 & 4 \end{bmatrix}}{2} = \frac{1}{2}\begin{bmatrix} 1 & -2 \\ -1 & 4 \end{bmatrix}$$

A.5 Derivative of a Matrix

The derivative of a matrix is obtained by taking the derivative of each of its elements, i.e.,

$$\frac{d}{dx}\mathbf{A} = \begin{bmatrix} \frac{da_{11}}{dx} & \frac{da_{12}}{dx} \\ \frac{da_{21}}{dx} & \frac{da_{22}}{dx} \end{bmatrix} \tag{A.21}$$

A.6 Transpose of a Matrix

The transpose of a matrix is obtained by changing the order of the columns and rows. This procedure is employed to multiply the shape functions, their derivatives, and/or variables within the integral expressions to obtain a compatible matrix. For example, the mass matrix term

$$\mathbf{M} = \int_A N_i N_j dA, \qquad 1 \le (i,j) \le m \tag{A.22}$$

is actually

$$\mathbf{M} = \int_A \mathbf{N}^T \mathbf{N}\, dA \tag{A.23}$$

where N^T is the transpose of **N**. Otherwise, we would have **N** × **N**, which is non-conformable. The transpose of a row matrix, written **A**, $\mathbf{A}^T$, becomes a column matrix; i.e., the transpose $\mathbf{A} = [a_1\ a_2\ a_3]$ is

$$\mathbf{A}^T = \begin{bmatrix} a_1 \\ a_2 \\ a_3 \end{bmatrix}$$

If **A** is symmetric, then $\mathbf{A}^T = \mathbf{A}$.

B. UNITS

Quantity	Symbol	SI	English	Conversion
Area	A	m^2	ft^2	$1\ m^2 = 10.7639\ ft^2$
Convection heat transfer coefficient	h	W/m^2C	$Btu/hr\ ft^2F$	$1\ W/m^2C = 0.1761\ Btu/hr\ ft^2F$
Density	ρ	kg/m^3	lb_m/ft^3	$1\ kg/m^3 = 0.06243\ lb_m/ft^3$
Energy (heat)	$\bar{q}$	kJ	Btu	1 kJ = 0.94783 Btu
Force	F	N	lb_f	$1\ N = 0.2248\ lb_f$
Gravity	g	m/s^2	ft/s^2	$9.8\ m/s^2 = 32.2\ ft/s^2$
Heat flux (per unit area)	q/A	W/m^2	$Btu/hr\ ft^2$	$1\ W/m^2 = 0.317\ Btu/hr\ ft^2$

Quantity	Symbol	SI	English	Conversion
Heat flux (unit per length)	q/L	W/m	$Btu/hr\ ft$	$1\ W/m = 1.0403\ Btu/hr\ ft$
Heat generation (per unit volume)	Q	W/m^3	$Btu/hr\ ft^3$	$1\ W/m^3 = 0.096623\ Btu/hr\ ft^3$
Length	L	m	ft	1 m = 3.2808 ft
Mass	m	kg	lb_m	$1\ kg = 2.20462\ lb_m$
Pressure	p	N/m^2	lb_f/in^2	$1\ N/m^2 = 1.45038\ x\ 10^{-4} lb_f/in^2$
Specific heat	c_p	kJ/kgC	Btu/lb_mF	$1\ kJ/kgC = 0.23884\ Btu/lb_mF$
Thermal conductivity	k	W/mC	$Btu/hr\ ftF$	$1\ W/mC = 0.5778\ Btu/hr\ ft^2F$
Thermal diffusivity	$\alpha(k/\rho c_p)$	m^2/s	ft^2/s	$1\ m^2/s = 10.7639\ ft^2/s$
Velocity	v	m/s	ft/s	$1\ m/s = 3.2808\ ft/s$
Viscosity: Dynamic	m	$kg/m\ s$	$lb_m/ft\ s$	$1\ kg/m = 0.672\ lb_m/ft/s$
Viscosity: Kinematic	$\nu(\mu/\rho)$	m^2/s	ft^2/s	$1\ m^2/s = 10.7639\ ft^2/s$
Volume	V	m^3	ft^3	$1\ m^3 = 35.3134\ ft^3$

C. THERMOPHYSICAL PROPERTIES OF SOME COMMON MATERIALS

Material	Melting Point (K)	ρ (kg/m^3)	c_p (J/kg K)	K (W/m K)	$\alpha \times 10^6$ (m^2/s)*
Aluminum: Pure	933	2701	903	237	97.1
Aluminum: 2024-T6	775	2770	875	177	73.0
Boron	2573	2500	1107	27.0	9.76
Carbon: Diamond	—	3500	509	300	—
Carbon: Amorphous	1500	1950	—	1.6	—
Copper: Pure	1358	8933	385	401	117
Copper: Bronze	1293	8800	420	52	14
Gold	1336	19300	129	317	127
Iron: Pure	1810	7870	447	80.2	23.1
Iron: Carbon Steel (AISI 1010)	—	7832	434	63.9	18.8
Iron: Stainless Steel (AISI 304)	1670	7900	477	14.9	3.95
Lead	601	113450	129	35.3	24.1
Nickel: Pure	1728	8900	944	90.7	23.0
Nickel: Inconel (X-750)	1665	8510	439	11.7	3.1
Platinum	2045	21450	133	71.6	25.1
Silver	1235	10500	235	429	174
Tin	505	7310	228	66.6	40.1
Tungsten	3660	19300	132	174	68.3
Zinc	693	7140	389	116	41.8
Zirconium	2125	6570	278	22.7	12.4

* At 300 K.

D. DIMENSIONLESS GROUPS[a]

Biot number	$Bi = hL / k$
Eckert number	$Ec = u^2 / c_p \Delta T$
Fourier number	$Fo = \alpha t / L^2$
Friction factor	$f = \tau_w / (1/2 \rho u^2)$
Froude number	$Fr = u^2 / gl$
Grashof number	$Gr = \beta g L^3 \Delta T / \nu^2$
Mach number (for perfect gas)	$M = u / u_{sound} = u / (\gamma \bar{R} T / \bar{M})^{1/2}$
Nusselt number	$Nu = hL / \kappa$
Péclet number	$Pe = (\mathrm{Re})(\mathrm{Pr})$
Prandtl number	$\mathrm{Pr} = \mu c_p / \kappa$
Rayleigh number	$Ra = (Gr)(\mathrm{Pr})$
Reynolds number	$\mathrm{Re} = uL / \nu = \rho u L / \mu = \dot{m} L / \mu$
Schmidt number	$Sc = \nu / D = \mu / \rho D$
Stanton number	$St = (Nu)(\mathrm{Re})(\mathrm{Pr}) = h / \rho c_p u$
Stefan or Jakob number	Ste or $\mathrm{Ja} = c_p \Delta T / \Delta h$
Strouhal number	$Sr = \nu L / u$

[a] The symbol L in the dimensionless groups stands for generic length and is defined according to the particular geometry being described; i.e., it may be diameter, hydraulic diameter, plate length, etc.

Physical Constants

Stefan–Boltzmann constant	$\sigma = 5.67051 \, x \, 10^{-8} W / (m^2 K^4)$
Universal gas constant	$\bar{R} = 8.31441 \, x \, 10^3 J / kmol \, K$

E. COMPUTER PROGRAMS

A set of computer codes for solving one-, two-, and three-dimensional (1-D, 2-D, and 3-D) problems is available on the Web (located at: femcodes.nscee.edu), along with source listings for several MAPLE, MATHCAD, and MATLAB files. In addition, COMSOL files are included that use COMSOL Viewer, which is a limited, free version of COMSOL 3.1 that lets users examine specific data sets. The FORTRAN codes are written in FORTRAN 95 and consist of source and executable versions, including example data sets. All of the codes and examples run on a PC. Minimum PC systems requirements include the following:

1. Intel Pentium 4 or equivalent processor
2. Microsoft Windows 95/98/2000/XP

3. 256 MB RAM (for Windows and COMSOL)
4. 40 MB of available disk space
5. 16-bit graphics adaptor and display to support 256 simultaneous colors at 640 × 480 resolution; 24-bit color at 800 × 600 (or greater) resolution

In addition, there are C++ and JAVA versions of the 2-D codes available on the Web. The C++ version runs under Windows and allows the user to directly input boundary conditions using mouse drag-and-drop routines; results are instantly produced as display contours along with an adapted (refined) mesh. The JAVA version permits the user to input boundary conditions directly into the code on the Web site and instantly displays contours along with an adapted mesh. Both codes use bilinear quadrilateral elements along with local mesh refinement. If users require assistance or have questions regarding access to the software, they should contact personnel identified in the Web site.

It is recommended that a subdirectory be created on the hard drive of the PC for storing and modifying the codes. Copying the files from the Web into the subdirectory can be accomplished using Windows drag and drop. Once the files are copied, the computer codes can be edited and run on the PC. Most of the codes require data files that are read to establish a mesh (preprocessing) and boundary conditions, then create an output file for displaying the results (post-processing).

E.1 Source Codes

The source codes permit examination of the programming structure, and allow changes to be made to the codes as desired. Any suitable FORTRAN compiler contains an editor that allows the user to examine the source listing and change lines as desired. This can also be accomplished using a word processor, but be sure to save the code with a .FOR extension when finished. Once the codes are changed and updated, they will need to be recompiled to produce an executable version. The code listings in MAPLE, MATHCAD, and MATLAB will all execute within their program kernels; i.e., the user must have working versions of these software packages to run these files. Similarly, the user must have COMSOL Viewer (free) to run the COMSOL files.

A review of the FORTRAN source programs will reveal a great deal of similarity among the 1-D, 2-D, and 3-D code listings. The 2-D and 3-D FEM versions were built on the 1-D FEM code structure. These FORTRAN codes all assume only one degree of freedom per node (1 DOF/node). It is left to the user to alter the codes accordingly to accommodate more than 1 DOF per node (as in structural problems involving u,v deflections per node, etc.).

The compiled versions of the computer codes (extension .EXE) were made using Compaq Visual FORTRAN (version 6.5), developed by Compaq Computer Corporation. This compiler works easily and provides good diagnostics—and permits C++ coding to be embedded within the FORTRAN. The FORTRAN codes were also compiled using Lahey and Ryan–McFarland FORTRAN compilers, and produced equally good executable codes. Any changes to the FORTRAN codes contained on the Web will require both compiling and linking—the user must have a FORTRAN compiler resident on the PC. The FORTRAN codes can also be uploaded onto larger class machines, if desired (e.g., workstations, mainframes, etc.).

The programs and data files are listed in seven subdirectories and grouped for convenience to permit association of files with one another. The principal subdirectories are as follows:

Subdirectory 1-D:

MESH-1D.FOR*; MESH-1D.EXE
FEM-1D.FOR; FEM-1D.EXE
PLOTFEM.EXE (graphics display)
EX301.DAT; EX302.DAT; EX305.DAT; EX306.DAT
LIN-1D.DAT; QUAD-1D.DAT; CUBIC-1D.DAT

Subdirectory 2-D:

MESH-2D.FOR and MESH-2D.EXE
FEM-2D.FOR and FEM-2D.EXE
PLOTFEM.EXE
EX405.DAT; EX503.DAT; EX504.DAT
EX901.DAT; EX902.DAT; EX903.DAT
EXS1.DAT; EXS2.DAT; EXS3.DAT; EXS4.DAT

Subdirectory 3-D:

FEM-3D.FOR; FEM-3D.EXE
HEX-3D.DAT; TET-3D.DAT
EX705.DAT

Subdirectory MAPLE:

EXMPL301; EXMPL302; EXMPL305; EXMPL306
EXMPL405; EXMPL503; EXMPL504
EXMPL705

Subdirectory MATHCAD:

EX301.MCD; EX302.MCD; EX305.MCD; EX306.MCD
EX405.MCD; EX503.MCD; EX504.MCD
EX705.MCD

* The .FOR extension denotes source code listing in FORTRAN; the .DAT extensions are data sets; the .EXE files are executable versions of the codes.

Subdirectory MATLAB:

M-files: Linear1D; Quadratic1D; Lintriangle2D; Quadtriangle2D; Linquadrilateral2D; Quadquadrilateral2D; Lintetra3D
EXMTL301; EXMTL302; EXMTL305; EXMTL306
EXMTL405; EXMTL503; EXMTL504
EXMTL801; EXMTL901; EXMTL902; EXMTL903
EXMTL705

Subdirectory COMSOL:

EX301.FL; EX302.FL; EX305.FL; EX306.FL
EX405.FL; EX503.FL; EX504.FL; EX705.FL
EXS1.FL; EXS2.FL; EXS3.FL; EXS4.FL
EX801.FL; EX901.FL; EX902.FL; EX903.FL
EX1001.FL; EX1002.FL
PROB1001.FL; PROB1002.FL; PROB1003.FL

PLOTFEM.EXE is problem dependent with the total number of nodes limited to 101. In addition, the FORTRAN codes produce output files that can be read by TECPLOT, which is a well-known post-processing graphics program used by many engineers and scientists.

Information regarding flowcharts and subroutine descriptions of the FORTRAN 1-D and 2-D mesh and FEM codes can be obtained from the Web site: femcodes.nscee.edu.

Index

H

I

Q

R

S

T